# THE BEGINNINGS OF GREEK MATHEMATICS

# SYNTHESE HISTORICAL LIBRARY

## TEXTS AND STUDIES IN THE HISTORY OF LOGIC AND PHILOSOPHY

VOLUME 17

ÁRPÁD SZABÓ

*Mathematical Institute, Hungarian Academy of Sciences*

# THE BEGINNINGS
# OF
# GREEK MATHEMATICS

D. REIDEL PUBLISHING COMPANY

DORDRECHT : HOLLAND / BOSTON : U.S.A.

Library of Congress Cataloging in Publication Data

Szabó, Árpád 1913–
    The beginnings of Greek mathematics.

        (Synthese Historical Library; v. 17.)
        Translation of Anfänge griechischen Mathematik.
        Includes bibliographical references.
        1. Mathematics, Greek. I. Title. II. Series

QA 22. S 9713        510'.0938        78–7452
ISBN 90–277–0819–3

---

Published as a coedition of
D. Reidel Publishing Company,
P.O. Box 17, Dordrecht, Holland

and

Akadémiai Kiadó, Budapest, Hungary


Sold and distributed in the U.S.A., Canada, and Mexico
by D. Reidel Publishing Company, Inc.
Lincoln Building, 160 Old Derby Street, Hingham, Mass. 02043, U.S.A.


Translated from the German

*Anfänge der Griechischen Mathematik*

by A. M. Ungar

Printed in Hungary

TO THE MEMORY OF MY FRIEND
IMRE LAKATOS

(1922—1974)

Late Professor of Logic, University of London

# CONTENTS

# PREFACE TO THE ENGLISH EDITION

When this book was first published, more than five years ago, I added an appendix on How the Pythagoreans discovered Proposition II.5 of the 'Elements'. I hoped that this appendix, although different in some ways from the rest of the book, would serve to illustrate the kind of research which needs to be undertaken, if we are to acquire a new understanding of the historical development of Greek mathematics. It should perhaps be mentioned that this book is not intended to be an introduction to Greek mathematics for the general reader; its aim is to bring the problems associated with the early history of deductive science to the attention of classical scholars, and historians and philosophers of science.

I should like to conclude by thanking my translator, Mr. A. M. Ungar, who worked hard to produce something more than a mechanical translation. Much of his work was carried out during the year which I spent at Stanford as a fellow of the Center for Advanced Study in the Behavioral Sciences. This enabled me to supervise the work of translation as it progressed. I am happy to express my gratitude to the Center for providing me with this opportunity.

*Árpád Szabó*

# NOTE ON REFERENCES

The following books are frequently referred to in the notes.
Unless otherwise stated, the editions are those given below.

Burkert, W. *Weisheit und Wissenschaft, Studien zu Pythagoras, Philo-laos und Platon*, Nuremberg 1962.

Diels, H. & Kranz, W. *Fragmente der Vorsokratiker*, 8th edn, Berlin 1956.

Heath, T. L. *Euclid's Elements*, 3 vols. Dover Publications 1956.

*Mémoires scientifiques*, ed. J. L. Heiberg & H. G. Zeuthen, Toulouse–Paris: vol. 1, 1912; vol. 2, 1913; vol. 3, 1915.

Proclus, *Diadochi in primum Euclidis Elementorum Librum commentarii.* ed. G. Friedlein, Lipsiae 1873.

Thaer, C. *Die Elemente von Euclid*, in Ostwald's *Klassiker der Exacten Wissenschaften*, Leipzig 1933–7.

van der Waerden, B. L. *Science Awakening* (Noordhoff, Groningen, Holland, 1954; Science Editions, New York, 1963). *Erwachende Wissenschaft, ägyptische, babylonische und griechische Mathematik* (Basel–Stuttgart 1956)

*Zur Geschichte der griechischen Mathematik*, ed. O. Becker, Wissenschaftliche Buchgesellschaft, Darmstadt 1965.

# CHRONOLOGICAL TABLE

These investigations into the origins of Greek mathematics are concerned with the pre-Euclidean period. Individual scientific discoveries cannot be dated exactly within this period, but the following table should help the reader to start out with a rough idea of the chronology. I have tried to give the likely dates of those persons mentioned in the text. (Only the dates referring to Plato and his circle are absolutely certain.) In the right-hand column, I have mentioned some mathematical discoveries which can be associated with the person in question.

### 6th Century

Thales (circa 639—546)

Because of the tradition which attributes new mathematical discoveries to Thales, the concept of '*angle*' must have been known at this time. The elaboration of this concept seems to have been a new achievement of the Greeks.

Anaximenes (circa 560—528)
Pythagoras (circa 510)
Parmenides (a little younger than Pythagoras)
Epicharmus (flourished about 500)

The theory of odd and even was already generally known.

### 5th Century

Zeno (a pupil of Parmenides)
Hippasus of Metapontum (a pupil of Pythagoras)

His experiments with bronze *diskoi* further verified the ratios

|  |  |
|---|---|
|  | associated with consonances. I have therefore come to the conclusion that the original experiments with a monochord and measuring rod, to which the concept of *diastema* is due, must have taken place earlier. |
| Oenopides (an older contemporary of Hippocrates of Chios) | The first three postulates of Euclid. |
| Hippocrates of Chios (active in Athens around 430) | The construction of a mean proportioned to two straight lines (*Elements*, Book VI, 13) was already known at this time. |

## 4th Century

|  |  |
|---|---|
| Archytas (about the same age as Plato) | Doubling the (volume of a) cube. |
| Plato (427—347) | |
| Eudoxus (a younger contemporary of Plato) | The fifth book of the *Elements* |
| Aristotle (384—322) | |
| Eudemus (a pupil of Aristotle) | |
| Autolycus | |
| Euclid (around 300) | Compiled the *Elements* |

# INTRODUCTION

This book is entitled *The Beginnings of Greek Mathematics*. The reader should not, however, expect a systematic and comprehensive account of the oldest period of Greek science. The investigations presented here revise and extend the many works which, in the course of the last few years, have sought to throw light from different viewpoints on the early history of Greek mathematics. However, I did not collect together these investigations supposing that a definitive new picture of the early history of Greek science could be presented on the basis of the historical conclusions reached here. On the contrary, I believe that today we are still at the beginning of the difficult work which one day will lead to a completely new picture of the historical development of Greek mathematics. To this task it is hoped that the present book may make a contribution which, although modest, nonetheless opens up some new avenues.

First of all, I must give an account of what is new in this book.

*

In the first place I would like to discuss the method used, for undoubtedly it is this which distinguishes the book from most of its predecessors. I want to illustrate the difference straight away with an example, and have chosen for this purpose van der Waerden's book[1] which at present is the most widely read and, quite deservedly, the most highly thought of work on Greek mathematics. When the German edition of this book appeared more than ten years ago, I wrote in a review:[2]

---

[1] *Erwachende Wissenschaft, ägyptische, babylonische und griechische Mathematik*, translated from Dutch by H. Habicht with additions by the author, Basel–Stuttgart 1956. (The first Dutch edition appeared in 1950 and the first English edition in 1954.)

[2] *Acta Scientiarum Mathematicarum* (Szeged, Hungary) **18** (1957) pp. 140–1.

"I expect that all further historical research in this field will, for a long
time to come, be based on van der Waerden's work." The time which
has passed since then has only served to bear out this prediction.
I should also emphasize that without van der Waerden's pioneering
work, it would hardly have been possible to begin my own researches.
(Furthermore, I do not believe that the undisputed merits of his work
are made obsolete or in any way replaced by what is new in this book.)
But however highly I esteem van der Waerden, I disagree in one essential respect with the method he employed.

One of the greatest merits of his book lies in the fact that it "is based
on actual study of the sources". This was an indispensable requirement.
As the author himself emphasized: "So many statements in books on
the history of mathematics have been copied from other books uncritically and without studying the sources. Thus there are many tales in
circulation which pass for universally known truths." A glaring example of this is given, and then follows the passage against which I would
like to argue from a methodological viewpoint. We are interested only
in those parts of the following quotation which concern Greek
mathematics, nonetheless the passage is given here in full for the sake
of completeness.[3]

> To avoid such errors, I have checked all the conclusions which I
> found in modern writers. This is not as difficult as might appear,
> even if, as in my case, one cannot read either the Egyptian characters
> or cuneiform symbols and one is not a classical philologist. For
> reliable translations are obtainable of nearly all texts. For example,
> Neugebauer has translated and published all mathematical cuneiform
> texts. The Egyptian mathematical texts have all been translated
> into English or German. Plato, Euclid,[4] Archimedes, . . . of all these
> good translations exist in French, German and English. Only in a
> few doubtful cases it became necessary to consult the Greek text.

[3] B. L. van der Waerden, *Science Awakening* (Noordhoff, Groningen, Holland,
1954, Science Editions, New York, 1963) p. 6.

[4] Let me take this opportunity to note something not directly relevant to
the contents of this passage. The correct transcription of this Greek name is
the one used throughout by van der Waerden, namely Eukleides. I however,
will always use the English form Euclid, so as to distinguish the author of the
*Elements* from the Megarean philosopher of the same name (Eukleides), with
whom he was often confused even in ancient times.

In Part 1 of this book a short passage from Plato (*Theaetetus*, 1470–1488) which is not particularly difficult to understand from the linguistic point of view is discussed in detail. This is because it seems to contain some interesting facts about two Greek mathematicians–Theodorus and Theaetetus. I am, it is true, a classical philologist, nonetheless, I have read the passage in question not only in the original, but also in a multitude of German, English, French, Italian and even Hungarian translations. Before proceeding further, I want to recount my experiences with these. To my surprise, most translations I looked at (with some unimportant exceptions) were accurate on the whole. Even more astonishing was the fact that by and large they rendered the mathematical content of the passage faithfully. (Although they also contained some obvious mistakes[5] which lead to an erroneous interpretation of the text.) It only became fully apparent to me that these translations were completely unreliable after I had studied in detail the interpretation of this passage, which regrettably has become a tradition among philologists and historians of the last hundred years. A translation can be 'accurate on the whole' and still give rise to a completely wrong interpretation of the text. Let me give a small example of this.

In the passage from Plato mentioned above, the technical term δύναμις occurs more than once. If this word is translated as *power* (as it usually is), this is simply a mistake. It is true that the mathematical content of the passage will still be comprehensible, but the incorrect use of this word inevitably suggests erroneous ideas about Greek mathematics. In classical times the Greeks did not as yet have the concept of *power*. Furthermore the Greek term *dynamis* has nothing to do with the Latin notion of *potentia* (ability or possibility). If, on the other hand, the word is translated as 'square', and this translation is justified by saying (as Heath[6] does) that the particular 'power' found in Greek

---

[5] Apelts' admirable translation should be mentioned as an exception to this: *Platons Dialog Theätet*, translated into German with annotations by O. Apelts (Leipzig 1911).

[6] T. L. Heath, *Euclid's Elements* (Dover Publications, 1956, Vol. 3, p. 11): '*Commensurable in square* is in the Greek δυνάμει σύμμετρος. In earlier translations (e.g. Williamson's) δυνάμει has been translated '*in power*', but as the particular *power* represented by δύναμις in Greek geometry is *square*, I have thought it best to use the latter word throughout.'

**2** Szabó

geometry is the square, then this is still misleading for it implies the following two ideas, both of which are patently false:

*(a)* The Greek word *dynamis* literally means 'power'. It is translated as 'square', because in practice *dynamis* is always the second power.

*(b)* If in mathematics the word *dynamis* is related to 'power', then perhaps κύβος could at least in principle also be construed as a kind of *dynamis*.

Finally, if the word is translated as 'square' without any further explanation, then the translation is undoubtedly correct; but not much is gained for the history of mathematics. The term δύναμις, correctly translated as 'square', really becomes a key to understanding this whole passage of Plato only if one also knows that this geometrical term did not originate in Plato's time. Even in the first half of the 4th century B.C. it had been handed down from an earlier time. Furthermore one should remember that this scientific term could not have been coined without the prior discovery of some important mathematical facts. In other words one has to be clear about the etymology of this expression as a technical term in Greek mathematics. Only then does its correct translation become informative as far as understanding the text is concerned. If one knows this, the translation of this piece appears in a new and totally different light. Many questions which were previously discussed in great detail (and could still not be answered satisfactorily), suddenly become insignificant. Indeed they seem to be red herrings. In their place other problems emerge which could not even have been thought of before the discovery of the real meaning of the term.

At this moment, however, I just want to sketch the method followed in this book, not give the results obtained by its use. So it should be emphasized that as far as the history of mathematics is concerned, translations of the source materials are frequently unreliable, even when they are philologically excellent. A good example of this is the passage from the *Theaetetus* mentioned above. This piece, as far as I have been able to check in the rather extensive technical literature, was never considered problematic by philologists or historians. The whole passage could be translated almost without objection. It didn't matter whether the expression δύναμις was translated as 'power', 'square' or 'side of a square', since everyone knew that in the text it referred to squares or to the sides of these squares. Furthermore this knowledge ensured that the mathematical content of the passage was correctly understood (at least in broad outline). One might have thought it just philological pedantry to take a lot of trouble over this single word. Yet the correct understanding of this word would have been important not for philology, but for the history of mathematics. It would have avoided a lot of mistakes in reconstructing the early history of Greek mathematics. It is perhaps worthwhile to recall here at least briefly the change which the interpretation of this expression in the history of Greek mathematics has undergone over the last hundred years.

As far as I know, Tannery was the first to notice that the meaning of the term δύναμις in this passage from Plato was in any way problematic. On the one hand it was clear that for some obscure reason this expression must have its usual meaning of 'square' here. On the other hand the same word in the very same passage seems to mean 'square root' as well. This idea is justified in as much as the passage in question deals not only with squares, but also with their sides. Nevertheless had the text been analysed carefully enough, it would have been immediately apparent that there is no justification for supposing that the word δύναμις is ambiguous. Tannery, however, instead of making a careful linguistic analysis of the text, chose the wrong way in his paper of 1876[7] and conjectured that mathematical terminology had not yet been fixed in Plato's time; so δύναμις could have meant 'square root' *(racine carrée)* as well as 'square' *(carré)*.

This erroneous conjecture of Tannery survived for a long time in the history of science, although he himself realised his mistake eight years later. It was simply impossible for one and the same word to have had two meanings at the same time. Unfortunately, however, he failed to find the correct interpretation of the passage this time as well. On the contrary, in this later work (of 1884) he proposed[8] that the difficulties of interpretation be done away with by substituting δυναμένη for δύναμις throughout the text of the *Theaetetus*. This proposal by Tannery is of course a heavy-handed and completely unnecessary adulteration of the

---

[7] P. Tannery, 'Le nombre nuptial dans Platon', *Revue Philosophique* 1 (1876) 170–88 (*Mémoires scientifiques*, ed. J. L. Heiberg & H. G. Zeuthen, Toulouse–Paris 1912, Vol. 1, pp. 28–38). The incidental remarks cited in the text are to be found in *Mém. Scient.* 1, 33, n. 2.

[8] P. Tannery, 'Sur la Langue mathématique de Platon', *Annales de la Faculté des Lettres de Bordeaux*, 1 (1884) 95–105 (*Mém. Scient.* 2, 91–104).

2*

text, which is worthy of an amateur. But I still consider this second attempt of his noteworthy for the following two reasons; firstly, because he clearly rejected his previous erroneous view about the ambiguity of the mathematical term δύναμις, and secondly, because his proposal to alter the text is clear proof that he believed the usual interpretations of this piece to be unsatisfactory, even though its mathematical content was correctly understood by and large.

In 1889 Tannery made another vain attempt to deny his earlier view.[9] But his erroneous idea had been accepted by the best scholars in the field, even though he himself had already realised its inadequacy. Consequently it was just his attack on the authority of the text that was rejected.[10]

It is not surprising that Tannery's correct realization that δύναμις could not mean 'square root' (or 'side of a square' as well) did not prevail, for he made no serious effort to elucidate the process by which concepts evolve, and whose result in this case was that a square could be referred to as *dynamis*. Instead, in the long run he just created confusion with two startling and incompatible ideas. Unfortunately these are still current, so I give them here in some detail.

(1) In his paper of 1884, Tannery wanted to explain the fact that δύναμις means the side of a square in Definition 4, Book X of the *Elements*, as follows:[11]

---

[9] P. Tannery, 'L'hypothèse géométrique du Menon de Platon', *Archiv f. Gesch. der Philosophie* **2** (1889) 509–14 *(Mém. Scient.* **2** 400–6). It is worth quoting the following from this work: 'Je citais même, comme exemple typique, le passage de Théétète, où δύναμις est employée dans le sens de *racine carrée*, tandis que dans La République ... le même mot signifie au contraire *carré*. Mais depuis, la poursuite de mes études sur les variations qu'a pu subir la langue mathématique des Grecs, *m'a conduit à des conclusions tout à fait opposées* et je n'hésite plus désormais etc.'

[10] Cf. T. L. Heath, *Euclid's Elements*, Vol. 2, p. 288; *A History of Greek Mathematics*, Oxford 1921, Vol. 1, p. 209, n. 2: also B. L. van der Waerden, *Science Awakening*, pp. 142, 166 etc. On page 142 of this last reference, one finds '... it is not necessary, as Tannery does, to replace the word δύναμις by δυναμένη creating'.

[11] See n. 8 above. Tannery of course is also solely to blame for van der Waerden's use (*Science Awakening*, pp. 142, 156) of such expressions as: δύναμις *(impulse, force)*, the '*generation* of mean proportionals', 'the *generation* of ... geometric, arithmetic and harmonic means' etc. by reason of his strange paraphrase *pouvoir une aire* (sic).

"Soit un carré dont l'aire soit déterminée, de *trois pieds* par exemple, le côté de ce carré est, dans la langue mathematique classique, La δυναμένη (la ligne qui peut) cette aire de trois pieds. *Pouvoir* une aire *(δύνασθαί τι χωρίον)* c'est de même, pour une ligne droite limitée, être telle que le carré construit sur elle ait précisément cette aire."

I must confess that it remains a mystery to me how one could possibly 'explain' δυναμένη and δύνασθαι by the equally obscure phrases '*la ligne qui peut*' and '*pouvoir une aire*' respectively. Nonetheless this fine 'explanation' which the master-historian of science gave did not lack adherents. It would be easy to show that this explanation of his was carried further, when the mathematical term δύναμις was construed in German as '*generating force*', and the term δυναμένη as '*that which generates*'. Clearly the line segment concerned is called δυναμένη (that which generates), because being the side of a square it *generates* a square. So much for the evolution of concepts ! (The Greeks could never have used the verb δύνασθαι to express this kind of 'generation'.) So Tannery's one idea led to these kinds of mistakes.

(2) His second attempted explanation in 1902 had even more unfortunate consequences. It must be presumed that by now he had forgotten what had been correctly established by him in 1884, namely that the technical term δύναμις could not mean 'square-root' or 'side of a square'. Otherwise he could never have pursued the train of thought which I want to recall here. In this later work Tannery sketched the method of approximation which the Greeks used to calculate the square-roots of non-square numbers (i. e. surds). He concluded his words on this topic with the following remark:[12] " . . . en exprimant de plus en plus près la valeur de cette moyenne, si l'on ne peut la construire que géométriquement, si elle n'existe qu'en *puissance*, non en *acte*, pour employer le langage des Grecs."

For the moment the only things that interest us in this quotation are the two words in italics. To appreciate their significance, one should follow Tannery's own advice and translate them back into Greek. He believed that when he spoke of *puissance* and *acte*, he was using a 'quasi-Greek language of geometry', and his ideas can be better under-

[12] P. Tannery, 'Du rôle de la musique grecque dans le développement de la mathématique pure', *Bibliotheca mathematica*, **3** (1902) 102, 161 — 75. *(Mém. Scient.*, **3** pp. 68–9. The quotation is taken from p. 82.)

stood if one uses the original Greek words, namely δύναμις for *puissance* and ἐνέργεια, ἐντελέχεια for *acte*. In other words, Tannery is saying that δύναμις means the 'irrational square-root of a number' in Greek, because such a number could be constructed *(puissance)* geometrically, but did not exist as *acte (ἐντελέχεια)*. In the first place, this conjecture of his is totally foreign to the way in which the early Greeks thought about mathematics. It is just an unsuccessful construction of his own. In the second place, the idea is wrong on two counts. (a) δύναμις never meant 'irrational square-root' in Greek; and (b) the origin of the mathematical term δύναμις has nothing to do with the Aristotelian contrast between *dynamis* and *entelechia*. (Aristotle's *dynamis* is not the same as *dynamis* in geometry.)

Here I attempt to give an authentic account of the early history of Greek mathematics, not just in the sense that I have consulted the original texts, but also because I have endeavoured to elucidate the new concepts of Greek mathematics *in statu nascendi*. Even in its earliest period a whole series of new concepts were introduced into Greek mathematics; concepts which do not correspond to anything in pre-Greek mathematics (or at least are not known to do so). Unfortunately no mathematical texts have survived from the period during which most of these evolved.

Yet since these same concepts also occur in the extant Greek mathematical texts which originated much later, I believe it is possible to elucidate at least in part the development of mathematical thought during this earlier period by reconstructing the history of these concepts.

For example, there is no doubt that the axiomatic foundations of mathematics laid down by Euclid in the *Elements* are the culmination of a lengthy pre-Euclidean development. Previous historians only drew upon Aristotle, who lived just before Euclid, or at best on Plato to explain this process. They never even contemplated the idea that Greek 'axiomatics' could have originated in a still earlier time. Vague 'interactions' between the Pythagoreans and Eleatics had also been conjectured (by Tannery, for example, and by Rey — one of his followers), although no one had determined precisely what these 'interactions' comprised, nor which side might have gained the most from them. The idea that the founding of this science on definitions and axioms is attributable to the Eleatic influence has not been considered previously. Yet this is what I hope to prove in the sequel, mainly by analysing the history of concepts.

As far as I am concerned, a special significance is attached to language in this analysis. I intend to investigate Euclid's mathematical language from a historical point of view, for this technical language provides living proof of the developmental process which started long before the texts themselves came into being. For example, with the help of linguistic analysis it can be shown that all the technical terms of the geometrical theory of proportions have their origins in music. Indeed I will demonstrate that the concepts of 'ratio' *(diastema* or *logos)* and 'proportion' *(analogia,* i.e. 'sameness of ratio'), and even the expressions for operations on ratios, were all developed from considerations about and experiments in the theory of music. From this I conclude that the musical theory of proportions must have developed up to a certain stage before the geometrical theory. For example, 'similarity of rectilinear figures' could only be defined as 'sameness of ratios between the corresponding sides' after the theory of music had made the concepts of 'ratio' *(logos)* and 'sameness of ratio' *(analogia)* available to geometry. In this case linguistic analysis helps us to reveal a historical connection between music and geometry which the literary sources only hint at. Although they report that music and geometry were twin disciplines of the Pythagoreans, only a historical investigation into terminology discloses that the theory of music must have preceded the development of geometry, at least in the creation of its fundamental concepts.

I should emphasize, however, that in this work I am not pursuing the history of terminology for its own sake. Although for the most part philological methods are used in this book, these are designed to contribute towards mathematical and historical understanding, not to philology itself. I am convinced that the significance of many important facts about science in antiquity simply cannot be appreciated either historically or mathematically without using the philological precision I have attempted here. Let me illustrate this with an example.

Reading Thaer's[13] outstanding translation of the *Elements,* one encounters in Book X the notion of 'straight lines commensurable in

[13] C. Thaer, 'Die Elemente von Euklid', in Ostwald's *Klassiker der Exacten Wissenschaften* (Leipzig 1933–7).

square'. This translates the corresponding Greek phrase quite correctly and its mathematical meaning can easily be understood. Yet if one studies only the modern translation and not the Greek text, the belief that it faithfully renders the original might mislead one into thinking that in antiquity there were two different theories of 'straight lines rational in square' which led to the same result by different methods. The one seems to stem from questions about the existence of a geometric mean between straight lines (or numbers); the other deals with the 'squaring of straight lines', not with the notion of geometric mean. But this idea becomes untenable as soon as one realises that these two phrases, 'straight lines rational in square' and 'squaring of straight lines', both translate the Greek expression εὐθεῖαι δυνάμει σύμμετροι. Also, it should be remembered that producing a *dynamis* by its very nature presupposes the use of the geometric mean (cf. pp. 97–8, 174–81).

Therefore it cannot be emphasized too strongly that research into the history of ancient mathematics is impossible without a very careful and thorough investigation of its language. It should not be forgotten that mathematical thought and language were still very closely linked at that time. The mathematical symbols which we all rely on nowadays, could express practically nothing, they did not even exist, so words had to be used. Furthermore these words, even the ones which later acquired special mathematical meanings, were drawn mostly from everyday language or from the language of philosophy. So the characteristics of ancient mathematical thought, which sometimes provide a marked contrast to later ideas about mathematics, are only accessible to us through language.

*

Having sketched at least in outline the methodological novelties to be found in this book, it remains for me to indicate briefly the actual results of my investigations and the extent to which a different picture of the origins of Greek mathematics emerges from them. As I emphasized above, a definitive new picture of early Greek science cannot as yet be painted. Nonetheless, I believe that these investigations can help to change significantly our conceptions of the development of pre-Euclidean mathematics. To characterize this new perspective in a few lines, I will start by describing the fundamental work which tried more than 30 years ago to mark out the boundaries within which our attempts at reconstructing pre-Euclidean mathematics have taken place.

In his paper, 'The Theory of Odd and Even', Becker showed that an ancient Pythagorean *mathema* can most likely be recovered in its original form from the *Elements*.[14] This is the oldest example of deductive knowledge in Greek science, or at least the oldest surviving example. It is important to us for two reasons. The first is that it can be dated fairly accurately. A fragment of Epicharmus clearly indicates that he knew this theory.[15] Considering that he flourished around 500 B.C., the theory of odd and even must date from at least the beginning of the fifth century. However, the second reason is more important. It is that this theory clearly culminated in the proof of the incommensurability of the diagonal and sides of a square. It seems that linear incommensurability, at least in the special case of the sides and diagonal of a square, as well as the theory of odd and even must have already been known to Greek mathematicians at the beginning of the fifth century. (The opposing view which maintains that knowledge of irrational quantities dates only "from the middle of the fifth century"[16], is simply a tradition deriving from modern efforts to fix the construction of the theory of irrationals around 400 B.C. Since one wants to place the 'further development' of the theory during Plato's younger years, the 'first case' of this discovery (namely the diagonal and sides of a square) has to be dated as late as possible, i.e. no earlier than the first half of the fifth century.)

In connection with this discovery, Becker suggested four stages of historical development. He thought that these could be traced in the changes which took place in arithmetic and the theory of proportions

[14] O. Becker, 'Quellen und Studien zur Geschichte der Mathematik', *Astronomie und Physik B*, **3** (1934) 533–53. Reprinted in the collection *Zur Geschichte der griechischen Mathematik* ed. O. Becker (Wissenschaftliche Buchgesellschaft, Darmstadt 1965).

[15] Cf. Also van der Waerden's *Science Awaking* (2nd English edition, Science Editions, New York 1963) pp. 109–10. (The passage in question was unfortunately omitted from the German edition.) On the interpretation of the Epicharmus-fragment, see also K. Reinhardt *Parmenides und die Geschichte der griechischen Philosophie*, (2nd edition, Frankfort 1959) pp. 120 and 138.

[16] K. Gaiser, *Platons ungeschriebene Lehre* (Stuttgart 1963), p. 471, n.

during the fifth and the first half of the fourth century. This dating is noteworthy because it has been retained with minor modifications to the present day. I will give the four stages here, together with those of their features which Becker thought most important.

(1) The early Pythagorean stage: According to Becker the most significant achievement of this period was the theory of odd and even (although it did not as yet include the theorem that any number may be decomposed uniquely into prime factors). Distinct from this was a theory of rational proportions (the proof of the incommensurability of the sides and diagonal of a unit square) which arose from musical and geometric questions.

(2) The stage of Theodorus: This period saw the general definition of incommensurability by means of *antanairesis* of magnitudes, also the proof of the irrationality of $\sqrt{n}$ (for $n = 3, \ldots, 17$ and not a perfect square), and a general theory of proportions classified according to kind (straight lines, numbers, etc.).

(3) The stage of Theaetetus: At this stage *antanairesis* was applied to number theory; the 'Euclidean algorithm' was discovered, and it was proved that any number can be decomposed uniquely into prime factors *(Elements* VII. 30).

(4) The stage of Eudoxus: This saw a general and abstract theory of proportions uniformly applicable to ratios of all kinds. Also sameness of ratio was defined in terms of the 'Eudoxus cut' *(Elements* V, definitions 5 and 7), and the axiom about the multiplication of magnitudes (otherwise known as the axiom of Archimedes) was formulated *(Elements* V, definition 4).

As one can see, Becker was simply interested in giving the chronology of arithmetic and the theory of proportions. Geometry is considered only incidentally. Furthermore, the problem of incommensurability, which was a geometric problem both in the minds of the ancients and in its origin, is not just a matter of arithmetic. An immediate drawback of this attitude is that the most important geometer of the fifth century B.C., Hippocrates of Chios, cannot be placed correctly in this chronology. He must have lived before Theodorus and Theaetetus, yet it is scarcely possible to believe that Hippocrates had no knowledge of the mathematical facts which Becker associates with their stages.

Furthermore, Becker's four stages completely ignore the historical problem connected with laying the foundations of mathematics and its systematic construction. They leave open the question as to when Greek mathematicians were first able to lay the foundations of their subject on definitions and axioms. Becker himself dealt with this question in an earlier work which was to some extent superseded by his paper on the theory of odd and even.[17] At that time (1927), however, Becker still had the view that 'Plato was the first to be clearly aware of the vigorous methodical procedure for constructing the elements of mathematics.'[18] Bearing in mind these earlier ideas of Becker (which were formulated under the influence of Zeuthen's conjecture on the matter), one is inclined to supplement his four stages with a fifth – namely, the attempt to systematically construct mathematics first became possible under Plato's influence. Hence it took place directly after the fourth stage, or perhaps during it.

I believe that Becker's chronology has to be extensively revised in the light of my own investigations.

First of all, the most likely date of the earliest attempt to lay the foundations of mathematics and construct it in a deductive manner lies far back in the time of Becker's first stage. Following Becker, we might call this the Platonic stage. I believe I can show that the theory of odd and even presupposed the basic definitions of Euclidean arithmetic. Not only does the oldest Pythagorean *mathema* go back to the first half, if not to the beginning, of the fifth century B.C. but this is also the time at which the theoretical foundations of arithmetic were laid. Only the theoretical foundations of geometry are perhaps later, since Euclid's first three postulates are attributable to Oenopides (around the middle of the fifth century).

Becker's second and third stages are omitted in my reconstruction. In view of the interpretation proposed in Part 1 of this book for the relevant passage of Plato, it cannot be shown that Theodorus or Theaetetus (as they appear in Plato) made any new contributions to mathematics. On the contrary, since the concept of *dynamis* is used in this passage, I have come to the conclusion that the discoveries which were once attributed to the characters Theodorus and Theaetetus in Plato's dialogue date in fact from pre-Platonic times and in all likelihood antedate Hippocrates of Chios.

[17] See chapter 3.30, section II.
[18] O. Becker, *Mathematische Existenz* (Halle 1927), p. 250, n. 2.

I have nothing to say about Becker's fourth stage, since my investigations are not directly concerned with the mathematical discoveries which modern research attributes to Eudoxus.

Instead of Becker's chronology, the following investigations enable us to divide up the development of early Greek mathematics into different stages. These will be briefly outlined here.

The oldest stage in the development of Greek mathematics which is still accessible to us is the musical theory of proportion. All the technical terms of the later general theory originated in the musical one. (This stage is dealt with in chapters 2.3–2.19.) I believe it would be a hopeless undertaking to try and pin down this stage as having taken place in some particular century or half-century. Even if the fragments are included, our oldest texts touching on musical questions scarcely antedate the age of Plato. Yet those terms from the theory of music which had already been taken over into geometry at the time of Hippocrates of Chios must have been coined much earlier. It seems to me much more important that within this stage two periods can be clearly distinguished. In the earlier period experiments with the monochord took place (although these did not as yet involve a measuring rod). Also, some important terms in the theory of music originated at this time, namely *diplasion diastema* (2:1), *hemiolion diastema* ($1\frac{1}{2} = 3{:}2$) and *epitriton diastema* ($1\ 1/3 = 4{:}3$). These of course were later carried over into the theory of proportions. Finally the method of the 'Euclidean algorithm' (successive subtraction, *antanairesis* or *anthyphairesis*) was developed during this period. The later period was characterized by the introduction of the measuring rod which was divided into twelve intervals. This brought into being the new musical/mathematical concept of *logos* (the relation between two numbers).

At the next stage the musical theory of proportions was applied and extended to arithmetic, particularly to the geometrized arithmetic of 'plane numbers', 'similar plane numbers' etc. (See many of the propositions in Books VII, VIII and IX of the *Elements*.) There are two remarks to make in this connection. Euclidean arithmetic is predominantly of musical origin not just because, following a tradition developed in the theory of music, it uses straight lines (originally 'sections of a string') to symbolize numbers, but also because it uses the method of successive subtraction which was developed originally in the theory of music. However, the theory of odd and even clearly derives from an 'arithmetic of counting stones' *(ψῆφοι)*, which did not originally contain the method of successive subtraction. It seems that only the arithmetic of the theory of proportions (just for numbers, of course) has developed from the musical theory of proportions. Actually, this theory seems to have come after the theory of odd and even. But this is simply a conjecture on my part and not a proven fact. In my opinion it is extremely unlikely that Book VII of the *Elements* goes back to Archytas, for the problems with which it deals are very closely linked to the theory of incommensurability. They seem indeed to prepare the way for this theory. Yet quadratic incommensurability (the general theory of *dynameis*, not just the particular case of the diagonal of a square) must have been known well before the time of Archytas, and before the time of Hippocrates of Chios as well.

The third stage comprises the application of the theory of proportions to geometry. This led to the precise definition of *similarity of rectilinear figures* ('sameness of the ratios of the corresponding sides') and subsequently to the construction by geometric methods of the mean proportional. All this, of course, took place at the time of the early Pythagoreans. The construction of the mean proportional led directly to the discovery of linear incommensurability. This construction (already well known to Hippocrates of Chios) implied the extension of the notion of 'ratio' to arbitrary quantities, whether the Greeks were aware of it or not. It is especially noteworthy that the discovery of incommensurability is due to a problem which arose originally in the theory of music. Moreover it is interesting that Tannery had already conjectured that "The problem of the irrationality of $\sqrt{2}$ could just as well have originated in the theory of music as in geometry."[19]

The discovery of incommensurability prepared the way for a further stage in the development of mathematics which is treated in Part 3 of this book under the heading, 'The construction of mathematics within a deductive framework'. To prove in an unobjectionable way that incommensurability really existed, it became necessary to develop a new technique of proof and to lay the theoretical foundations of deductive mathematics. According to my reconstruction, these developments took place under the influence of the Eleatics.

[19] Du rôle de la musique grecque dans le développement de la mathématique pure. (*Mém. Scient.* **3**, 68—89 especially pp. 83-9). Tannery's conjecture was also taken up by Becker (*Zur Geschichte der griechischen Mathematik*, p. 143).

As one can see, my 'chronology' fails to answer many questions about dates. I cannot pin-point any more accurately the time at which the theory of proportions was first applied to arithmetic and then to geometry. Indeed the idea that it was applied to arithmetic first and only later to geometry is just a conjecture based on the fact that the theory of proportions of numbers is closer to the original musical theory than the theory of proportions of arbitrary geometrical quantities. Hence this 'chronology' serves only as an attempt to reconstruct in their likely order the succession of *problem situations* which led from 'simpler' to 'more complicated' knowledge.

It still remains for me to mention a problem which is not dealt with in this book, although a treatment of it could rightly be expected. One might expect that in a book entitled *The Beginnings of Greek Mathematics* the pre-history of this subject would at least be touched upon. Modern research has shown that Greek mathematicians took over a lot of well developed mathematics from pre-Greek times. A well-known example of this is what is usually called Pythagoras' theorem. This could not have been discovered first by Pythagoras, because it had already been learned from much older, Babylonian texts. Similarly it can be shown that other portions of Greek mathematical knowledge are of oriental descent.

Nevertheless I have deliberately omitted a discussion of what was borrowed from pre-Greek mathematics, mainly because in my view we still do not know enough about Greek mathematics itself to be able to treat this problem with any success. As an illustration of this, let me mention here the problem concerning 'the geometrical algebra of the Greeks'.[20]

It was Zeuthen who noticed that there are some interesting geometric propositions in Books II and VI of the *Elements* which would normally be written out as algebraic formulae. From this he concluded that they dealt with 'algebraic propositions in geometric clothing', or as he put it – *with geometric algebra*.

However, he could not furnish satisfactory answers to such questions as whether the early Greeks actually had an algebraic theory which was

---

[20] Cf. what Neugebauer has to say about the following in 'Studien zur Geschichte der antiken Algebra, III', *Quellen und Studien zur Geschichte der Mathematik, Astronomie und Physik B*, **3** (1936) 245.

later geometrized, how this theory might have been arrived at, and what the role of 'geometric algebra' might have been in early Greek science. On the other hand Neugebauer, the discoverer of 'Babylonian algebra', found Zeuthen's 'discovery' very opportune. According to him, the reason the Greeks 'geometrized algebra' is " . . . on the one hand that after the discovery of irrational quantities, they demanded that the validity of mathematics be secured by switching from the domain of rational numbers to the domain of ratios of arbitrary quantities; and on the other hand that it then became necessary to *reformulate the results of pre-Greek 'algebraic' algebra in 'geometric' algebra.*"[21]

Ever since then, Euclid's 'geometric algebra' has been regarded as a borrowing from Babylonian science or as a geometrical version of Babylonian algebra. Only a thorough examination of Zeuthen's theory about 'algebraic propositions in geometric clothing' would allow one to decide the extent to which the above is a historically sound reconstruction. But until we have obtained a better understanding of Greek mathematics itself, and in particular have answered the question as to whether these propositions of 'geometric algebra' really deal with problems originating in algebra and not in geometry, I think it best to put aside all questions concerning the origins of 'Greek geometric algebra'. (But see the Appendix to this book, pp. 332 ff.)

Moreover, one must be careful not to misunderstand the phrase *"switching from the domain of rational numbers to the domain of ratios of arbitrary quantities"*. If one understands it as saying that after the discovery of linear incommensurability it became necessary to extend the concept of 'ratio' (which hitherto had been applied only to numbers) to incommensurable quantities as well, then there is no objection to Neugebauer's words. If on the other hand one understands by 'switching' some kind of 'additional geometrization', then I have some decided objections to raise. In the first place incommensurability was a geometric problem to the Greeks right from the start. In the second place it is simply incorrect to suppose that the discovery of incommensurability shook the Greeks' 'faith in numbers' in any way. They knew very well that even though a number could not be assigned to the length of the diagonal of a square, for example, nonetheless this same diagonal could be expressed numerically in terms of the square constructed on it.

[21] Ibid., p. 250.

It was not necessary to 'abandon the domain of numbers' after this discovery. The explanation of the genesis of the mathematical concept *dynamis* given in this book throws some new light on the above topic as well.

In conclusion I would like to list here those of my papers on which I have relied heavily in this new discussion of early Greek mathematics.

'Wie ist die Mathematik zu einer deduktiven Wissenschaft geworden?' *Acta ant. Acad. Sci. hung.* (Budapest) **4** (1956) 109—51.

'*Deiknymi*, als mathematischer Terminus für *beweisen*', *Maia* N. S. **10** (1958) 106—31.

'Die Grundlagen in der frühgriechischen Mathematik' *Studi Italiani di Filologia Classica* (Florence) **30** (1958) 1—51.

'Der älteste Versuch einer definitorisch-axiomatischen Grundlegung der Mathematik', *Osiris* (Bruge, Belgium) **14** (1962) 308—69.

'The Transformation of Mathematics into Deductive Science and the Beginnings of its Foundation of Definitions and Axioms', *Scripta Mathematica* (New York) **37**, 27–49 and 113–39.

'Anfänge des Euklidischen Axiomensystems'. *Archive for History of Exact Sciences* (Springer-Verlag) **1** (1960), 37—106.

'Ein Beleg für die voreudoxische Proportionen Lehre?'(Aristoteles: *Topic* Θ 3, p. 158b 29–35), *Archiv f. Begriffsgeschichte* (Bonn) **9** (1964) 151–71.

'Der Ursprung des "Euklidischen Verfahrens" ' *Math. Ann.* **150** (1963) 203–17.

'Die frühgriechische Proportionenlehre', *Archive for History of Exact Sciences* (Springer-Verlag) **2** (1965) 197–270.

'Der mathematische Begriff *dynamis*', *Maia* N. S. **15** (1963) 219–56.

Theaitetos und das Problem der Irrationalität in der griechischen Mathematikgeschichte. *Acta. ant. Acad. Sci. hung.* (Budapest) **14** (1966) 303–58.

# 1. THE EARLY HISTORY
## OF THE THEORY OF IRRATIONALS

### 1.1 CURRENT VIEWS
### OF THE THEORY'S DEVELOPMENT

In what follows I will give an example of the method I intend to use for investigating the history of early Greek mathematics. I believe that amongst other things this method sheds new light on the development of the theory of irrationals. Let me begin by recalling what has been thought up to now about the history of this theory.

Recognizing the existence of irrationals was an outstanding achievement of early Greek mathematics. Yet historical research has been unable to give a satisfactory account of how this discovery came about. The only thing that seems to be more or less established on the basis of previous investigations is "that Greek mathematicians knew of the existence of irrational quantities (or incommensurable ratios) from the middle of the fifth century".[1] If one surveys the relevant literature of the last fifty years, what stands out is that the circumstances surrounding this discovery remain obscure.

The prime example of irrationality in ancient texts is always the lengths of the diagonal and sides of a square.[2] Hence it used to be thought that the discovery was 'undoubtedly' suggested by the diagonal of a square.[3] Recently, however, historians have been more inclined to attribute the discovery to Hippasus of Metapontum and claim that he was led to it (in the fifth century B.C.) *by considering the dodecahed-*

---

[1] K. Gaiser, *Platons ungeschriebene Lehre*, Stuttgart 1963, p. 471 n.

[2] Aristotle, *Metaphysics*, 983a19ff. and 1053a14ff. cf. also T. L. Heath, *Mathematics in Aristotle*, Oxford 1949, p. 2: "The incommensurable is mentioned over and over again, but the only case is that of the diagonal of a square in relation to its side etc."

[3] K. von Fritz in *Realencyclopädie der Klassischen Altertumswiss.*, ed. Wissowa, Kroll, *et. al.* A1813; see also W. Burkert, *Weisheit und Wissenschaft, Studien zu Pythagoras, Philolaos und Platon*, Nuremberg 1962, p. 435 n.

*ron*.[4] This latter view represents a compromise between conflicting accounts handed down from antiquity.[5]

On the one hand, it is known that the pentagram was an emblem of the Pythagoreans. On the other hand, a late source credits Hippasus with working on the penta-dodecahedron. He was supposedly the first person to disclose the properties of this solid and is said to have perished at sea for this irreligious act.[6] Now it is argued that the incommensurability of the side and diagonal of a regular pentagon (i.e. of the faces of a penta-dodecahedron) is easily seen. Furthermore, in some ancient accounts the very mention of mathematical irrationality was held to be a 'terrible betrayal of the teaching of Pythagoras' amounting to a 'scandal'. This historical reconstruction (Hippasus' shocking discovery — his 'treason' and the divine retribution for it) supports the ancient tradition.

In recent years, however, convincing arguments have been advanced against the credibility of this tradition. I mention only two of these here.

Reidemeister has pointed out "that there is no mention of any scandal in the various texts of Plato and Aristotle which deal with irrationals, even though its effects must still have been felt at that time".[7] This leads one to think that the story of discovery and betrayal is perhaps only a later legend arising from the ambiguity of the word ἄρρητος. Then the explanation and open mention of mathematical irrationality becomes rather a 'terrible breach of a holy tradition'.[8] In the language of mystical religious writings (particularly those of the Neo-Pythagoreans) ἄρρητα meant the 'carefully guarded secret teachings which were dangerous to the uninitiated'. A non-mathematician might easily

---

[4] K. von Fritz, 'The Discovery of Incommensurability by Hippasus of Metapontum', *Ann. Math.* **46** (1945). See also S. Heller, 'Die Entdeckung der stetigen Teilung durch die Pythagoreer', *Abhandlungen der Deutschen Akademie der Wiss. zu Berlin, Klasse f. Math., Physik und Technik*, 1958; reprinted in *Zur Geschichte der griechischen Mathematik*, pp. 319–54.

[5] Burkert, *Weisheit und Wissenschaft*, p. 435.

[6] Iamblichus, *On the Philosophy of Pythagoras* (ed. Deubner) 52.3–5, cf. Heller, 'Die Entdeckung' (n. 4, above), p. 6.

[7] K. Reidemeister, *Das Exakte Denken der Griechen*, Hamburg 1949, p. 30.

[8] Burkert, *Weisheit und Wissenschaft*, p. 437.

have thought that such a secret also existed in the ἄρρητον of mathematics.

In view of these dubious parts of the tradition, Burkert comes to the conclusion[9] that "The only certainty about the discovery of irrationality is that Theodorus of Cyrene proved that $\sqrt{n}$ (for $n = 3, \ldots, 17$ and not a perfect square) is irrational. Furthermore, the irrationality of $\sqrt{2}$ was known earlier."

Therefore, in this view, a "new epoch" in the history of the theory begins with Theodorus. Whether the square or the dodecahedron first led to the discovery of irrationals, subsequent development of the theory is linked to the name of Theodorus. It had to be he who proved further cases of irrationality for numbers between $\sqrt{3}$ and $\sqrt{17}$. This remarkable dating of events is due to Vogt, who attempted more than fifty years ago to reconstruct the early history of the theory in the following three stages.[10]

(1) The younger Pythagoreans discovered and proved (before 410 B.C.) that the lengths of the sides and diagonal of a square are incommensurable, and in so doing, they gave approximations to their ratio (διάμετρος ῥητή and διάμετρος ἄρρητος).

(2) Theodorus of Cyrene (around 410–390) posed the inverse problem of squaring, i.e. he discovered the irrationality of square roots in general and proved this by generalizing the methods of the Pythagoreans (σύμμετρον and οὐ σύμμετρον; ῥητόν and ἄρρητον).

(3) Theaetetus of Athens (around 390–370) laid the foundations of a general theory of quadratic irrationals, and classified them (ῥητὸν μήκει and ῥητὸν δυνάμει or ἄλογον: μέση, ἐκ δύο ὀνομάτων, ἀποτομή).

As one can see, special significance attaches to the names of Theodorus and Theaetetus in Vogt's outline. Theodorus is supposed to have discovered 'the irrationality of square roots in general,' and Theaetetus to have 'laid the foundations of a general theory of quadratic irrationals'. (No one would claim, however, that these words are sufficient to distinguish clearly between the contributions of Theodorus and

---

[9] Ibid., p. 439.
[10] H. Vogt in *Bibliotheca Mathematica, Zschr. f. Gesch. d. math. Wiss.*, **3** Series, 10 (1909/10) 97–155 and **14** (1914/15) 9–29. See also C. Thaer, 'Antike Mathematik, 1906–1930' in C. Bursian, *Jahresberichte über die Fortschritte der klassischen Altertumswissenschaft*, Jahrgang 1943, Vol 283, Leipzig.

Theaetetus.) This reconstruction of the history of irrationals is still found today in textbooks on Greek mathematics. Tannery[11] argued for it even earlier, and it was doubtless suggested by Plato's dialogue *Theaetetus* or rather by the mathematical part of this dialogue (147D–148B). As Vogt wrote,[12] "The mathematical part of the *Theaetetus* is the birth certificate of irrationals, drawn up by a contemporary."

This interpretation of the text led to even more far-reaching conclusions in later research, especially where Theaetetus himself was concerned, for this same passage of Plato is the most important source of information about the mathematical achievements of the young Theaetetus. However, the so-called 'Theaetetus problem' is not our main concern here.[12a] The passage has to be examined from an entirely different viewpoint to see first whether it really supports the idea that "Theodorus was the first to pose the inverse problem of squaring", or that "he discovered the irrationality of square roots in general", and then whether it allows a different reconstruction of the early history of the theory.

## 1.2 THE CONCEPT OF 'DYNAMIS'

In this next section we will inspect more closely the text itself, together with a translation. But since my translation, as well as my interpretation of the text, is based primarily on the definition of a single word, I want to preface the translation with some remarks about this definition in the hope that these will make it easier to understand.

The mathematical term $\delta\acute{v}\nu\alpha\mu\iota\varsigma$ occurs more than once in the passage, and the corresponding verb $\delta\acute{v}\nu\alpha\sigma\vartheta\alpha\iota$ occurs there as well.[13] It is my opinion that unless one understands this term correctly, one cannot give a satisfactory interpretation of this passage from Plato, or recon-

---

[11] 'Sur la langue mathématique de Platon', *Mém. Scient.* **2**, 91–104.

[12] *Bibliotheca Mathematica*, **10** (1909/10) 131.

[12a] For the so-called 'Theaetetus problem' cf. my paper, 'Theaitetos und das Problem der Irrationalität in der griechischen Mathematikgeschichte", *Acta ant. Acad. Sci. hung.* (Budapest) **14** (1966) 303–58.

[13] The verb $\delta\acute{v}\nu\alpha\sigma\vartheta\alpha\iota$ occurs again in the Greek text this time in its *non-technical, everyday sense: $\acute{o}$ $\mathring{\alpha}\varrho\iota\vartheta\mu\grave{o}\varsigma$ $\delta\upsilon\nu\acute{\alpha}\mu\varepsilon\nu o\varsigma$ $\mathring{\iota}\sigma o\varsigma$ $\mathring{\iota}\sigma\acute{\alpha}\varkappa\iota\varsigma$ $\gamma\acute{\iota}\gamma\nu\varepsilon\sigma\vartheta\alpha\iota$, 'the number which *can* be obtained as the product of two equal factors'.

struct the development of the theory of irrationals. It seems remarkable that so little attention was paid to it in earlier research. I was finally able to establish the following facts about this term.[14]

δύναμις is frequently translated as 'power,' but this translation is misleading – in fact it is downright *false*, for Greek mathematicians did not as yet have a general notion of 'power'.[15] The Platonic term αὔξη[16] comes close to our concept of 'power' but does not correspond to it, since there was only a second and third αὔξη in Greek (see κατὰ τρίτην αὔξην in Plato).

Furthermore, such later terms as δυναμοδύναμις (a⁴), δυναμόκυβος (a⁵) and κυβόκυβος (a⁶) which occur in Diophantus,[17] show that δύναμις cannot mean 'power'. Since κύβος undoubtedly means 'cubic', the word δύναμις can only mean 'square' in mathematical contexts. (This applies to Diophantus' usage of the word as well.)

More important than just establishing that δύναμις (in its technical sense) means square is to explain how this usage comes about, i.e. to show how a word which meant something entirely different (force, capacity or power) in everyday language, came to have this unusual meaning in the specialized language of mathematics.

It is not just the noun δύναμις, but also the verb δύνασθαι which has a special meaning in mathematics. The latter means 'to have the value' 'to be worth' or 'to amount to', and is used only in connection with

---

[14] See Á. Szabó, 'Der mathematische Begriff δύναμις und das sog. geometrische Mittel' *Maia*, N. S. 15 (1963) pp. 219–56. I am indebted to Professor François Lasserre for bringing to my attention the paper by Baerthlein in *Rhein. Mus.* 108 (1965) 35ff. There is much useful philological material in this work, nonetheless, its author fails to recognize that δύναμις as a mathematical term ('value of the square' or 'square') and Aristotle's δύναμις (as opposed to *energeia* or *entelecheia*) have absolutely nothing to do with each other.

[15] For example Heiberg in his Latin translation of the *Elementa* (Lipsiae 1888–1916) renders Definition X. 2 as 'Rectae potentia commensurabiles etc.' (εὐθεῖαι δυνάμει σύμμετροι); also Timpanaro-Cardini *(Pitagorici*, Vol. 2, Florence, 1962, p. 77) translates a part of the Platonic text in question as 'Teodoro qui, ci aveva construito delle figure relative alle *potenze* ecc.'

[16] *Republic*, Book IX, 587d.

[17] Cf. T. L. Heath, *A History of Greek Mathematics*, Oxford 1921, Vol. 2, pp. 457–8.

*squaring.*[18] It should be noticed that in mathematics the word δύνασθαι
was used originally about *transformations of surfaces in the plane.*
A *rectangle* was transformed into a *square* of the same area and the side
of the resulting square was described as follows — this *straight line
(εὐθεῖα) when squared* [i.e. if one were to construct a square on the
straight line in question] *has the same value ( ἴσον δύναται ) as the previous
rectangle (τῷ περιεχομένῳ ὑπό...).*[19] It should also be pointed out that
this verb was only used about the transformation of a *rectangle* into a
square of the same area and in this context it indicated the value of
that area. It would have been impossible for example, to express the
value of the square of a triangle using this verb. There is no evidence of
any such usage in the language of ancient mathematics. (Moreover, a
triangle was always transformed first into a rectangle of the same area,
and then the rectangle was transformed into a square. This can be seen
from propositions I.42 and II.14 of the *Elements.* So it seems that
δύνασθαι was only used in the latter step, when it expressed *the value of
the square* of the rectangle.) Therefore, this particular transformation
must have been considered especially important for some reason. There
is no other way to account for the fact that δύνασθαι and δύναμις were
used to express 'sameness of value' only in the *special case of the rectangle and square.*

In my opinion the mathematical expression δύνασθαι must have been
derived from the language of finance. Just as when converting one kind
of money into another, the *value* was expressed by this word,[20] so in
geometry, when transforming one figure into another, the same word
expressed the *value of a rectangle when squared.* Furthermore, just as

[18] Cf. F. Rudio, *Der Bericht des Simplicius über die Quadraturen des Antiphon
und Hippokrates*, Leipzig 1907, p. 139 (see the index under δύνασθαι); also T.
L. Heath, (ed.), *Archimedes*, Dover reprint, p. clxi.

[19] Cf. T. L. Heath, (ed.) *Archimedes*, Dover reprint, p. clxi: "The verb δύνασθαι
(with or without ἴσον) has the sense of being and, when δύνασθαι is used alone,
it is followed by δυνάμει ἴσα the accusative; thus *the square (on a straight line)
is equal to the rectangle contained by* . . . is: *(εὐθεῖα) ἴσον δύναται τῷ περιεχομένῳ
ὑπό* . . .

[20] Compare this with Xenophon (*Anabasis* 1.5. 6): "the σίγλος [an Asiatic
coin, the Hebrew shekel] *amounts to, is worth* or *has the value of (δύναται)* 7½
Attic obols"; or Demostenes (34.23) "the stater of Cyzicus *had a value (ἐδύνατο)*
of 28 drachmas".

the term δύναμις meant 'value' in general[21] in the language of finance, so in geometry it came to mean 'the value of the square of a rectangle', then 'the value of the square' and finally just 'square'.

It should be emphasized, however, that these meanings apply only to mathematical language. Neither δύνασθαι nor δύναμις could ever take on these meanings in everyday Greek. In this instance the special usage of mathematicians did not lead to a corresponding everyday usage. (A square was never called δύναμις in everyday Greek.) It is important to point this out because we will see in Part 2 of this book that the expression ἀνὰ λόγον (or ἀνάλογον) is also of mathematical origin. It was coined for a particular mathematical purpose and meant 'equal (when taken) in *logoi*' or 'in the same ratio'. Unlike δύναμις it had originally no meaning outside mathematics, so it could subsequently be taken over into everyday language with the extended meaning of 'similar'. In geometry 'similarity of rectilinear figures' was defined as ἀναλογία (i.e. 'having sides in the same ratio to each other'), and when 'similarity' came to be expressed by ἀναλογία outside mathematics as well, the technical term was being used to express a very common notion, well known in everyday life. δύναμις, on the other hand, never acquired an everyday meaning of this kind, because 'the transformation of a rectangle into a square of the same area' never figured as prominently in everyday experience (if it occurred there at all) as it did in the pure science of geometry. So let me emphasize once more that the mathematical concept δύναμις and its derivation from the language of finance *have nothing to do with everyday life*. δύναμις and δύνασθαι had *the meanings explained above, only in theoretical geometry*.

Before turning to the text itself *(Theaetetus, 147c–148b)*, I would like to make one further remark. In the part of the *Theaetetus* which we are going to discuss, δύναμις is sometimes translated as 'the side of a square' or as 'square root'. I have shown in a previous work[22] that this translation is completely unfounded. The expression δύναμις in mathematics never meant anything except '*value of a square*' or '*square*'. That this word has been translated as 'side of a square' or 'square-root', only goes

---

[21] For example in Plutarch (*Lycurgus* 9 and *Solon* 15): "he gave only *a small value* to money": δύναμιν ὀλίγην τῷ νομίσματι ἔδωκεν.

[22] See n. 14 above.

to show how superficially and arbitrarily the mathematical part of the *Theaetetus* has been treated in the literature.[23]

This is enough about the word δύναμις for now. I will have to return to it when I come to give a detailed interpretation of the text.

### 1.3 THE MATHEMATIC PART OF THE 'THEAETETUS'

*Theaetetus*

Ῥᾴδιον, ὦ Σώκρατες, νῦν γε οὕτω φαίνεται· ἀτὰρ κινδυνεύεις ἐρωτᾶν οἷον καὶ αὐτοῖς ἡμῖν ἔναγχος εἰσῆλθε διαλεγομένοις, ἐμοί τε καὶ τῷ σῷ ὁμωνύμῳ τούτῳ Σωκράτει.

It seems easy when you put it like that, Socrates. For you are asking something like what your namesake, the other Socrates, and myself were discussing here recently.

*Socrates*

Τὸ ποῖον δή, ὦ Θεαίτητε;

What was that, Theaetetus?

*Theaetetus*

Περὶ δυνάμεών τι ἡμῖν Θεόδωρος ὅδε ἔγραφε, τῆς τε τρίποδος πέρι καὶ πεντέποδος [ἀποφαίνων] ὅτι μήκει οὐ σύμμετροι τῇ ποδιαίᾳ, καὶ οὕτω κατὰ μίαν ἑκάστην προαιρούμενος μέχρι τῆς ἑπτακαιδεκάποδος· ἐν δὲ ταύτῃ πως ἐνέσχετο· ἡμῖν οὖν εἰσῆλθέ τι τοιοῦτον, ἐπειδὴ ἄπειροι τὸ πλῆθος αἱ δυνάμεις ἐφαίνοντο, πειραθῆναι συλλαβεῖν εἰς ἕν, ὅτῳ πάσας ταύτας προσαγορεύσομεν τὰς δυνάμεις.

Theodorus was drawing some squares *(περὶ δυνάμεων)* to demonstrate to us that *the lengths of the sides* of those having an area of three or five square feet *are not commensurable* (μήκει οὐ σύμμετροι) with the length of the sides of a unit square. He discussed the case of each square *(ἑκάστην* scil. δύναμιν) individually until he reached the case of the square with area seventeen square feet. Then for some reason he stopped. Now it occurred to us that since

---

[23] I am pleased to report that van der Waerden has accepted this suggestion of mine in *Zur Geschichte der griechischen Mathematik*, p. 254.

there are infinitely many squares (δυνάμεις), we should try to group these under one name by which we could refer to all squared rectangles (δυνάμεις).

*Socrates*

'Η καὶ ηὗρετέ τι τοιοῦτον;

And did you manage to do this?

*Theaetetus*

Ἔμοιγε δοκοῦμεν· σκόπει δὲ καὶ σύ.

Yes, I think so, but see for yourself.

*Socrates*

λέγε.

Go on.

*Theaetetus*

Τὸν ἀριθμὸν πάντα δίχα διελάβομεν· τὸν μὲν δυνάμενον ἴσον ἰσάκις γίγνεσθαι, τῷ τετραγώνῳ τὸ σχῆμα ἀπεικάσαντες τετράγωνόν τε καὶ ἰσόπλευρον προσείπομεν.

We divided all numbers into two groups. Those which can be obtained as the product of two equal factors, we compared in shape to a square and called them (equilateral) *square numbers*.[24]

*Socrates*

καὶ εὖ γε.

Very good.

*Theaetetus*

---

[24] Since numbers wich can be decomposed into 'two equal factors' (ἰσάκις ἴσος) are simply called τετράγωνος ἀριθμός in the *Elements* (Definition VII. 9), I used to think that Plato's term 'equilateral square number' (τετράγωνός τε καὶ ἰσόπλευρος) was pleonastic. Prof. H. Cherniss, however, has made me realise that this conclusion is somewhat hasty. The literal meaning of τετράγωνον is 'quadrangular'. The restricted meaning 'square', which became accepted after Euclid, was in reality not as common, even in post-Platonic times, as I had tacitly assumed. Cherniss has correctly pointed out that evidence for this can be found in *Timaeus* 55b7; Aristotle, *Metaphysics* 1054b2; Heron, *Definitions* 100; Proclus; *Commentary on Euclid* (ed. Friedlein), 166. 10–1.

*Τὸν τοίνυν μεταξὺ τούτου, ὧν καὶ τὰ*
*τρία καὶ τὰ πέντε καὶ πᾶς ὃς ἀδύ-*
*νατος ἴσος ἰσάκις γενέσθαι, ἀλλ᾽ ἢ*
*πλείων ἐλαττονάκις ἢ ἐλάττων πλεο-*
*νάκις γίγνεται, μείζων δὲ καὶ ἐλάτ-*
*των ἀεὶ πλευρὰ αὐτὸν περιλαμβάνει,*
*τῷ προμήκει αὖ σχήματι ἀπεικάσαν-*
*τες προμήκη ἀριθμὸν ἐκαλέσαμεν.*

On the other hand, those numbers which lie between the previous ones (e.g. three, five and
in general any number which cannot be obtained as the product
of two equal factors but only as
the product of a greater and a
smaller or a smaller and a greater,
and so is always enclosed by a
greater and a lesser side)[25] we
compared to rectangles and called
them rectangular numbers.

### Socrates

*Κάλλιστα. ἀλλὰ τί τὸ μετὰ τοῦτο;*

Excellent, but what comes next?

### Theaetetus

*ὅσαι μὲν γραμμαὶ τὸν ἰσόπλευρον*
*καὶ ἐπίπεδον ἀριθμὸν τετραγωνί-*
*ζουσι, μῆκος ὡρισάμεθα, ὅσαι δὲ*

Now those straight lines whose
*square is an (equilateral) square
number,*[26] we denoted by *μῆκος*.

---

[25] In this context 'side' (*πλευρά*) means 'factor' as well, just as it does in Definition VII. 16 of the *Elements*. From this we can see that a product of two factors
was in general viewed as a *rectangle*. Moreover the term 'rectangular number'
(*ἑτερομήκης* or *προμήκης* in our text) does not occur in Euclid. He only uses
the concept of 'plane number' (*ἐπίπεδος ἀριθμός*) which can refer to *square
numbers* as well as *rectangular ones*. (Let me add another important point here.
I have concluded that the concepts 'plane number' and 'similar plane number'
must be old and in any event pre-Platonic, because Plato frequently uses *πλευρά*
to mean both 'the *side* of a parallelogram' and 'factor'. Now these concepts
('plane number' and 'similar plane number') are used particularly in Book VIII
of the *Elements* and this book is attributed to Archytas by van der Waerden
('Die Arithmetik der Pythagoreer', *Math. Ann.* **120** (1947–49) 127–53). Apart
from the fact that I disagree with him on this for other reasons, it seems to me
that the concepts 'similar plane and solid numbers' could not possibly have
been first introduced by Archytas.)

[26] The words in italics, although unambiguous, are somewhat brief and care-
lessly written. Their meaning is as follows: 'Those straight lines which when
squared (*τετραγωνίζουσιν*) form a figure corresponding to a square number

τὸν ἑτερομήκη, δυνάμεις, ὡς μήκει μὲν οὐ συμμέτρους ἐκείναις, τοῖς δ'ἐπιπέδοις ἃ δύνανται. Καὶ περὶ τὰ στερεὰ ἄλλο τοιοῦτον.

Those which yield a *rectangular number when squared*[27] we denoted as δυνάμεις, for although the latter are not commensurable with the former in length, nonetheless *the areas (ἐπιπέδοις) which they enclose when squared (ἃ δύνανται),*[28] are commensurable. We also tried to do something similar for solids.

*Socrates*

Ἀριστά γ' ἀνθρώπων, ὦ παῖδες· ὥστε μοι δοκεῖ Θεόδωρος οὐκ ἔνοχος τοῖς ψευδομαρτυρίοις ἔσεσθαι.

Well done, lads! I think that Theodorus was right (to praise you . . .) etc.

We will begin our interpretation of the text by showing that the mathematical portion of the *Theaetetus* and the rest of the dialogue form a coherent whole.

The above text begins with the young Theaetetus asserting that after the example just given, Socrates' question is *easy;* he is asking "something like what your namesake . . . and myself were discussing here recently". These words are the tenuous connection between the mathematical part and the conversation which precedes it. However, the mathematical part cannot be properly understood without taking into account what precedes it.

Socrates began by asking the young man 'What is knowledge?' (146c), and Theaetetus responded first by *enumerating* different kinds of knowledge. But Socrates was not satisfied with this answer. Instead of an *enumeration*, he demanded a comprehensive and general definition

---

we denoted by μῆκος . . .'. Furthermore the term for 'square number' is redundant here. For ἰσόπλευρος καὶ ἐπίπεδος ἀριθμός means literally 'equilateral and *plane* number'.

[27] Again the meaning of the Greek text requires that the verb τετραγωνίζουσιν be added. So the sentence is best translated as 'those straight lines on the other hand, which yield a rectangular number when squared (τετραγωνίζουσιν), we denoted by δυνάμεις . . .'.

[28] We should not forget that in mathematics the precise meaning of δύνασθαι was 'to have the value when squared' or 'to amount to when squared'.

of knowledge and immediately gave an illustration of what he wanted by means of an example. If one were asked 'What is clay?', then one could not answer by *enumerating* the kinds of clay used by each particular craftsman. One should rather try to give a general and comprehensive definition of clay (147aff). The young man replies to this example by saying that he now understands; Socrates' question is quite easy because it resembles something which he was discussing recently with his friend.

So we can see that the central point of the preceding conversation is to contrast 'enumeration of particular instances' with a 'comprehensive description (definition) of the same phenomenon'. (For this reason one should reject the view found among some historians that Plato is giving an example of a mathematical investigation by the young Theaetetus in the dialogue. They would say that the example is rather farfetched; it is supposed to serve as an introduction to a philosophical discussion, but is not well suited to the purpose etc., etc.)

According to the text, "Theodorus was drawing some squares . . . having an area of three of five square feet . . . He discussed . . . each square . . . until he reached . . . a square with area seventeen square feet." So we can see that in this section 'the enumeration of individual instances' consists of squares *(δυνάμεις)*, just as previously Theaetetus had enumerated different kinds of knowledge and Socrates had ˙enumerated the kinds of clay used by different craftsmen.

Before investigating more fully what Theodorus might have intended by enumerating *squares*, I must make a further remark about the term *δύναμις*. Thus far I have given its precise definition (as a mathematical term), but have not indicated how this bears on the history of mathematics. I will proceed to do this now, since in my opinion it helps us to understand Plato's text better.

1.4 THE USAGE AND CHRONOLOGY OF 'DYNAMIS'

The term *δύναμις* is used in the text without any explanation, as though it were completely familiar and could be understood unambiguously by the reader. From this we can infer that the concept *δύναμις* ('the value of the square of a rectangle' or 'square') cannot have originated at the time of Plato. Indeed one can cite other passages from

Plato which attest to the same fact. Of these I shall mention here only *Politicus* 266 a5–b7, a passage which will be discussed in detail later from another point of view. In it the concept δύναμις is used in a rather complicated play on words, which would hardly have been possible if Plato had not been able to assume that the word (and its technical meaning) was well known, and familiar to the educated reader at least.

Exactly when in pre-Platonic times the word *dynamis* was coined, we do not know at present. There is no text earlier than the writings of Plato in which it occurs in its technical sense. The most we can do is to appeal to Eudemus' account (as transmitted by Simplicius) of the squaring of lines by Hippocrates of Chios. It begins as follows:[29] "He prepared a foundation for himself and erected on it first a proposition which served his purpose, namely that similar segments of a circle have the same ratio to each other as the *squares of their bases*" (καὶ αἱ βάσεις αὐτῶν δυνάμει).

This question may be considered as evidence for the pre-Platonic use of the mathematical term *dynamis*. I should mention in passing that when I established the precise meanings of δύναμις and δύνασθαι in mathematics, I had in mind (apart from Archimedes) mainly the usage of Simplicius–Eudemus (or rather, of Hippocrates).

There is, of course, abundant evidence of the same sort for these two expressions being used in the senses explained above in the whole of later mathematical literature. An interesting example of this is the formulation given by Athenaeus of Pythagoras' theorem.[30] τριγώνου ὀρθογωνίου ἡ τὴν ὀρθὴν γωνίαν ὑποτείνουσα ἴσον δύναται ταῖς περιεχούσαις. "The square of the hypotenuse of a right-angled triangle *is equal to* (ἴσον δύναται) the sum of the (squares of) the other two sides."

It might be concluded from this that δύνασθαι and δύναμις were common technical terms, used constantly in Greek mathematics since before Plato. But this assertion becomes questionable when the *Elements* are considered. δύνασθαι in the sense of '*to have the value when squared*', does not occur in Euclid. He only uses the derived terms δυνάμει σύμμετροι or ἀσύμμετροι (cf. Book X, definition 2) which will

---

[29] Cf. O. Becker, *Grundlagen der Mathematik in geschichtlicher Entwicklung*, Freiburg–München 1954, p. 29; also his article in *Quellen und Studien zur Geschichte der Math.* etc. B, **3** (1936) 417.

[30] Athenaeus X 418f.

have to be considered later. Furthermore a *square* is never denoted by δύναμις in his writings.

This fact is noticeable because one can easily find a proposition of Euclid in which the term δύναμις would at least have suggested itself. I am referring to Proposition 9, Book X. The second part of this proposition deals with "squares on straight lines incommensurable in length" *(τὰ ἀπὸ τῶν μήκει ἀσυμμέτρων εὐθειῶν τετράγωνα)* which "do not have the ratio to each other, of a square number to a square number". In the *Theaetetus* this kind of τετράγωνα is called δύναμις. According to Theaetetus, Theodorus sketched δυνάμεις of area three, five, etc., square feet, for his pupils, i.e. he constructed τετράγωνα which did not have the ratio to each other of a square number to a square number, and then showed that the sides of these squares were incommensurable in length with the side of the unit square. So undoubtedly Euclid could have used δυνάμεις to denote these squares, but he never does. In Euclid, a *square* is always called τετράγωνον; I believe that this is because Euclid always strove for the greatest possible generality. The terms ἑτερόμηκες, ῥόμβος and ῥομβοειδές (cf. Book I, definition 22) also do not occur in his propositions; instead he always speaks of *parallelograms*. So just as ἑτερόμηκες, ῥόμβος and ῥομβοειδές are special cases of the parallelogram, δύναμις (according to the previous definition) is a special case of τετράγωνον. Hence all *squares* are τετράγωνα, but those which have the same area as certain rectangles, are called δυνάμεις.

## 1.5 'TETRAGONISMOS'

These speculations about τετράγωνον and δύναμις also suggest a plausible conjecture about the *relative chronology* of the two concepts, namely that δύναμις (the value of the square of a rectangle) must be of later origin than τετράγωνον σχῆμα, the other usual term for *square* in Greek geometry. To coin a phrase like τετράγωνον σχῆμα, one need only to have distinguished between the most important kinds of parallelograms (rectangles, squares, etc.). On the other hand, the concept of δύναμις presupposes that one already knows how to transform a rectangle into a square of the same area. In other words, this piece of knowledge must have originated at the same time as the mathematical concept *dynamis*.

This conclusion follows immediately from the simple facts established about the meaning of *dynamis*, and it leads to still further observations which in my opinion are of paramount importance not only for understanding our text from Plato, but also for the whole history of early Greek mathematics.

The question we are interested in is how the 'transforming of a rectangle into a square of the same area' was expressed in Greek. The correct technical name for it was τετραγωνίζειν, or the noun τετραγωνισμός. It is no accident that *this other important term* (*τετρα-γωνίζειν*) *occurs right by δυνάμεις in the text which we are discussing*, since the *dynamis* was obtained by means of *tetragonismos*. The two concepts are very closely linked. Furthermore, there is an important statement about *tetragonismos* by Aristotle which enables us to be more precise about the *relative chronology* suggested above. Before discussing it, however, I want to make a brief remark about English usage.

The Greek words τετραγωνίζειν and τετραγωνισμός are usually translated by 'squaring'. However, this translation can easily be misleading. 'Squaring' can sometimes mean 'raising a number to the second power' or 'constructing a square on a given line segment', whereas the Greeks, even when they were not just talking about τετραγωνισμός τοῦ κύκλου always understood by τετραγωνισμός the 'transforming of a *rectangle* into a square of the same area'. (Of course *tetragonismos* could also lead from other rectilinear figures to a square of the same area, by way of *a rectangle*.) Hence I have consistently tried to avoid the use of the word 'squaring' in this work.

In the *Metaphysics* Aristotle says the following about *tetragonismos*:[31] "What is τετραγονίζειν? It is the finding of a mean proportional" (*τί ἐστι τετραγωνίζειν ... μέσης εὕρεσις*). The meaning of this assertion and its context are explained as follows in the accompanying commentary by Ross.[32] "The definition, *the squaring of a rectangle is the finding of a geometrical mean between the sides*, is an abbreviated form of the syllogism: *a rectangle can be squared because a mean can be found between its sides*."

[31] *Metaphysics* 996b 18–21.
[32] *Aristotle's Metaphysics*, Oxford 1924, Vol. 1, p. 229.

Aristotle himself also explains the same idea in more detail, in another work.[33] "Definitions are usually like conclusions. For example, *what is tetragonismos? The construction of a square equal in area to a rectangle (ἑτερόμηκες).* This kind of definition is a conclusion. But he who maintains that *tetragonismos* is the finding of a mean proportional, also specifies the rationale behind it." *(ὁ δὲ λέγων ὅτι ἐστὶν ὁ τετραγωνισμὸς μέσης εὕρεσις, τοῦ πράγματος λέγει τὸ αἴτιον).*

These statements by Aristotle are important because they show us that his contemporaries (if they were well versed in the mathematics of the time) must have been aware of the impossibility of transforming a rectangle into a square of the same area without constructing a mean proportional between two given line segments.[34] So the *dynamis* (the value of the square of a rectangle) is obtained by constructing a mean proportional between two sides of the rectangle, for this is just what *tetragonismos* (the transformation of a rectangle into a square of the same area) comprises. Thus my previous conjecture to the effect that *dynamis* and *tetragonismos* originated at the same time inevitably leads to the conclusion that the creation of the concept of *dynamis* must have coincided with the discovery of how to *construct a mean proportional between any two line segments.*

To understand why this entirely new concept evolved at a particular time in the development of Greek mathematics, one has to look more closely at the problem of the mean proportional.

### 1.6 THE MEAN PROPORTIONAL

Euclid discusses the construction of a mean proportional between any two line segments in Proposition 13, Book VI (of the *Elements*). It is nowhere indicated in this book that the construction finds its most important application in *tetragonismos*. This in itself is not particularly surprising. More interesting is the fact that the same construction is discussed somewhere else in the *Elements*, namely in Proposition 14 of Book II. The problem with which this latter proposition deals is the

[33] *De Anima* II. 2. 413a 13–20.

[34] On this see chapter 1.6, 'The Mean Proportional', as well as the appendix to this book.

most general form of *tetragonismos*. It is 'to obtain from a given rectilinear figure, a square having the same area'. Since this is accomplished
by first constructing a rectangle of the same area and then finding a
mean proportional between two of its sides, one would expect the construction of a mean proportional to be fully discussed in this proposition. In fact the construction of the desired line segment (the side of the
square in question) in Proposition II.14 is basically the same as the
construction of a mean proportional between any two given line segments in Proposition VI.13.[34a] Even more noteworthy is the fact that
no use is made of proportions in Proposition II.14. It is never stated
that the line segment which provides a solution to the problem, is a
mean proportional. Instead Euclid gives an entirely different account
of why it can be used to construct the required square. Indeed the proof
of Proposition II.14 gives the impression that knowledge of the mean
proportional is perhaps unnecessary for carrying out *tetragonismos*.

Heiberg has discussed both these propositions together with the
passages from Aristotle quoted in the previous section, and has come
to what is clearly the correct conclusion.[35]

He argues that the passages from Aristotle unequivocally attest to
the fact that the transformation of a rectangle into a square of the
same area was *originally* accomplished by constructing a mean proportional. This original method is given in Proposition VI.13. The alternative account offered in Proposition II.14 must be the original in a *later
disguise*. It seems that Euclid, or whichever one of his predecessors gave
the propositions in Book II their final form, wanted to avoid using the
theory of proportions. Hence a new proof that a rectangle can be
transformed into a square of the same area was supplied, which did not
mention that the side of the square obtained was a mean proportional
between the two sides of the original rectangle.

[34a] Correction: compare this assertion with the appendix to this book. The
above discussion was written before I came into possession of the information
contained therein.

[35] *Mathematisches bei Aristoteles*, Abhandlungen zur Geschichte der math.
Wissenschaften, No. 18, Leipzig 1904, p. 20. See also T. L. Heath, *Mathematics
in Aristotle*, Oxford 1949, pp. 191–3. One should not forget that the construction
of the mean proportional (i.e. Proposition VI. 13) must already have been
known to Hippocrates of Chios, cf. n. 37 below.

Let me add that in my opinion Heiberg was quite right in defending the view that Proposition VI.13 antedates Proposition II.14. Certainly no one as yet has refuted his arguments for this view, or even cast doubt on them. Furthermore there are other arguments which can be advanced in support of his position. For as has been rightly observed: "The geometric construction of a mean proportional [i.e. Proposition VI.13] was completely familiar to Archytas and the Pythagoreans, and it must already have been known to Hippocrates . . . " (on this topic, see the references cited in n. 37 below). Consequently the μέσης εὕρεσις of which Aristotle speaks (in *Metaphysics* 996b 18–21 and *De Anima* II2, 413a 13–20), must also have been known at the time of Hippocrates. I do not see why these passages from Aristotle cannot 'be connected' with Greek mathematics of around 400 B.C. (perhaps only because of an obsolete view held by Tannery's school, according to which the theory of irrationals first became current shortly before 400 B.C.).

At the moment our interest in the problem of the mean proportional is confined to investigating the extent to which the discovery of a method for constructing a mean proportional between any two line segments must of necessity have led to the new mathematical concept of *dynamis*. My conjecture is that the creation of the concept *dynamis* must be related to this discovery.

First let us remember that the mean proportional was a much discussed problem in Greek arithmetic as well. For example, in Book VIII of the *Elements*, Proposition 11 states that there is exactly one mean proportional number between any two square numbers. The next proposition (VIII.12) establishes that there are two mean proportional numbers between two cube numbers. Of greater interest to us are the two propositions concerning the existence of mean proportional numbers. These are: *Proposition VIII.18* "There exists a mean proportional number between any two *similar plane numbers*" (δύο ὁμοίων ἐπιπέδων ἀριθμῶν) and *Proposition VIII.20* "If there exists a mean proportional number between two numbers, then these two are *similar plane numbers*".

To understand these two propositions, one must have grasped the interesting concepts of *plane number* and *similar plane number* which occur in early Greek mathematics. The definition given in the *Elements* is as follows: *Definition VII.16* "If two numbers are multiplied together to form a third, then the resulting number is called a *plane number (ἐπίπεδος ἀριϑμός)*, and the other two are said to be its *sides* (factors, *πλευραί)"*.

Any number which can be written as the product of two factors and hence can be represented geometrically by a parallelogram (usually a rectangle), falls under this definition. For example 6 is a plane number which can be represented geometrically by a rectangle with sides *(pleurai)* of length 1 and 6 or of length 2 and 3, since $2 \cdot 3 = 6 = 1 \cdot 6$. (Of course square numbers can also be represented in rectangular form. For $4 = 1 \cdot 4$ as well as $2 \cdot 2$. Similarly $9 = 1 \cdot 9$ and $9 = 3 \cdot 3$. Hence the squares $2 \cdot 2$ and $3 \cdot 3$ can serve as *dynameis*, i.e. as the values which the rectangles with sides 1 and 4, and sides 1 and 9, respectively, take when squared.) One need only think how common the term *πλευραί* (meaning factor) is in mathematics, to realise at once that the concept of plane number must be very old.

Euclid's other difinition is still more interesting. I am referring to *Definition VII.22 "Similar plane numbers (ὅμοιοι ἐπίπεδοι ἀριϑμοί)* are those whose sides *(πλευραί)* are in the same ratio."[36]

All square numbers (1, 4, 9 . . . , since they can always be written as the product of two equal factors and so have sides in the same ratio, i. e. $1:1 = 2:2 = 3:3 . . .$) are similar plane numbers in this definition, as are such pairs of numbers as 2 and 8 and 3 and 12. Each member of the pair can be written as a product of two factors (1 · 2, 2 · 4, and 1 · 3, 2 · 6 respectively) and the ratios of these factors are the same (i. e. $1:2 = 2:4$ and $1:3 = 2:6$).[36a] Clearly they are called *similar plane numbers* because numbers of this kind can be pictured as *similar rectangles*. Carrying this out for the example above, we get Fig. 1.

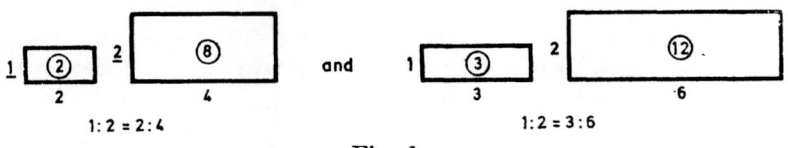

1:2 = 2:4          1:2 = 3:6

Fig. 1

[36] I have preferred to use 'in the same ratio' as a translation of *ἀνάλογον*, instead of the more usual 'proportional'. For an explanation of this mathematical term, see pp. 148—57 ff.

[36a] On the notion of 'similar plane number' cf. also T. L. Heath, *Euclid's Elements*, Vol. 2, p. 193: "Theon of Smyrna remarks (p. 36. 12) that, among plane numbers, *all* squares are similar, while of *ἑτερομήκεις* those are similar *whose sides, that is, the numbers containing them, are proportional.*"

4*

Proposition VIII.18, which was quoted above, states that between any two similar plane numbers there always exists a mean proportional number. For example, there exists a mean proportional number between *three* and *twelve*. In this case the required number is *six*, since $3:6 = 6:12$. It is worthwhile using this simple example to illustrate the difference between ancient and modern ways of thinking. Today the mean proportional would be obtained by multiplying the two numbers together and taking the square root of their product ($6 = \sqrt{3 \cdot 12}$). In antiquity, however, the numbers were first divided into the factors which had led them to be classified as *similar plane numbers* in the first place (here $3 = 1 \cdot 3$ and $12 = 2 \cdot 6$; of course if 12 were divided into factors other than $2 \cdot 6$, it would *not* be similar to 3). Then the mean proportional was obtained by multiplying two different 'sides' (factors) which were not similar together ($3 \cdot 2 = 6 = 1 \cdot 6$).

The above example could also be understood in the following way. We are given a *number-rectangle* of sides 3 and 12 which is to be transformed into a *square* of the same area. The feasibility of this depends on whether the two sides of the rectangle are similar plane numbers.

If the two sides of the rectangle can be decomposed into factors in such a way that the ratios of the factors are the same (i. e. if a smaller rectangle can be constructed from each side of the original one, and these smaller ones are similar to each other), then the problem can be solved. Proposition VIII.20 above is the converse of Proposition VIII.18. It states that "if a mean proportional number exists between two numbers, then these two are *similar plane numbers*". So a mean proportional number can only be found between two similar plane numbers.

It is now easy to see that at the time when Propositions VIII.18 and 20 were known, but Proposition VI.13 (the construction of a mean proportional between any two line segments) was not, an arbitrary rectangle could not be transformed into a square of the same area.[37] This transformation could only be carried out in the case of rectangles whose

---

[37] In my opinion Propositions VIII. 18 and 20 are older than Proposition VI. 13. Concerning the latter, I agree with van der Waerden who says that "The geometrical construction of the mean proportional (Prop. VI. 13) was quite familiar to Archytas and the Pythagoreans. Furthermore, it must already have been known to Hippocrates of Chios." (*Zur Geschichte der griechischen Mathematik*, p. 225, n. 28.) On the other hand, I cannot accept his dating of Book VIII. His view is that "The problems of Book VIII are very closely

sides were similar plane numbers. As we would say today, the transformation could be carried out only if the area of the rectangle was a perfect square. If, however, a rectangle with sides of, say, length one and three say given, it could not be transformed into a square of the same area unless one knew how to construct a mean proportional by geometric methods. The sides can only be decomposed into two factors in one way, namely $1 = 1 \cdot 1$ and $3 = 1 \cdot 3$ and since $1:1 \neq 1:3$, the lengths of the sides are not similar plane numbers.

When subsequently a geometric method of constructing a mean proportional between *any* two line segments was found, it became possible to transform *any* rectangle into a square of the same area. Thenceforth there was no longer any need to worry about whether the rectangle concerned had sides which were similar plane numbers or not. Instead a mean proportional was constructed by geometric methods which used similar right-angle triangles (as in the proof of Proposition VI.13).

Of course, as soon as the possibility of this construction became known, the question as to *what exactly the sides of such a square (i. e. of a square equal in area to a rectangle whose sides were not similar plane numbers)* were, must also have been raised. By Proposition VIII.20, a mean proportional *number* only existed between similar plane numbers. The newly discovered fact that a mean proportional also existed between two numbers which were not similar could only be reconciled with this proposition by introducing the notion of *linear incommensurability*. As the Greeks interpreted it, the mean proportional between two numbers which were not similar plane numbers was not a number.

The question is still left open as to whether (and if so, how) the linear incommensurability of a mean proportional between two such line segments could be rigorously proved. But the above considerations are important because they show that the problem of transforming a rectangle into a square of the same area, which in its most general form

---

linked to the theory of irrationals which according to a well-founded opinion (sic), *did not exist until shortly before 400."* The author of this 'well-founded opinion' is none other than Vogt who uncritically adopted Tannery's views and misinterpreted the mathematical section of the *Theaetetus* as "the birth certificate of irrationals". Furthermore, I remain unconvinced by the arguments which purport to show that Archytas was the author of Book VIII.

is equivalent to the problem of finding a mean proportional between any two line segments, led to the problem of linear incommensurability.

My conjecture is that the discovery of how to construct a mean proportional between any two line segments also *prompted the introduction of the new mathematical concept of 'dynamis'*. Of course I cannot produce any documentary evidence to support this conjecture, because no Greek mathematical texts have survived from the pre-Platonic times during which the concept originated. Nonetheless, I believe that it can be established by a study of the word *dynamis* itself.

The mathematical term *dynamis* denotes by definition an *area* — 'the value of the square of a rectangle'. The problem remains as to why it was found necessary to use this curious new term for a square which had the same area as some rectangle, and also what particular properties of the square led to its receiving this extraordinary name.

I can only think that as long as only those rectangles whose sides were similar plane numbers were transformed into squares of the same area, there was no reason to introduce a new name for the resulting squares. These were perfectly ordinary τετράγωνα σχήματα, since their sides were whole numbers. Matters changed, however, when a new *general* method (the geometrical construction of a mean proportional) was discovered. Any rectangle could now be 'squared', for a mean proportional always existed between any two line segments. Hence, as Aristotle emphasized, *tetragonismos* is equivalent to finding the μέση. In addition the Greeks must have been aware that the squares obtained in this way were frequently such that their sides were not numbers. (A mean proportional number only existed between two similar plane numbers. The mean proportional between two numbers which were not of this kind was itself not a number.) Thus squares having the same area as rectangles, whose sides were not similar plane numbers, must have aroused considerable interest. The lengths of their sides could not be assigned a number, even though their areas could be calculated exactly. This may have occasioned the surprising use of *dynamis* to denote their area.

This amounts to saying that the concept of *dynamis* was introduced at the same time it was discovered that linearly incommensurable line segments exist, whose squares *(dynameis)* can nevertheless be measured.

## 1.7 THE MATHEMATICS LECTURE DELIVERED BY THEODORUS

The above considerations have led us to conjecture that the creation of the mathematical concept *dynamis* is linked to the discovery of linear incommensurability. But this concept is, by all appearences, rather old. In any case, it must antedate Plato, since he uses it naturally and without feeling any need to explain it further.

In this case, however, it cannot be maintained that "the mathematical part of the *Theaetetus* is the birth certificate of irrationals" or that "Theodorus discovered the irrationality of square roots in general" (Vogt). It would perhaps be better to emphasize that Plato never actually says that Theodorus showed his pupils something new.[38] The impression he gives is rather that Theodorus was telling the Athenian youths something which must have been common knowledge amongst the mathematicians of the time.

In fact, we shall soon see that hardly any new mathematical discoveries can be attributed to Theodorus, at least on the basis of this passage from Plato. To prove this we have to return to a detailed interpretation of the text quoted above.

According to Plato, "Theodorus was drawing some squares *(dynameis)* ... having an area of three of five square feet .... . He discussed each square *(ἑκάστην* scil. *δύναμιν)* individually until he reached a square with area seventeen square feet, when for some reason he stopped."

Previous commentators have invariably asked how Theodorus constructed those squares; more particularly, how he constructed their sides, which he wanted his pupils to see here linearly incommensurable. The answer given was either to say that it did not much matter how those sides were constructed,[39] or to adopt Anderhub's interesting attempt at reconstructing the method used.[40] Anderhub's idea was to

---

[38] K. von Fritz, *Ann. Math.* 46 (1945) 244. See also the paper by Wasserstein in *Classical Quarterly*, N. S. 8 (1958) 165–79: "It is at least conceivable that Plato means no more than that Theodorus was demonstrating not a new discovery of his own, but something which though known to professional mathematicians might be new and interesting to his young hearers."

[39] B. L. van der Waerden, *Erwachende Wissenschaft*, p. 235.

[40] H. J. Anderhub, *"Joco-Seria, Aus den Papieren eines reisenden Kaufmannes,"* Kalle-Werke edition, Wiesbaden 1941.

construct first a right-angle, isosceles triangle whose hypotenuse had the length $\sqrt{2}$, then to draw another right-angle triangle with hypotenuse $\sqrt{3}$ and whose other two sides had the lengths 1 and $\sqrt{2}$. He continued this as in Fig. 2, until sixteen such triangles had been constructed; and the hypotenuse of these provided a geometric representation of $\sqrt{2}$, $\sqrt{3}$, ..., $\sqrt{17}$.

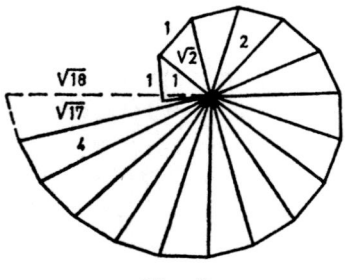

Fig. 2

This interesting reconstruction gains some additional plausibility from the fact that it seems to explain why Theodorus stopped at $\sqrt{17}$. Nonetheless I claim that *historically* it is simply misleading. My reasons are as follows.

(1) This attempt to ascertain how Theodorus might have constructed the squares in question does not proceed from the concept of *dynamis* which is the key to understanding the text. It simply ignores the fact that here is an interesting mathematical concept whose origin as well as meaning has to be understood before one can begin to interpret this passage from Plato.

(2) Plato's words undoubtedly imply that Theodorus must first have produced these *dynameis* somehow and then directed his pupils' attention to their *sides*. On the other hand, Anderhub's construction yields the sides straight away. The whole construction is based on the tacit (and erroneous) assumption that *dynamis* is actually the *side* of a square.

(3) Plato does not indicate how Theodorus constructed the *dynameis*, so this construction must clearly have been both simple and well known by the standards of those times. Anderhub, however, places the construction itself in the foreground, although by all appearences it was only of secondary importance to Theodorus.

(4) Instead of taking the trouble to understand the ancient mathematical concept of *dynamis*, Anderhub without a second thought identifies it with our modern notion of $\sqrt{3}, \sqrt{5}, \ldots \sqrt{17}$.

(5) It is even more unfortunate that Anderhub also tried to explain why Theodorus stopped at the *dynamis* of area 17 square feet, or rather at $\sqrt{17}$ (although this is surely not of central importance in the context of Plato's text). This fact also contributes towards obscuring the distinction between a '*dynamis* of area 17 square feet' and the modern notion of '$\sqrt{17}$'.

For these reasons I believe that Anderhub's reconstruction is historically inadequate.

Of course I do not dispute that there are a number of different ways in which one might obtain squares having as areas whole numbers between 3 and 17. Nevertheless I believe that the best way of explaining the text is to start with a definite and (so it seems to me) very simple conjecture, namely that Theodorus *obtained these squares (dynameis) by transforming certain rectangles into squares of the same area, and that he used Propositions VI.13 and 17 of the Elements to do this.* (This possibility, of course, has frequently been considered by other researchers.)[41] Apart from the fact that, as we shall soon see, this conjecture greatly facilitates further interpretation, there are at least three clues in the text itself which point to its being correct. These are as follows:

(1) Theaetetus speaks of *dynameis* and, as we already know, the word *dynamis* means 'the value of the square of a rectangle'. Furthermore *dynameis* were originally obtained by transforming rectangles into squares of the same area.

(2) The correct geometrical term for 'transforming a rectangle into a square of the same area' is '*tetragonizein*' in Greek. As noted above, it cannot be accidental that this term also occurs in Theaetetus' short speech. Furthermore in my opinion the use of this word by Theaetetus suggests that the discussion between the two youths (Theaetetus and the 'young Socrates') was preceded by a *tetragonismos* which was carried out by Theodorus.

(3) The *dynameis* investigated by Theodorus are whole numbers between 3 and 17. Theodorus seems to have started out with these

---

[41] B. L. van der Waerden, *Erwachende Wissenschaft*, p. 235. The proposition referred to there is II. 14 which is equivalent to VI. 13.

whole numbers and then to have constructed *dynameis* whose areas corresponded to them. If one recalls Euclid's definition (VIII.16) of *plane number* which was discussed above, it seems clear that within the context of the old Pythagorean arithmetic, these whole numbers were represented geometrically by rectangles.

Thus I believe that the further interpretation of the text can be based on this conjecture. Of course Theodorus would not have had to transform all number rectangles between 3 and 17 into squares. A single example would suffice to demonstrate the general method. Perhaps he took the rectangle of area 3 and showed how it could be transformed into a square of the same area by constructing a mean proportional between its two sides. This same method works for all the remaining cases. The construction of all *dynameis* between 3 and 17 could have been imagined simply by analogy with the construction of a single one. So, when the text states that each of these squares was considered individually *(κατὰ μίαν ἑκάστην προαιρούμενος)*, it should not be taken to mean that Theodorus actually constructed each one of them. I shall have more to say about this matter later. For the moment I want to recall two points which modern researchers have made about the alleged 'mathematical discoveries' of Theodorus.

(1) Theodorus is supposed to have been the first to discover and prove the irrationality of square roots between $\sqrt{3}$ and $\sqrt{17}$.[42]

(2) Nevertheless he is only supposed to have recognized the irrationality of the particular cases $\sqrt{3}, \sqrt{5}, \sqrt{6}, \ldots, \sqrt{17}$. Subsequently "Theaetetus, who was still quite young at the time, grasped the general concept of irrationality and so laid the groundwork for a general theory of irrationals."[43] Up to now Theodorus has never been given credit for knowing a comprehensive definition of irrationality. As Heath wrote,[44] "It does not appear, however, that he reached any definition of a surd in general or proved any general proposition about all surds."

The impartial reader, knowing what was said above about the concept of *dynamis*, will from the outset receive the second point at least with a certain scepticism. It seems unlikely that Theodorus discovered only the *particular cases* $\sqrt{3}, \sqrt{5}, \sqrt{6}, \ldots, \sqrt{17}$ of irrationality, since he

[42] Cf. H. Vogt, op. cit. (n. 10 above), p. 97.
[43] E. Frank, *Plato und die sog. Pythagoreer*, Halle 1923, p. 228–9.
[44] T. L. Heath, *A History of Greek Mathematics*, Oxford 1921, Vol. 1, p. 203.

used freely not only the concept of *dynamis* but also the technical term μήκει ἀσύμμετροι which is found in Book X of the *Elements*. Perhaps Theaetetus' words just mean that for some reason he stopped his individual demonstrations when he reached 17, rather than that he was only able to construct *dynameis* up to 17. Once a single rectangle with sides which are not similar plane numbers has been transformed into a square of the same area by constructing a mean proportional between its two sides, there seems to be no reason why all such rectangles cannot be transformed into squares. Furthermore, if it is known that the side of the square obtained is incommensurable in length, there seems to be no reason why this result cannot be generalized to all such squares.

The first point, however, is just as questionable when one thinks of the sense in which it is maintained that Theodorus *proved* the irrationality of $\sqrt{3}$, $\sqrt{5}$, . . . , $\sqrt{17}$. It is assumed, quite without justification, that only the irrationality of $\sqrt{2}$ (the linear incommensurability of the diagonal of a square with its side) could be proved before Theodorus. He is supposed to have distinguished himself not just by recognizing that $\sqrt{3}$, $\sqrt{5}$, . . . , $\sqrt{17}$ were irrational, but also by being able to prove this fact. If this assumption is accepted, the 'proofs of Theodorus' become very important. Hence a lot of trouble has been taken to discover these proofs, i.e. "to find a method of proving the irrationality of $\sqrt{3}$, $\sqrt{5}$, . . . , $\sqrt{17}$, which is analogous to other methods used in Greek mathematics".[45]

I do not wish to deny that Theodorus *could prove* the mathematical assertions which he made in front of his pupils. Indeed, shortly I shall try to show what kind of proofs he might have given. Nonetheless, I must emphasize that no such proofs are mentioned in Plato's account. Vogt already observed quite correctly that "Plato nowhere tells us Theodorus' method of proof. He does not even hint at it."[46] Furthermore the proper technical term for mathematical proof, the verb δείκνυμι, does not occur in Theaetetus' speech. (That Theodorus nevertheless at least indicated some kind of proof, can be inferred from the verb ἀποφαίνω. This word often took on the meaning of 'prove' in everyday speech, especially in mathematical contexts.) But an interpretation which places the proofs of Theodorus in the foreground distorts the

[45] See Vogt (n. 10 above).
[46] Ibid.

meaning of Plato's text, for the point of the text is something other than these. Although I will try to reconstruct the steps of Theodorus' proof, it is not my intention by doing so to attach as much importance to them as is usually done in the history of mathematics. (In other words, I do not believe that Theodorus had to *invent* the proof which I shall try to reconstruct here. Such a proof might already have existed in the mathematical tradition, just as the concept of *dynamis* was handed down to Theodorus and not invented by him.) So let us consider what Theodorus' method of proof might have been, even though this was not of primary importance to Plato.

According to Theaetetus, "Theodorus was drawing some *dynameis* to demonstrate to us that the lengths of the sides of those having an area of three of five square feet, are not commensurable with the length of the sides of a unit square. He discussed the case of each *dynamis* individually until he reached the case of a square with area seventeen square feet ... ". The question now is how this was most simply demonstrated or 'proved', in the schools of the time. I believe that the following two steps suffice:

(1) The rectangle numbers 3 ( $= 1 \cdot 3$), 5 ( $= 1 \cdot 5$), ..., 17 ( $= 1 \cdot 17$), or at least one illustrative example of them, were transformed into *dynameis* by use of Proposition VI.13. Of course this proposition (and perhaps Proposition VI.17 which states that a:b = b:c implies ac = b², as well) would have to *be proved correctly*. In the last analysis it is only the proof which shows that the *dynameis* produced have the same areas as the rectangles concerned. Then the sides of those squares will be *mean proportionals* between the two sides of the corresponding rectangles.

(2) The proof that the sides of the individual *dynameis* were linearly incommensurable might have used Propositions VIII.18 and 20 of the *Elements*. According to these propositions a mean proportional *number* exists only between two similar plane numbers. Since the sides of the rectangle concerned were not similar plane numbers, the sides of the corresponding *dynameis* could not be *numbers*, i.e. magnitudes commensurable with the unit length.

This 'proof' needed to have been thorough and detailed in only one respect. Each *dynamis* between 3 and 17 must have been considered individually (κατὰ μίαν ἑκάστην προαιρούμενος as the text says) to show that the sides of each rectangle, having the same area as the individual *dynameis*, could not be similar plane numbers. In other words, Theodorus took all whole numbers between 3 and 17 (excluding of course the perfect squares 4, 9, and 16) and showed that none of these could be decomposed into two factors which were similar plane numbers. In this way he made the linear incommensurability of the sides of the *dynameis* being discussed seem sufficiently plausible to his pupils. The sides of these *dynameis* are mean proportionals between plane numbers which are not similar, and mean proportional *numbers* only exist between similar plane numbers.

The above, however, is not a rigorous proof by any means. Notice that in the text not only are the concepts of *dynamis* and *μήκει οὐ σύμμετροι* used freely (furthermore Theaetetus does not explain what a *dynamis* is, or how Theodorus obtained these *dynameis*), but in addition the Athenian youths seem to have understood their master's teaching immediately and to have grasped it with astonishing ease. The proofs of Theodorus, even supposing that they were carried out in full, do not seem to have interested the youths very much. Instead of worrying about the proofs, they immediately wanted to start on something completely different, using the results which had been obtained.

## 1.8 THE MATHEMATICAL DISCOVERIES
### OF PLATO'S THEAETETUS

In the light of the above, there seems to be no sound historical justification for the idea that a new chapter in the history of the theory of irrationals began with Theodorus. In any event this idea is not justified by that part of Plato's *Theaetetus* which is the sole reference given for it. It is clear that neither the creation of the mathematical concept of *dynamis* nor the method of constructing individual *dynameis* can be attributed to Theodorus; so the question remains as to what his epoch-making new discoveries were. We concluded from the text that Theodorus *was able to prove* the linear incommensurability of the sides of *dynameis* between 3 and 17. But it is misleading to attach epoch-making importance to these proofs (which are hardly even mentioned by Plato, and anyway were probably quite simple and closely connected with the construction of the mean proportional) and even more so to suspect that Theodorus *could only give proofs for the particular cases between* $\sqrt{3}$ *and* $\sqrt{17}$. The following considerations make one realise immediately how unfounded this suspicion is.

The theory that Theodorus could only prove particular cases of linear incommensurability has to be supplemented by another one, namely that Theaetetus could handle all possible cases. This, however, does not follow from Plato's text. Theaetetus does not report any proofs of his own. It is only fanciful to discuss any 'proofs of Theaetetus' in connection with this passage from Plato, when the text itself offers no evidence for them.

This brings me to the second point in my revision of the interpretation of Plato. It is my contention that, on the basis of Plato's text, new mathematical discoveries can no more be attributed to Theaetetus than they can to Theodorus of Cyrene. This is the view for which I want to argue in the following interpretation.

If Theodorus' purpose was *not to arrive at the proofs*, then the question remains as to why he enumerated all *dynameis* between 3 and 17 and proved linear incommensurability for twelve of them. In my opinion, this question can only be answered if one does not lose sight of the relationship between the mathematical section and the rest of the *Theaetetus* (as suggested in chapter 1.3 above).

One should not forget the way in which Theaetetus was reminded of his mathematical experience. Socrates had just told him that enumerating one by one the kinds of clay used by different craftsmen did not provide a correct answer to the question 'What is clay?' Bearing in mind those things which are common to all kinds of clay, one should rather give as a definition, 'Clay is earth mixed with water'. Thus Socrates' example resembles the mathematical part of the dialogue, in that the enumeration of individual cases is followed by the collecting together of these same cases.

Having sketched Theodorus' enumeration, Theaetetus says "Now it occurred to us that since there are infinitely many *dynameis*, *we should try to group these under one heading by which we could refer to all 'dynameis'.*"

Thus the enumeration of individual *dynameis* and the somewhat detailed proofs given by Theodorus seem to be a *preparation* for the comprehensive classification which was then given by Theaetetus and his friend. The question is whether Theodorus was deliberately preparing the way for this from the outset, perhaps even asking them for it, or whether the two youths, on their own initiative, spontaneously thought of it, as the text ("Now it occurred to us ... ") seems to imply. This matter will be taken up later. For the moment I want to point out a small difference between Socrates' example on the one hand, and the mathematical efforts of Theodorus and his pupils on the other. In Socrates' example a correct *definition* of clay was sought only after the enumeration of the different kinds of clay, whereas Theaetetus and his friend were just seeking a vantage point from which they could classify all *dynameis*.

Before going on to investigate in more detail whether, or to what extent, it can be said that Theodorus' pupils were acting independently of their teacher, we shall take a closer look at what they actually did.

Theaetetus and his friend started with *numbers*. As the text says, "We divided all numbers into two groups ... ". This is easy to understand, since Theodorus started out from numbers as well. He first transformed the number rectangles between 3 and 17 into squares, and then directed his pupils' attention to the sides of these squares.

Furthermore, in drawing a distinction between *square* and *rectangular* numbers, the pupils were following a distinction which must at least have been hinted at by their teacher. Theodorus could only prove that the sides of the *dynameis* 3, 5, 6 ... were linearly incommensurable. He must either have passed over 4, 9, and 16 in silence, or have explicitly emphasized that these cases differed from the remaining ones in that the sides of the rectangular numbers concerned were similar plane numbers.

After dividing all numbers into two groups (square and rectangular), the youths turned their attention to the *dynameis* which served to illustrate these numbers. First of all they called the sides of those *dynameis* which corresponded to square numbers 'lengths' *(μῆκος)*. The reason for this is obvious. Theodorus had previously shown that the sides of *dynameis* with area 3, 5, etc. square-feet were 'not measurable in *length*' *(μήκει οὐ σύμμετροι* linearly incommensurable). So the sides of the other squares (with area 4, 9 or 16 square feet) were called 'lengths' *(μῆκος)*, because unlike the dynameis 3, 5, etc. they were 'measurable in length' (i.e. linearly commensurable).

At most, one could object that this nomenclature is somewhat inexact and imprecise. The sides of those *dynameis* corresponding to square numbers should not really be called *μῆκος*, but rather *μήκει σύμμετρος*.

Euclid also uses the precise term (μήκει σύμμετρος or ἀσύμμετρος), for example in Definition 3 of Book X. Moreover, the correct form of the term occurs in Theaetetus' speech as well (μήκει οὐ σύμμετροι, 147d 4–5). Anyway, although Theaetetus' inexact use of this mathematical term is annoying, it should not surprise us too much, for there is another striking terminological inconsistency in Theaetetus' speech. It does not affect his meaning, however, any more than the preceding example. In 147e7 he uses τετράγωνός τε καὶ ἰσόπλευρος to mean *square number* (which Euclid calls τετράγωνος ἀριθμός). By itself, this way of describing square numbers could be interpreted as simply an antiquated form of the usual term, but in 148a6 Theaetetus uses ἰσόπλευρος καὶ ἐπίπεδος to mean the same thing. The word ἐπίπεδος is not only completely superfluous in this context, but almost troublesome as well. ἐπίπεδος ἀριθμός means not only square number, but also rectangular number (ἑτερομήκης). Furthermore, it occurs in a passage which deals only with plane numbers (i. e. numbers which can be decomposed into two factors). Theaetetus does not mention solid numbers (στερεοὶ ἀριθμοί) until later, and then only in passing. It seems that we have to put up with these inexactitudes in mathematical terminology on the part of Theaetetus.

There is another interesting and important fact about Theaetetus' unusual use of words, which should be established here. It was rightly emphasized in earlier research that there is a close connection between the word used by Theaetetus (μῆκος) and the correct technical term for the same notion (μήκει σύμμετρος) used in Book X of the "Elements"[47]. This observation has to be supplemented by a *relative* chronology of the two terms. It hardly seems possible that the precise technical term μήκει σύμμετρος was coined after Theaetetus (in Plato's dialogue) had just used the word μῆκος for the same concept. A moment's thought makes one realize that this order of events is extremely unlikely. If the word μῆκος by itself was used earlier for the concept of linear commensurability, this vague and by no means self-explanatory description could never have evolved into the completely clear and precise term μήκει σύμμετρος. Clearly these terms developed in the following three steps:

(1) When linear *incommensurability* was discovered, the term μήκει ἀσύμμετρος was coined for it.

---

[47] B. L. van der Waerden, *Erwachende Wissenschaft*, p. 234.

(2) Only then did it become necessary to give a name to the usual case of linear commensurability. To contrast it with incommensurability, the precise and exact term μήκει σύμμετρος was originally used. (*Symmetros* of course is the opposite of *asymmetros*. The dative *mekei* remains the same in both cases and serves to qualify the adjective.)

(3) Theaetetus' curious description of linear commensurability *(mekos)* must be a later abbreviation of the precise term μήκει σύμμετρος.

We are faced with the interesting fact that an abbreviation *(μῆκος)* which celarly postdates the precise term *(μήκει σύμμετρος)* occurs in Plato, whereas the earlier term is preserved in Euclid's later text. Furthermore, it should be noted that as, far as I know, Theaetetus' way of describing linear commensurability is unknown in ancient mathematical literature. Usually it is called *mekei symmetros* (except in the passage immediately following our part of the *Theaetetus*, 148b6–7 which deals once more with περὶ τοῦ μήκους καὶ τῆς δυνάμεως). Obviously this is a case in which Theaetetus has somewhat unnecessarily and arbitrarily abbreviated a clear, unambiguous and precise term.

Theaetetus' careless and inexact use of words in the last part of his speech is even more troublesome, however (although I should stress that once again the meaning of the text remains completely unambiguous). Having described the sides of those *dynameis* corresponding to square numbers as '*commensurable in length*', the two youths turned their attention to those other *dynameis* which corresponded to rectangular numbers. The sides of these they called simply δύναμις, *because these latter straight lines, although incommensurable with the others in length, are nonetheless commensurable with them in the area which they enclose when squared.* This explanation clearly tells us how to understand the peculiar name δύναμις. The correct term of course is not simply δύναμις, but rather δυνάμει σύμμετρος (i.e. commensurable with respect to the square constructed on them), just as previously the term *mekos symmetros* would have had to be used instead of *mekos* if one were speaking precisely. Furthermore, in this case as well, the correct and precise form of the term is to be found in Book X of the *Elements*.

Yet again the question has to be raised as to the relative chronology of these two ways *(dynamis* and *dynamei symmetros)* of describing a straight line incommensurable in length but measurable in square.

In my opinion, the possibility that such a line segment was originally described only by the word *dynamis*, and that this vague term later evolved into the precise and unambiguous term *dynamei symmetros*, has to be decisively rejected, for we already know the clear and precise meaning of the word *dynamis*. In geometry it describes an area and means 'the value of the square of a rectangle' (more generally 'the value of a square' or 'square'), never anything else. To use this same word on its own, as Theaetetus does, in the sense of *dynamei symmetros* (i.e. to apply it to a *straight line*, not an area) is such an inexact use of mathematical terminology that it almost leads to error and compels one to guess the true meaning of the term from its context.

But let us not be too hard on the young man for what seems to be characteristic imprecision in his use of mathematical terminology. At the moment it is more important for us to decide on the relative chronology of the two terms. Of course in this case, as in the previous one, the exact and precise term δυνάμει σύμμετρος (found in Book X of the *Elements*) must be the older of the two. The vague term δύναμις must have come later and could never have been a correct technical term. Once again Theaetetus has superfluously and arbitrarily abbreviated a completely clear and unambiguous scientific term.

### 1.9 THE 'INDEPENDENCE' OF THEAETETUS

We now want to look more closely at the question of how far Theaetetus and his friend can be said to have acted *independently* of their teacher Theodorus. In particular we want to start out by investigating the extent to which the classification of numbers proposed by them was new. Although it has been maintained on the basis of our passage from Plato that Theaetetus also stated and proved some mathematical theorems, we cannot investigate this claim at the moment. According to Plato, Theodorus' pupils wanted to *give all 'dynameis' a name* which would pick out the feature they had in common. So our first task will be to look at their 'descriptions' or 'definitions'.

The youths commenced their 'independent work' by dividing all numbers into two groups, *square* and *rectangular numbers*. Apart from

Frank,[48] no one has been willing to credit Theaetetus with inventing this classification. We need not cite the arguments against this view here. Anyone who is just a little better versed in Greek mathematics will prefer Vogt's opinion. He wrote on this question,[49] "The first step, the classification of numbers as being either square or rectangular as well as the association of numbers with figures, was *not a personal achievement of Theaetetus*. It was devised by the Pythagoreans and was common knowledge amongst number theorists of that time."

But if Theaetetus did not invent this classification, he could only have been responsible for the two concepts 'commensurable in length' and 'commensurable in square'. I have already indicated that the abbreviations *(mekos* and *dynamis)* which he used for these concepts must be later forms of the correct and precise terms, i. e. the expressions used by Theaetetus presuppose the earlier existence of the precise terms. But for the sake of completeness, we should consider the concepts themselves, not just their names. So let us ask whether these two *concepts* could have originated with Plato's *Theaetetus*.

Theaetetus' own words imply that the concept 'commensurable in length' *(μήκει σύμμετρος)* was not created independently by him. According to him, Theodorus had shown that *dynameis* of area 3 or 5 square feet were not commensurable in length *(μήκει οὐ σύμμετροι)*. From this, of course, it follows that Theodorus must have known that the sides of the other kind of square (those with area 4, 9, 16, etc. square feet) were commensurable in length. Hence neither the concept nor the name *μήκει σύμμετρος* could have originated with Theaetetus. He may have used this name on his own, but the concept was already known at least to his teacher. Furthermore, as I have shown above, the young man abbreviated the precise and correct name.

---

[48] E. Frank, op. cit. (n. 42 above), pp. 258–9: "The antithesis of *square and rectangular numbers*, according to Speusippus, derives from the Pythagorean table of opposites. The fact that it does not really date back to the early Pythagoreans but only to Plato's time, is proved by this piece of mathematical work which, as Plato himself says in the *Theaetetus, was the discovery of this mathematician . . .*". [Translation] On the other hand, Frajese *(Periodico di Mathematiche* Serie IV, **44** (1966) 4) has very properly observed that "il giovinetto attribuisce a se stesso (e al collega) nientemeno che la ripartizione dei numeri in quadrati e rettangolari, con la nomenclatura relativa: *cosa inconcepibile*".

[49] Op. cit. (n. 10 above), p. 113.

5*

This leaves the question of whether the concept δυνάμει σύμμετρος (commensurable in square) originated with Theaetetus. On the face of it, this seems rather unlikely. Theodorus showed his pupils that the sides of *dynameis* with area 3, 5, etc. square feet were not commensurable in length; yet he also knew that the *dynameis* he was considering had areas of 3, 5, etc. square feet. It would be astonishing if this did not prompt him into thinking that those straight lines whose *linear* incommensurability he had just demonstrated to his pupils could most easily be measured by the *squares* constructed on them.

Moreover I have also made a case for the conjecture that Theaetetus' inexact description *(dynamis)* of a length measurable only in square must be a later form of the precise term *(dynamei symmetros)*. All this speaks for the fact that the notion of 'quadratic commensurability' could hardly have been a new creation of Theaetetus.

In fact, I believe there are *other* texts of Plato from which it can be proved conclusively that the notion of quadratic commensurability *(δυνάμει σύμμετρος)*, like the concept of *dynamis* itself, originated in pre-Platonic times. Let me digress here and give this proof.

*

The first piece of evidence for the pre-Platonic existence of this concept is a curious and rather complicated pun in Plato's *Politicus* (266 a5–b7), which is hardly intelligible without a knowledge of contemporary Greek mathematics. Some philologists who have studied this section of Plato[50] have correctly recognized that the pun depends on the ambiguity of the word *dynamis*. In everyday speech *dynamis* meant 'ability', whereas in geometry it meant 'square.' I want to go one step further than this and maintain that knowledge of quadratic commensurability was also an indispensable prerequisite for understanding the play on words. If knowledge of this concept had not been common and widespread, Plato could never have thought of making such a pun. Let us now look at the pun itself.

The problem which the 'Eleatic Stranger' and his interlocutor would like to solve in Plato's *Politicus*, is one of classification, namely how to

[50] Cf. L. Campbell, *The Sophistes and Politicus of Plato*, Oxford 1867, pp. 30–3; H. Leisegang, *Die Platondeutung der Gegenwart*, Karlsruhe 1929; as well as Apelt's notes to his translation of this passage.

distinguish between men and pigs. The simplest way would be to say that men walk on two feet, whereas pigs walk on *four feet*. But this is too simple. The 'Eleatic Stranger' jokingly suggests a somewhat more learned method of classification. His interlocutor is the same 'young Socrates' who appears in the *Theaetetus* as a pupil of the mathematician Theodorus. (Incidentally it is also mentioned in this dialogue that Theaetetus and the 'young Socrates' shared a keen interest in geometry, 266a6–7.) For this reason the Stranger proposes that the distinction between men and pigs indicated above should be made *according to the diagonal, and the diagonal of (the square constructed on) the previous diagonal. (τῇ διαμέτρῳ δήπου καὶ πάλιν τῇ τῆς διαμέτρου διαμέτρῳ.)*

Naturally the young man is at first bewildered by this curious suggestion and no more understands what the Stranger is really driving at than we do ourselves.

However, we can make it easier for ourselves to understand the pun which follows, if we keep in mind the sentence "Man is (distinguished) by his ability to walk *(κατὰ δύναμιν . . . εἰς τὴν πορείαν)* on two feet." The phrase 'ability (to walk on) two feet' would most likely be translated back into the everday Greek of Plato's time by δυνάμει δίπους. Now a geometer would immediately think of the diagonal of the unit square on hearing this expression, for this linearly incommensurable straight line was also denoted by δυνάμει δίπους, i. e. 'two feet when measured by the area of the square constructed on it'. The pun is in fact made possible by the ambiguity of the expressions δύναμις ('ability' and 'square') and δίπους ('two legged' and 'measuring two feet'). This is why the Stranger jokingly suggests that man should be classified by the diagonal (or more precisely, by the diagonal of the unit square). We have succeeded in making sense of the first part of the pun, but the joke goes further. We still have to look more closely at the second part of the pun, to see how the Eleatic Stranger wanted to characterize the four-leggedness (four-footedness) of pigs 'by means of a diagonal'.

His suggestion is that one should construct a square on the diagonal of the unit square (see Fig. 3). This diagonal, it was jokingly suggested, yielded the numerical value of our human δύναμις (ability) (to walk on) two feet. Then one should ask how many feet the diagonal of this second square has. Now this question cannot be answered in the way that a geometer of Plato's time would have correctly answered it, for a geometer would have said that, since this second diagonal is a *linearly com-*

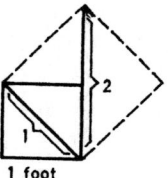

Fig. 3

*1* The diagonal δυνάμει δίπους which is two feet when squared. *2* The diagonal μήκει δίπους (two feet in length) which measures twice two feet when squared *(κατὰ δύναμιν δυοῖν γέ ἐστι ποδοῖν δὶς πεφυκυῖα)*.

*mensurable* quantity, it should be described as having a *length of two feet (μήκει δίπους)*. On the other hand, the Eleatic Stranger (just for the sake of the joke, of course) insists that this second diagonal is to be measured 'by its square' (δυνάμει or 'by its ability' if the word is taken to have its everyday meaning). Thus he can maintain that 'measured by its square, it has twice two feet' *(κατὰ δύναμιν . . . δυοῖν γέ ἐστι ποδοῖν δὶς πεφυκυῖα)*. This insistence on measuring a line by the square constructed on it is akin to our use of the square root sign in cases where it is completely superfluous, e.g. when we write the numbers 5 and 6 as $\sqrt{25}$ and $\sqrt{36}$ respectively.

The pun we have just discussed makes us realise that measuring straight lines incommensurable in length, by the squares constructed on them, in other words quadratic commensurability (which Theaetetus is supposed to have discovered, according to the orthodox view), must have been quite a familiar concept at the time when the pun was made. Plato must have counted on his readers understanding the complicated play on words, especially since it is given in a very concise form and without further explanation. This, however, would hardly have been possible if the concept of the quadratic commensurability of linearly incommensurable straight lines had just been discovered by Theaetetus, for the whole play on words depends on this notion. To me it seems more likely that knowledge of quadratic commensurability was common in Plato's time, and widespread at least among mathematicians. It was a notion about which schoolboys could boast in mathematics classes and with which philosophers could make puns in their discussions.

The second piece of evidence for the pre-Platonic existence of the concept of quadratic commensurability (applied to certain linearly incommensurable straight lines) is to be found in Plato's *Republic*, Book VIII 546c4–5. Since this passage will be treated in full later (cf. pp. 89 ff.), I do not want to go into much detail here. Let me just say the following. The number 50 is described in this passage as the '*square on the (unexpressable) diagonal of five*'. By this Plato means that if a length of five units is assigned to the sides of a square, then the diagonal of that square is expressed by the number which we would call $\sqrt{50}$. In such a case, however, the Greeks did not wish to assign a number to the *length* of the diagonal (for, according to them, the length of a linearly incommensurable straight line was not a number). They preferred to say that the square on the diagonal of a square whose sides had length five, is 50. It is therefore clear that 50 could only be described in this way, because it was known that *linearly incommensurable straight lines existed which were nonetheless measurable in square*.

I believe that I have conclusively proved the existence of the concept of quadratic commensurability of linearly incommensurable straight lines in pre-Platonic times. Although the correct technical term for this concept *(δυνάμει σύμμετρος)* does not occur in either of the references I have cited, the casual way in which a pun is made about it in the *Politicus* and the curious expression 'the square on the diagonal of five' (instead of 50) which is used in the *Republic* shows that it must already have been current in Plato's time. Hence this concept cannot have been introduced by Theaetetus.

### 1.10 A GLANCE AT SOME RIVAL THEORIES

Thus far my interpretation of Plato has led to the following conclusions.

It appears that Theodorus did not intend to teach his pupils about new scientific results of his own when he gave the lecture on mathematics sketched by Plato. The lecture seems rather to have just been an exercise. Theodorus exhibited various *dynameis* which had areas between 3 and 17 square-feet. In so doing, he made his students aware of the fact that most of these had sides which were incommensurable in length. Although no explicit mention is made of what Theodorus hoped to accomplish by this enumeration (indeed this question cannot be definitively answered on the basis of the interpretation given up to now),[51] one can at least say that his demonstration gave the youths food for thought. They immediately came to the conclusion that there were infinitely many *dynameis* of which Theodorus had only enumerated a few, and they wanted to classify all of them.

One's first reaction is naturally to suppose that the youths extended in some way the teachings of Theodorus which may well have consisted of scientific achievements of an earlier period. It might be that the classification of *dynameis* was a piece of independent work carried out by Theodorus' pupils.

This idea has to be abandoned, however, as soon as it is realised that three very important concepts *('square and rectangular numbers'*, *'measurable in length'* and *'measurable in square')* which not only were necessary for this classification, but also can be viewed in part as resulting from it, existed long before Plato's time. 'Commensurable in length' and 'commensurable in *dynamis*' can be said to result from the classification, because both concepts presuppose that one has already learnt to distinguish (classify) the sides of individual *dynameis* (the values of the squares of rectangles). If all numbers are represented as rectangles (plane numbers) and these number rectangles are transformed into squares of the same area *(dynameis)*, then the sides of the squares obtained are of two kinds, those which are *'measurable in length'* and those which are only *'measurable in square'*. If Theaetetus had invented or introduced these two concepts, he would indeed have had to be credited with the classification of *dynameis*. However, since it has been shown above that both concepts must have been known long before Plato's time, Theaetetus could not have discovered this classification; at most he could only have rediscovered it, in the way that talented students of mathematics occasionally rediscover the earlier results of their predecessors.

These conclusions by no means exhaust my interpretation of the mathematical part of the *Theaetetus*. Indeed I will return later to the question of how this passage was understood in antiquity. But first I want to raise three points which are based on the conclusions reached up to now. These contradict some views which are found

---

[51] I shall try to answer this question in the next chapter.

today in the best and most authoritative works on Greek mathematics.[52]

(1) First of all I cannot agree with the interpretation which asserts that exact definitions of the concepts 'measurable', 'measurable in square', 'rational', and 'irrational' are due to Theaetetus. For in the dialogue he only talks about the first two of these, namely 'measurable' (or more precisely 'linearly commensurable') and 'measurable in square' (or 'commensurable by the square constructed on it'). Furthermore, both of these concepts undoubtedly antedate Plato and Theaetetus only refers to them in an inexact, imprecise and almost misleading way (i.e. he uses *mekos* and *dynamis* instead of the correct terms *mekei symmetros* and *dynamei symmetros*).

(2) In addition I cannot agree with the view that the mathematical passage concludes with Theaetetus stating 'very briefly but nonetheless clearly' the following theorem: "Straight lines which form a square whose area is a whole but not a square number, have no common measure with the unit of length."

This theorem can indeed be reconstructed from Theaetetus' words, but there is nothing in the dialogue to suggest that he intended to assert a theorem. It has been *extracted* from the following part of the text. "Those straight lines which yield a rectangular number when squared, we denoted by *dynameis*. For although these are not commensurable in length, nonetheless the areas which they enclose when squared, are commensurable." (Cf. p. 43 above.)

The arbitrary way in which this theorem has been extracted from the words just quoted should also be mentioned. The same words could be used with more justice to reconstruct the following, more complete, theorem: "Straight lines whose squares are whole but not square numbers, have no common measure with the unit of length, but they can be measured by their squares." It should be emphasized

---

[52] On the following cf. van der Waerden's *Erwachende Wissenschaft*, pp. 233ff. and pp. 271ff. as well as his paper 'Arithmetik der Pythagoreer' in the collection *Zur Geschichte der griechischen Mathematik*, pp. 203–54. I should also mention here that after I had finished writing these lines, Professor H. Cherniss sent me the following communication in connection with the paper of mine mentioned in n. 12a above: "I had myself argued in *Plato as Mathematician* that the passage (Theaetetus 147c–148d) does not indicate anything about a *discovery* or a new *demonstration*." I hope that the above discussion will serve to confirm this thesis of Professor Cherniss, of which I was previously unaware.

once more, however, that there is absolutely nothing in Plato's text about any theorem of Theaetetus.

(3) Not only do I have to reject individual assertions like these, but I also believe that some other historical reconstructions, which up to now were more or less plausible, will have to be reexamined in the light of the development given above. Let me give just one example of this here.

Van der Waerden, who today is the most prominent historian of Greek mathematics, published in 1947/49 an important paper[53] which was the "result of a fruitful discussion with Reidemeister". In this work he says "there was not just *one* theory of squared rational straight lines,[54] but two which led to the same results by quite different methods".

Van der Waerden referred to the one theory as $A\varrho$, since it was supposed to go back to Archytas; the other one was denoted by $\Theta\varepsilon$, because its founder was supposed to be Plato's Theaetetus. First of all, I should emphasize that I used to find this historical reconstruction acceptable. In particular I found the way in which the two theories were distinguished especially enlightening. Only after I had come to understand the mathematical concept of *dynamis* better, did I first begin to have doubts about it. I would like to give a brief account of these doubts here, so let us see how van der Waerden distinguished between the theories $A\varrho$ and $\Theta\varepsilon$. About $A\varrho$ he says "it proceeds from the question of the existence of a *geometric mean* between two lines or numbers, and its method is based on the arithmetic of geometric series and the related theory of similar plane and solid numbers". On the other hand, $\Theta\varepsilon$ "does *not use the concept of geometric mean,* but employs the *squaring of straight lines, etc."*

So the two theories are distinguished by whether or not they use geometric means. Before trying to test the validity of this observation, the modern term 'squaring of straight lines' should be translated back into Greek.

If I am not mistaken, both 'square rational straight lines' and 'squaring of straight lines' correspond simply to the Greek concepts

---

[53] See the previous footnote.

[54] "Squared rational straight lines" translates εὐθεῖαι δυνάμει σύμμετροι cf. Definition X. 2 of the *Elements*.

*dynamis* and *dynamei symmetros*.[55] Furthermore, these concepts are used by Plato's Theaetetus, as well as by the so-called Theory $\Theta\varepsilon$, But the following closely related facts about *dynamis* have already been established above:

(a) The scientific term *dynamis* means 'the value of the square of a rectangle'.

(b) The value of the square of a rectangle was obtained by *tetragonismos* (transforming it into a square of the same area).

(c) According to Aristotle, the nature of *tetragonismos* was the finding of the geometric mean.

Thus the concept *dynamis*, which is paraphrased as the '*squaring of straight lines*', cannot be separated from the concept of geometric mean. In other words, the concept of geometric mean is used by Theaetetus (in the dialogue) or by the Theory $\Theta\varepsilon$, just as much as it is by the Theory $A\varrho$ which is attributed to Archytas.

But this obliterates the distinction between the two theories of "squared rational straight lines" which "led to the same results by quite different methods". I no longer believe that there was more than one theory, and can see no argument which could be used to support such a historical reconstruction.


## 1.11 THE SO-CALLED 'THEAETETUS PROBLEM'

The investigation presented in Part 1 of this book deals with the early history of the Greek theory of irrationals. By the 'Theaetetus problem' is meant the question of whether Theaetetus really made a significant contribution to mathematics, and which scientific discoveries can be attributed to him. This problem is only of secondary importance to us. Nonetheless I feel compelled to consider it here at least in passing, because the passage from Plato discussed above, which in my opinion is an important historical source for the theory of irrationals (notice that this opinion does not depend in any way on the fact that the mathematicians Theodorus and Theaetetus are mentioned therein), also provides 'evidence of Theaetetus' role in the history of science'. Furthermore, the mathematical section of Plato's

[55] See n. 54 above.

dialogue cannot be satisfactorily interpreted without considering this whole complex of questions. Therefore, I have inserted here some remarks about the 'Theaetetus problem'.

I have already shown that there are no new mathematical concepts (or names for them) and still less any theorems which can be attributed to Plato's Theaetetus on the basis of the text we have been analyzing (*Theaetetus* 147c–148b).[56] We possess two ancient sources which seem to contradict this. They support a tradition according to which concrete discoveries connected with the theory of irrationals were attributed to Theaetetus even in antiquity. Since both these sources, a scholium to Proposition 9 in Book X of the *Elements*[57] and a report which probably stems from Pappus' commentary on Book X and survives only in an Arabic translation,[58] are very closely related to the dialogue *Theaetetus*, we will have to examine them both in detail. This will also enable us to throw light on the mathematical section of Plato's dialogue from a new angle.

An accurate translation of the scholium in question is as follows:[59]

"This theorem is a discovery of Theaetetus. It is also mentioned by Plato in the dialogue *Theaetetus*, except that only a special case is stated there, as opposed to the general formulation given here."

---

[56] The text from Plato which was quoted above concludes with the statement that Theaetetus and his friend ('young Socrates') attempted a *similar description of solid numbers*. No further details of this other 'discovery' are given in the text itself. Furthermore, modern researchers have not been able to infer anything more definite from this statement. Hence we need not concern ourselves with it here.

[57] Euclides, *Elementa* (ed. J. L. Heiberg, Lipsiae 1883–1916), Vol V, p. 450, line 16ff. (See n. 59, below, for the Greek text.)

[58] On this see T. L. Heath, *Euclid's Elements*, Vol. 3, pp 3–4. The text of this Arabic commentary, edited by G. Junge and W. Thompson was published with an English translation as Vol. 8 of the Harvard Semitic Series (Cambridge, Mass. 1930). Heath used an earlier edition of this commentary (edited by Woepcke, Paris 1855). The text which is quoted in English below, appears also in von Fritz's article on Theaetetus in *Realencyclopädie*.

[59] Τὸ θεώρημα τοῦτο Θεαιτήτειόν ἐστιν εὕρημα καὶ μέμνηται αὐτοῦ ὁ Πλάτων ἐν Θεαιτήτῳ, ἀλλ᾽ἐκεῖ μὲν μερικώτερον ἔγκειται, ἐνταῦθα δὲ καθόλου etc. (cf. n. 57 above).

To facilitate discussion of the scholium, Proposition X.9 is reproduced here as well:

"The squares on linearly commensurable straight lines have the same ratios to each other as square numbers have to square numbers; and squares which have the same ratios to each other as square numbers have to square numbers, must also have linearly commensurable sides; but squares on linearly incommensurable straight lines do not have the same ratios to each other as square numbers have to square numbers;
and squares which do not have the same ratios to each other as square numbers have to square numbers, cannot have linearly commensurable sides."

If this theorem is compared with the scholium quoted above, then the following can be established about the latter. Clearly the scholiast is referring to the passage from Plato which we have been discussing. He claims that Proposition X.9 is mentioned there. This assertion, of course, is inaccurate. Plato's Theaetetus did not state any real theorems, he only classified squares (and even then, he did not do so in a novel way) as follows: "The sides of squares whose areas are whole numbers, are of two kinds, those which are commensurable in *length*, and those which are only commensurable in *square*." The scholium is correct only in so far as Proposition X.9 contains the same 'classification of squares' which Theaetetus proposes in the dialogue. But the scholiast's remark clearly derives from Plato's text, hence it cannot be regarded as independent evidence of a mathematical discovery by the young Theaetetus. It is far more likely that Thaer is correct, when he writes:[60] "The tradition which attributes this proposition (X.9) to Theaetetus is suspect because it probably arose from a passage in Plato's dialogue of the same name. Both Vogt, and before him Hankel, concluded correctly from this passage that the proposition (X.9) was already known at least to Theodorus . . .". So it is clear that the scholiast wanted to credit Theaetetus with discovering Proposition X.9, just on the basis of this passage.

---

[60] C. Thaer, *Die Elemente von Euclid*, Part 4, Leipzig 1936, p. 108.

A similar remark holds true for the other source, the report mentioned above which survives only in an Arabic translation. This sketches Theaetetus' mathematical achievements in greater detail. For the sake of completeness, the most important part of it is reproduced here in an easily accessible English translation.[61] In the following quotation, I have italicized those words which clearly reveal that it was based on Plato's dialogue.

> ... the theory of irrational magnitudes had its origin in the school of Pythagoras. It was considerably developed by Theaetetus the Athenian, who gave proof, in this part of mathematics, as in others, of ability which has been justly admired. He was one of the most happily endowed of men, and gave himself up, with a fine enthusiasm, to the investigation of the truths contained in these sciences, *as Plato bears witness for him in the work which he called after his name. As for the exact distinction of the above-named magnitudes and the rigorous demonstration of the propositions to which this theory gave rise*, I believe that they were chiefly established by this mathematician, and later, the great Apollonius, whose genius touched the highest point of excellence in mathematics, added to these discoveries a number of remarkable theories after many efforts and much labour. *For Theaetetus had distinguished square roots* ('puissances' must be the δυνάμεις of the Platonic passage) *commensurable in length from those which are incommensurable*,[62] and had divided the well-known species of irrational lines after different means, assigning the 'medial' to geometry, the 'binomial' to arithmetic, and the 'apotome' to harmony, as is stated by Eudemus the Peripatetic, etc.

Actually this quotation is noteworthy only because of the last half of the last sentence, where it is maintained that according to Eudemus, the threefold division of irrational lines *(μέση, ἐκ δύο ὀνομάτων* and ἀποτομή) should also be attributed to Theaetetus. It must be left for future researchers to find out *whether an authentic and historically*

---

[61] See n. 58 above.

[62] The German translation by von Fritz (see n. 58 above) is superior here: " ...Theaitetes, welcher unterschieden hat *Quadrate* (dynameis), welche kommensurabel sind in der Länge von den nicht-kommensurablen etc."

*reliable tradition underlies this one point which is independent of Plato's dialogue.*

The two passages quoted above are interesting for the following reason. As stated above, the Arabic passage attributes the threefold division of irrational lines *(mese, ek dyo onomaton* and *apotome)* to Theaetetus. A critical examination of this claim lies outside the bounds of this investigation into the early history of the theory of irrationals. Both sources, however, relying on the mathematical section of the *Theaetetus,* also attribute scientific discoveries to Theaetetus which could not possibly be due to him. This can only be described as a false tradition. If these two reports did not betray the source of their information, there would not be any way of examining their credibility. However, since Plato's dialogue is explicitly mentioned as a source in both cases, their statements can immediately be checked. Without doubt Plato's dialogue has been misunderstood. From our interpretation of the text, it is clear that Theaetetus could neither have introduced any new definitions of irrational quantities, nor have stated any new mathematical theorems. The only remaining question is how this false tradition arose and whether the origins of the legend can be further explained.

In my opinion there is some serious textual evidence in favor of the false tradition. One could almost say that Plato misleads the reader (both ancient and modern) by the way in which he presents his material. Before inquiring into his motives for doing so, let us take a hard look at the evidence which seems at first sight to justify completely the false tradition.

Once before I posed the question of whether Theodorus, when he enumerated and explained the *dynameis* between 3 and 17 in front of his students, was from the outset deliberately preparing the way (perhaps even asking) for the comprehensive classification which they then carried out, or whether they accomplished this on their own. I believe that this question can be answered with certainty. Undoubtedly Theodorus must have expected this 'independent work' right from the start. Later I will attempt to establish this view more firmly, but for the moment I just want to point out that Plato's text, at least on a first reading, *gives very much the opposite impression.*

One just has to read Theaetetus' speech to get the impression that he is talking about his own scientific discovery in which at most his friend ('young Socrates') shared. There is nothing in the text to indicate that the ground was carefully laid for this discovery by Theodorus. It does *not* say that Theodorus challenged his students to find a way of classifying the *dynameis*, which he had shown them, from a common point of view. The text merely says that "for some reason Theodorus stopped" when he reached the *dynamis* 17. In this way Theaetetus emphasizes more strongly his own initiative. It is not just that at the beginning of his speech the young man says to Socrates, "You are asking something like what *your namesake the other Socrates and myself were discussing here recently*", but later on as well he emphasizes his own initiative in the matter, by the words "and now it occurred to *us*". Furthermore, 'us' certainly does not include Theodorus, just Theaetetus himself and 'young Socrates'.

Plato's Theaetetus is undoubtedly convinced that he is talking about his own scientific discovery. This cannot be an accident. It is clear that *Plato deliberately made Theaetetus speak in this way*. It is not surprising, therefore, that even in antiquity readers of this passage thought it had to be interpreted as meaning that Theaetetus was talking about his own independent discovery.

Now the question is what did Plato hope to accomplish by this curious portrayal. He could hardly have been ignorant of the fact that everything which the youth spoke of as his own scientific 'discovery' was known to Theodorus as well, since this was mathematical knowledge which had been in existence for a long time. One could argue that he wanted to deck out his 'hero' Theaetetus in borrowed plumes, but I think that this would be obviously false and misleading. It implies that he must have wanted to make Theaetetus appear ridiculous; and this is clearly impossible. On the other hand, 'Theaetetus actually decks himself out in borrowed plumes' when right at the beginning of his speech, he claims that he and his friend discovered the distinction between square and rectangular numbers, even though this discovery could not possibly have been due to them. Frajese[63] is correct, when he writes on this problem,

Questa presentazione del Teeteto platonico come vero *enfant prodige* spiega anche il fatto che il giovinetto attribuisca a se stesso (e al

[63] *Periodico di Mathematiche*, Serie IV, **44** (1966) 4.

collega) nientemeno che la ripartizione dei numeri in quadrati e rettangolari, con la nomenclatura relativa: cosa inconcepibile. Si tratta invece, secondo noi, della baldanza, e anche della scarsa conoscenza ad essa collegata, propria del giovanissimo che ritiene d'avere scoperto lì dove altri hanno trovato prima di lui . . .".

Not only does Theaetetus ingenuously depict old knowledge as 'his own discovery', but he is not even reproved for doing so. No one suggests to him that he is serving up as his own discovery something which must have already been known to his teacher. Instead of a reproof, the youth earns praise from Socrates; "Well done, lads! I think that Theodorus was right to praise you." Furthermore, these words are undoubtedly meant seriously. They should not be taken as mockery or irony.

Now in my opinion it is by no means easy to find out Plato's true intentions in this case, but there is no reason why it should be. The Mona Lisa's smile is not easy to understand. The worthy Philistine, for the most part, does not understand it at all. He just shakes his head, for ambiguities and subtle nuances are simply not accessible to him. But then there is no reason to expect that great artist Plato to be any less enigmatic than the creator of the Mona Lisa.

Perhaps we can come a little closer to Plato's true meaning by observing closely the way in which he portrays the young Theaetetus. At the beginning of the dialogue, Theaetetus is described in the loftiest tones as $\varkappa\alpha\lambda\acute{o}\varsigma\ \tau\varepsilon\ \varkappa\alpha\grave{\iota}\ \mathring{\alpha}\gamma\alpha\vartheta\acute{o}\varsigma$; he is brilliant and has great mathematical aptitude.[64] Never before has Theodorus come across such a youth. "He approaches learning and research like a stream of noiseless flowing oil, and yet he is very tranquil and modest about it." One could go on like this for a long time, just repeating the praise which is so liberally bestowed on the youth in Plato's dialogue.

One thing, however, is usually forgotten in modern discussions of Plato's portrayal of Theaetetus, namely that the portrait itself is not completely unambiguous. There are inconsistencies in it and if these are ignored, one runs the risk of misunderstanding Plato.

Let us begin by observing that Theaetetus is not only a gifted and mathematically very talented youth, but also a very *naive* one.

---

[64] Cf. B. L. von der Waerden, *Erwachende Wissenschaft*, p. 271ff.

6 Szabó

It would seem that sometimes he can only give false and even simple (not to mention stupid) answers to the questions which are asked him. For example, to Socrates' question "What is knowledge?", he who "approaches learning . . . like a stream of noiseless flowing oil" replies without any idea of how difficult and complicated this question really is, by giving first of all a surprisingly facile enumeration of the different kinds of knowledge. Socrates can only respond with concealed irony, *"You are very generous and liberal, my friend, for you were only asked for a single thing and see, you immediately give us many things of different kinds"* (146d). But the young man does not understand this irony and asks with touching simplicity, *"What do you mean by that, Socrates?"* Socrates, unable to answer such a naive question, prefers to abandon irony for the time being and says, "I don't mean anything by it, or perhaps I do . . .".

To think that simplicity and naivety are incompatible with brilliant talent is surely mistaken. As a matter of fact, it is just this unaffected naivety which lends especial charm to the gifted youth and makes him so sympathetic to both Plato and the reader. There is no question of Theaetetus being simple-minded or stupid. He immediately gives a conspicuous display of intelligence by his quick and accurate association of ideas. No sooner has Socrates explained the banal and over-simplified example of the different 'kinds of clay' to him, than the youth is reminded of a related problem on a much higher level, which lies in the field of mathematics. Socrates' 'kinds of clay' and Theodorus' *dynameis* lie rather far apart. An ordinary youth would not have seen the connection between the two so readily, but Theaetetus' capacity for abstract thought seems to have functioned quickly. In a flash he picks out what they have in common and at that moment he also establishes real contact with Socrates.

In answer to Socrates' question, the quiet and unassuming Theaetetus tells of his mathematical 'discovery'. Yet he is a little *nervous* and *excited* as he does so. We experience his excitement with him, as memory carried him back to the time at which he made his beautiful discovery. I am really amazed by the simplicity of the techniques which enable Plato to portray Theaetetus' mental state while he is speaking. This state is marked by *prolixity* on the one hand and *impatience* and *haste* on the other. As I have tried to point out above, Theaetetus sometimes uses mathematical expressions which are redundant and pleonastic, and at other times he uses ones which are excessively abbreviated, almost to the point of incoherence. Furthermore, it is by no means uninteresting to observe at which points he is long-winded and pleonastic, and at which points he is terse and concise.

He does not refer to the trivial concept of 'square' or that of 'square number' by the simple name τετράγωνος or τετράγωνός τε καὶ ἰσόπλευρος, but adds quite superfluously that it is also ἐπίπεδος. This overzealousness would be ludicrous in a trained mathematician, but Theaetetus is not a trained mathematician. He is just a young and gifted, but still inexperienced scholar who is anxious not to be misunderstood. He is striving for precision in his words, yet this endeavour which in itself is laudable, misleads him to a simple pleonasm in the simplest case of all.

But this minute exactness and striving for precision is abandoned as soon as he reaches the point which he supposes to be the most important – the correct description of sides of squares as falling into two classes, those which are *mekei symmetros* and those which are *dynamei symmetros*, which he considers his most significant 'discovery'. These are the very concepts which have to be explained with precision, for they are not so familiar to non-mathematicians as the trivial concept of *tetragonon*. Yet in explaining them, the youth falls victim to hasty impatience and the joy of discovery. From the grammatical point of view, he describes them in a completely incoherent way. The one is referred to by the singular *mekos*, the other by the plural *dynameis*.[65] It is hardly surprising that the content of these ungrammatical and imprecise expressions could only be understood in a general way. (As a matter of fact, insofar as they led one to believe that *dynamis* could be called a correct mathematical term as well as *dynamei symmetros*, they were not properly understood at all.)

---

[65] The inconsistent use of singular and plural (*μῆκος* and *δυνάμεις*) is surprising because a few lines later (148b 6–7) Theaetetus does not repeat this anacoluthon. There he speaks of περὶ τοῦ μήκους τε καὶ δυνάμεως. Moreover Prof. G. Calogero, who was kind enough to inform me of this inconsistency, has pointed out that Theaetetus' use of the expression ὁρίζεσθαι is ungrammatical. Let us consider the following example of Plato's use of language: τῷ ἀγαθῷ καὶ τῷ κακῷ τὴν ἀρετὴν ὁρίζεσθαι (to define virtue in terms of good and evil). It requires only two slight alterations to repair the grammar of our text and make it conform to this example. We need only change μῆκος to μήκει in 148b 7

If one now reflects on the subtle nuances of Plato's portrait, the following interpretation of the mathematical part of the *Theaetetus* becomes more plausible.

I need not emphasize once more that in reality Plato's Theaetetus did not discover anything new. Nonetheless Plato makes the youth speak as though he were discussing his 'own discovery'. In fact, Theaetetus seems to be of the opinion that he made a new scientific discovery by his own efforts, with at most some assistance from 'young Socrates'. Indeed he does not even give an account of the fact that the very concepts (square number, rectangular number, *mekei symmetros* and *dynamei symmetros*) which he uses to describe his alleged new discovery, just represent traditional, old ideas.

But the ease with which the gifted youth deceives himself, indicates that Theodorus scored a great pedagogic success. Like every true mathematics teacher, Theodorus probably wanted his students to learn to think for themselves. Since he was a skillful teacher, he seems to have prepared them well for this. Both his students succeeded brilliantly in giving the answer which must have been expected of them. On the other hand, in teaching his students to think for themselves, he also seems to have been very tactful. He asked no direct questions; nor did he just tell them to find a name for the *dynameis* which he had shown them – a name which reflected what they had in common. Such a demand was not necessary, for based on past experience, he could rely on his students to guess his unspoken intentions on their own. It was more important to him that the actual pedagogic strategy remained unnoticed. He succeeded so well in this, that it remained a mystery as to why he stopped his demonstration at the *dynamis* 17. All that Theaetetus says is, "For some reason he stopped at this *dynamis*". Clearly the teacher unobtrusively fell silent at the appropriate time as if unable to continue the train of thought

and replace δυνάμεις by δυνάμει in 148a1; then a complete paraphrase of the passage would be: ὅσαι μὲν γραμμαὶ τὸν ἰσόπλευρον καὶ ἐπίπεδον ἀριθμὸν τετραγωνίζουσι, μήκει (or συμμέτρους) ὡρισάμεθα, ὅσαι δὲ τὸν ἑτερομήκη (or ἀριθμὸν τετραγωνίζουσι), δυνάμει (or συμμέτρους ὡρισάμεθα) etc. Although this is not only an irreproachable, but also a very elegant paraphrase of the passage, I do not think that it is absolutely necessary to emend the traditional text. Without alteration, the meaning of the text is completely clear and unambiguous; and it remains so with the alterations.

any further, he let his students go on to find out *for themselves* what he had deliberately, carefully and yet inconspicuously prepared for them. Thus they found 'independently' the solution which their teacher had been expecting from them right from the start.

The above interpretation, however, might suggest that Theodorus would subsequently have had to clear up the youths' illusion. I am firmly convinced that nowadays no one would omit to do this, yet it seems to me that Theodorus was a better pedagogue for not doing so. Furthermore, I am not in the least bit surprised by the fact that Socrates was so pleased about the youths' discovery and did not find it necessary to disabuse them. After all, from his point of view, their little self-deception could only have been of *secondary importance*. In view of what one might call their independent rediscovery of some interesting mathematical facts, it would have been ludicrous and pedantic to inform them that these facts had already been discovered by others.

So now it does not seem very surprising that the later scholiast took Theaetetus' words about 'his own discovery' at face value.

## 1.12 THE DISCOVERY OF INCOMMENSURABILITY

The most important historical conclusions which may be drawn from the foregoing interpretation of Plato can be summarized as follows:

Theodorus of Cyrene did not begin a new chapter in the history of the Greek theory of irrationals. The text *Theaetetus* 147c–148b furnishes no support for the view that Theodorus or Theaetetus contributed something new to this theory, whether it be a concept, the name of a concept, a theorem or a proof. As a matter of fact the concept *dynamis* (the value of the square of a rectangle) which is used so casually in the same text shows that the theory of irrationals must have already been quite well developed in pre-Platonic times. The value of the square of a rectangle' cannot be separated from *tetragonismos* (the transformation of a rectangle into a square of the same area). Furthermore, according to Aristotle, the essence of *tetragonismos* consists in finding a mean proportional between two arbitrary line segments. Of course all three of these concepts *(dynamis, tetragonismos,*

and finding a mean proportional between two line segments) antedate Hippocrates of Chios, because the construction of a mean proportional (Proposition 13, *Elements*, Book X) must have been known
to him.[66]

As we can see, a historical explanation of the mathematical concept
of *dynamis* is sufficient to open up new possibilities for reconstructing
the historical development of the theory of irrationals. For this reason
I will discuss quite fully the most important and best-known terms
which refer to irrationals in Greek mathematics, using the same
method which has been demonstrated above for the example of the
concept *dynamis*.

*

The most common pair of concepts which are due to the discovery
of mathematical irrationality are called in Greek σύμμετρον and ἀσύμμετρον *(measurable* and *non-measurable).* As Euclid states in Definition 1, *Elements*, Book X. "Those magnitudes which are measured
by the same measure are said to be *commensurable (σύμμετρα μεγέθη)*
and those for which there is no common measure, are said to be
*incommensurable (ἀσύμμετρα δέ . . .)."*

Undoubtedly the concepts *commensurable* and *incommensurable*
originated within Greek mathematics itself. The existence of incommensurability could never have been discovered outside the scientific
and deductive framework of Greek mathematics. As Aristotle perti-
nently remarked:[67]

"...all begin by wondering that things should really be as they are...
Thus one wonders ... at the incommensurability of the diagonal
and side of a square. For at first everyone finds it astonishing that
there exists something which cannot be measured by the smallest
common measure. Finally however one comes to a different conclusion ... provided that one is informed about the matter. For
nothing would surprise a geometer more than if the diagonal should
suddenly become commensurable."

---

[66] See n. 37 above.
[67] *Metaphysics* A 2.983a13ff.

I think that the following point is worth emphasizing from this quotation. Aristotle is fully aware of the fact that incommensurability, by which he seems to mean only *linear* incommensurability (hence his example of the diagonal and side of a square), cannot be a naive concept obtained directly by reflection. "Everyone finds it astonishing that there exists something which cannot be measured by the smallest common measure . . .". This astonishment marks the beginning of the scientific way of thinking; at a naive pre-scientific stage, one would probably not even be aware of the *possibility* that two straight lines could be linearly incommensurable. Also a naive way of thinking, which is concerned first and foremost with practical matters, does not recognize the existence of incommensurable magnitudes. In practice a 'common measure' can ultimately be found even for the diagonal and side of a square. One need only choose a measure which for all *practical* purposes is sufficiently small to obscure the incommensurability of the two quantities. Thus incommensurability is a concept whose origin is *theoretical* rather than practical.

If Aristotle is right in saying that *"one wonders . . . at the incommensurability of the diagonal and side of a square"*, then it may also be conjectured that to begin with, people could not calmly and coolly accept this newly discovered and surprising fact without further ado. It seems much more likely that they initially made attempts to assign a precise ratio to the lengths of the two lines in question.

There is in fact a passage from Plato which, amongst other things, seems to bear this out. I am referring to the famous passage about the 'wedding number' in the *Republic*. (Book VIII, 546cff.) A full interpretation of this mathematically interesting passage would take us too far afield.[68] It suffices here, however, just to draw attention to the important mathematical term which makes its first appearance in this passage.

According to the literature,[69] "Plato calls the number 7, the *rational diagonal* belonging to the side 5." This assertion means that if one assigns a length of five units to the side of an arbitrary square, then

---

[68] On the interpretation of this passage see A. Ahlvers, 'Zahl und Klang bei Platon' (*Noctes Romanae, Forschungen über die Kultur der Antike*, ed. Professor V. Wili, No. 6), Bern–Stuttgart 1952.

[69] B. L. van der Waerden, *Erwachende Wissenschaft*, p. 206ff.

the length of the diagonal of that square is approximately seven units. It is easy to see this by using Pythagoras' theorem. If $a$ represents the side of the square and $d$ represents its diagonal, then the theorem states that $d^2 = 2a^2$. So for $a = 5$, $d^2 = 50$ and $d = \sqrt{50} \approx 7$. This explains Plato's terminology. For the moment, however, we are only interested in the expression 'rational diagonal' itself, not in the above-mentioned fact.

In fact the two mathematical terms used by Plato in this connection ($\dot{\varrho}\eta\tau\acute{o}\nu$ and $\ddot{\alpha}\varrho\varrho\eta\tau o\nu$) are usually translated as 'rational' and 'irrational'. However, it is worthwhile recalling that $\dot{\varrho}\eta\tau\acute{o}\nu$ originally meant 'that which can be expressed' and $\ddot{\alpha}\varrho\varrho\eta\tau o\nu$ meant 'that which cannot be expressed'. If one tries to explain the origin of the latter expression, the question arises as to why the diagonal of a square should be called $\ddot{\alpha}\varrho\varrho\eta\tau o\nu$. It can only be because, having assigned a *number* to the side of a given square, one wanted to do the same for its diagonal, and when it was realised that this was not possible, the diagonal in question received the name $\ddot{\alpha}\varrho\varrho\eta\tau o\nu$ (that which cannot be expressed).

The fact that in this passage Plato explicitly mentions not only $\delta\iota\acute{\alpha}\mu\varepsilon\tau\varrho o\varsigma$ $\ddot{\alpha}\varrho\varrho\eta\tau o\varsigma$ (the unexpressable diagonal), but also $\delta\iota\acute{\alpha}\mu\varepsilon\tau\varrho o\varsigma$ $\dot{\varrho}\eta\tau\acute{\eta}$ (the expressable diagonal), shows that $\ddot{\alpha}\varrho\varrho\eta\tau o\nu$ cannot be construed in this context as meaning something forbidden. Undoubtedly mystical-religious $\ddot{\alpha}\varrho\varrho\eta\tau\alpha$[70] were concerned with things which should not be expressed as well. Nonetheless the diagonal of a square was not called $\ddot{\alpha}\varrho\varrho\eta\tau o\nu$ for this reason, but just because a number could not be assigned to its length (assuming of course that the length of the sides of the same square had been assigned a number). The number 7, on the other hand, being approximately equal to the length of the diagonal of a square with side 5, can be described as $\delta\iota\acute{\alpha}\mu\varepsilon\tau\varrho o\varsigma$ $\dot{\varrho}\eta\tau\acute{\eta}$, the 'expressable (rational) diagonal', because it is actually a number. This short explanation belies the tradition which views the discovery and even more so the public discussion of mathematical irrationality as 'sacrilege'. It seems that the tradition is just a naive legend which sprang up later. This discovery was most probably never a 'scandal' to mathematicians.

Moreover the phrase in Plato (*Republic*, Book VIII, 546c4–5) which mentions both concepts (*expressable* and *unexpressable* diagonal) in

---

[70] Herodotus 5.83; Xenophon, *Hellenica* 6.3, 4; Euripides, *Helen* 13, 23, etc.

one breath, reads as follows: ἑκατὸν μὲν ἀριθμῶν ἀπὸ διαμέτρων ῥητῶν πεμπάδος, δεομένων ἑνὸς ἑκάστων, ἀρρήτων δὲ δυοῖν . . . .

Instead of a translation, we shall content ourselves in this case with an exact paraphrase of the text. As we know, the Greek words express the same number (4800) in two different ways, namely as '100 squares on the *expressable* diagonal of 5, each such square diminished by 1' (i.e. $100 \times (7^2-1) = 100 \times 48 = 4800$) and also as '100 squares on the *unexpressable* diagonal of 5, each such square diminished by 2' (i.e. $100 \times (50-2) = 100 \times 48 = 4800$).[71]

Thus the diagonal of a square whose sides have a length of five units can be described both as 'unexpressable' and as 'expressable'. The 'expressable' diagonal is 7 in this case. So we can say that in Greek 'expressable diagonal' is a locution for the approximate value of the length of the diagonal. Furthermore, the length of the unexpressable diagonal is written as $\sqrt{50}$ in modern notation. The Greeks, however, instead of using this notation, preferred to measure the square which could be constructed on the incommensurable straight line in question rather than its length. (We have already discussed this point fully in connection with the *Theaetetus*.) They then said that this diagonal, when 'measured by its square' (i.e. when 'squared'), was of 50 (square) units.

We know from Proclus[72] as well as from other ancient sources[73] that the Pythagoreans developed the method by which one can generate side and (expressable) diagonal numbers. The pair 5 and 7 mentioned by Plato form the third term of this infinite sequence of numbers. Of course the sequence can also be viewed as an ancient method of approximating $\sqrt{2}$. If one calculates the ratio of each diagonal to its side $(d : a)$ as a decimal fraction, then the interesting sequence obtained tends towards the limit $\sqrt{2}$.[74]

Not only are the Pythagoreans supposed to have discovered side (and diagonal) numbers, but nowadays it is generally agreed (and

[71] See n. 68 above. A. Ahlvers, op. cit., p. 12, n. 4: "In the language of mathematics, ἀριθμὸς ἀπό . . . meant *square of* . . .".

[72] Proclus Diadochus, *In Platonis Rem publ. Comm.* (Ed. W. Kroll, Lipsiae) 1901, II, 23, pp. 24–5.

[73] See the sources quoted by van der Waerden in *Erwachende Wissenschaft*, pp. 206ff.

[74] Cf. E. Stamates, *Euklidou Geometria*, Vol. 2, Athens 1953, pp. 9ff.

rightly so) that they also gave the first scientific proof of the incommensurability of the side and diagonal of a square.[75] This proof is given in the proposition to be found in the *Elements*, Book X, Appendix 27 (the reference here is to Heiberg's edition of the text; the proposition does not appear in Heath's translation of Euclid, but is summarized there in the introduction to Book X). It is based on a *reductio ad absurdum* argument which shows that if the side and diagonal of square are commensurable, then the same number must be both odd and even.

But it is worth noting that there is a slight difference in terminology between the text from Plato which has just been quoted and the Pythagorean theorem found in Euclid. Plato speaks of the 'unexpressable diagonal' *(διάμετρος ἄρρητος)* whereas in Euclid the diagonal of a square is said to be 'linearly incommensurable' with its side *(ἀσύμμετρός ἐστιν . . . μήκει)*. Of course both these expressions refer to the same fact which we would describe as the irrationality of $\sqrt{2}$ (if the side of the square is chosen to have unit length). Nonetheless a historical conjecture is suggested by the fact that the Greeks had two ways of describing this phenomenon, although it must be admitted that there is no further evidence to confirm the conjecture.

The conjecture is that 'unexpressable diagonal' is the older of the two descriptions. First of all the Greeks attempted to assign numbers to the diagonals of squares whose sides had also been assigned numbers. When they realised that this could not be done, they expressed this fact by describing the diagonals concerned as 'unexpressable quantities' *(ἄρρητα)*. Thus they were already very close to the creation of completely new mathematical concept; but the new concept finally made its appearance only after the expression 'incommensurable' *(ἀσύμμετρον)* had been coined to describe it (more precisely, only after the expression 'incommensurable *in length*', μήκει ἀσύμμετρον, had been coined).

The question remains as to how the problem of the diagonal of the square ever arose, and whether there is some way of reconstructing

---

[75] Cf. O. Becker, 'Die Lehre vom Geraden und Ungeraden im neunten Buch der Euklidischen Elemente', *Quellen und Studien zur Geschichte der Math.* etc. B, **3** (1936), pp. 533–53 (Reprinted in *Zur Geschichte der griechischen Mathematik*, ed. O. Becker, Darmstadt (1965).)

at least in part, the process of mathematical thought which led to the discovery of incommensurability. At first this sounds like an odd question. On the one hand, it cannot be answered with certainty on the basis of the facts which have been handed down to us. On the other hand, the problem of the diagonal of the square seems to be so simple as to make one believe at first sight that one would eventually arrive at it almost by oneself. So there seems to be no sense in searching out further historical connections when the matter being investigated is so simple and straightforward. Nevertheless, I believe that the following conjectures can claim some historical interest, even though they are just conjectures.

### 1.13 THE PROBLEM OF DOUBLING THE SQUARE

My conjecture is that the problem of the diagonal of the square arose in connection with *doubling the square*.

I have two pieces of evidence in favor of this view. Firstly it is known that the doubling problem occupied Greek mathematicians over a long period of time. To substantiate this claim, it suffices to note here the imaginative proposal made by Hippocrates of Chios for attacking the famous "Delian problem" (the doubling of the cube).[76] Although it is true that he was not able to carry through his own proposal, nonetheless soon after him Archytas of Taras found the first solution to this problem, a solution which made use of Hippocrates' proposal. Furthermore, after Archytas there were even more mathematicians who concerned themselves with this same problem and found other solutions to it which did not depend on Archytas' work. Thus it may be conjectured with some plausibility that in an earlier time (I mean of course a time before Hippocrates of Chios) *doubling the square* was just as interesting a problem to mathematicians as doubling the cube became at a later stage in the development of mathematics. One just has to think about this possibility to realise immediately that there is a famous literary passage which shows, amongst other things, exactly how the problem of

---

[76] Cf. O. Becker, *Das mathematische Denken der Antike*, Göttingen 1957, pp. 75 ff.

doubling the square was treated in early Greek mathematics instruction.[77] This brings me to my second piece of evidence.

As we know, there is a passage in Plato's *Meno* where Socrates asks a simple, uneducated slave to tell him how the area of a square whose sides are two feet long, can be doubled without altering its shape. So as not to be misunderstood, Socrates immediately draws the square whose sides are each supposed to represent a length of two feet. (Thus the square which he draws consists of four smaller squares.) He then asks how a square with twice the area can be constructed. Since Socrates explicitly mentioned that the sides of the original square are two feet long, the slave's first idea is that the required square must have sides which are twice this length, i.e. which are four feet long. Socrates' response is to draw this new square, by extending one side of the original square to twice its length and then constructing a square on this new side. It can be seen immediately from his drawing that the area of the original square has not been doubled, but quadrupled. The slave inevitably realises that his first attempt at an answer to Socrates' question was incorrect. If the length of the sides is doubled, the area obtained is four, not twice the original.

The slave now thinks as follows: a square whose area is twice that of the original will clearly have longer sides as well, so the sides of the square being sought, will have to be *longer than two feet*. However, they cannot be four feet long, because a square with four feet long sides has an area which is four times that of the original. Thus the required length must be *less than four feet*. Now three lies between two and four, so perhaps a square which has sides *three feet in length* will have twice the area of the original square. Once again Socrates responds with a diagram. He extends one side of the original square by a single unit and constructs a larger square on this new side. Again it is obvious from the diagram that the area has not been doubled, for the new square consists of nine smaller ones, whereas it should only consist of eight. Thus the slave is forced to admit that yet again his answer is false.

[77] O. Becker (*Grundlagen der Mathematik in geschichtlicher Entwicklung*, Freiburg–Munich 1954, p. 109) correctly described this passage from the *Meno*, as "a living portrait of a lesson in elementary geometry, as it would have been taught at that time".

The problem is finally resolved when Socrates draws a new diagram and shows that if the sides are doubled in length, then the original square can be quadrupled. (This is the construction which the slave first wanted to carry out.) Now, however, the *diagonals* of these four new squares also form a square (see Fig. 4) whose area is made up of four equal triangles. Since the original square is only made up of two such triangles, Socrates has succeeded in drawing a square whose area is twice that of the original, by means of the diagonals.

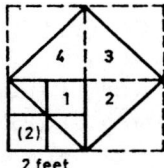

Fig. 4

The passage just summarized is especially instructive not only because it shows that the way in which a naive and uneducated person attempted to solve this problem was first of all to give a numerical value to the side of the required square, but also because of some very informative remarks made by Socrates. After the slave's second unsuccessful attempt, Socrates says "Now then, try and tell us exactly how long the required sides will be, and *if you do not want to express this as a number*, then just show us." (83e πειρῶ ἡμῖν εἰπεῖν ἀκριβῶς· καὶ εἰ μὴ βούλει ἀριθμεῖν, ἀλλὰ δεῖξον.) These words are a quiet hint to the informed reader that the length of the sides of the square with twice the original area (i.e. the length of the diagonal of the original square) cannot be given a numerical value, because it is an ἄρρητον.

Without a doubt the problem of doubling the square – the attempt to find the number which, when assigned to the length of the sides, would yield a square with twice the area – led to the problem of the *diagonal of the square* and thence to the problem of linear incommensurability. Indeed one would almost like to postulate a direct connection between the mathematical problem in the *Meno*, and the Pythagorean proof that the side and diagonal of a square are incommensurable which is found in Euclid (or rather in the appendix to Euclid cited above). On the one hand, it is indicated in the *Meno* that once a numerical value has been assigned to the side of a square, then the diagonal of that square cannot be expressed as a *number*. On the other hand, the Pythagorean proof is purely number-theoretic and takes the form: if the side and diagonal were commensurable, then the same number would have to be both even and odd, which is impossible.

I believe that the evidence just quoted (the long-standing interest in the problem of doubling the cube and the passage from the *Meno*) goes some way towards confirming my conjecture about the discovery of linear incommensurability. If we accept that the Greeks discovered the linear incommensurability of the side and diagonal of the square in their attempts to *double the square*, then there is a sense in which the usual interpretations of this discovery will have to be revised. Up to now the historical significance of the discovery of mathematical irrationality was interpreted in the following way.

The earliest Pythagoreans could be said to have "worshipped numbers".[78] For them, numbers were "the rock bottom of the entire universe", the world was formed "by imitation of numbers" and the heavens were "harmony and number". According to Aristotle (*Metaphysics* A5), they arrived at this view because they were so much occupied with mathematics. The old teaching, however, is supposed to have been shaken by the discovery that the side and diagonal of a square were incommensurable. The diagonal of a square having no common measure with its sides meant that if the length of the sides is taken as the unit, then the length of the diagonal cannot be measured, i.e. *it cannot be expressed either as a whole number or as a fraction.* Thus something which was "not a number" but just an ἄρρητον (something which could not be expressed) had been discovered. It is alleged that this disconcerting discovery led to what used to be called a *crisis in the foundations of Greek mathematics.*[79]

---

[78] Cf. B. L. van der Waerden, *Erwachende Wissenschaft*, p. 204.
[79] Cf. H. Hasse and H. Scholz, *Die Grundlagenkrisis der griechischen Mathematik*, Pan-Bücherei, Berlin 1928. For an opposing view, see B. L. van der Waerden, 'Zenon und die Grundlagenkrise der griechischen Mathematik', *Math. Ann.* 117 (1940) pp. 141–61.

Although the newer textbooks hardly mention this 'foundational crisis' any more, the historical interpretation based on such a 'crisis' (or at least on a 'critical turning-point') still survives, albeit in a disguised form. Van der Waerden, for example, writes that the Greeks "turned away from numbers" in the course of their mathematical development, just because of the discovery of incommensurability, and that subsequently a geometrization or geometric 'formulation' of mathematics took place.[80] In his own words:

> In the domain of numbers, the equation $x^2 = 2$ cannot be solved, not even in that of ratios of numbers. But it is solvable in the domain of segments; indeed the diagonal of the unit square is the solution. Consequently, in order to obtain exact solutions of quadratic equations, we have to pass from the domain of numbers to that of geometric magnitudes.

There are two points in the interpretation sketched above with which I disagree. Firstly I do not believe that the new discovery required a revision of the ancient Pythagorean doctrine ('everything is number'). Secondly I cannot accept the conjecture that the discovery of irrationality led the Greeks to favor geometry at the expense of arithmetic which up to then had been esteemed more highly. (In principle, arithmetic was always prized more highly than geometry by the Greeks; cf. Chapter 3.27 of this book.)

Let us first consider the extent to which the ancient Pythagorean doctrine 'everything is number' was shaken by the incommensurable. Even if the side of the square is chosen as the unit of length, it is misleading to say that the diagonal is really 'not a number'. According to the *Meno*, the Greeks originally sought the line segment which would enable them to double the area of a given square (whose sides had been assigned a numerical value). Thus the problem which led to the discovery of incommensurability was a *geometrical* one. They could only establish that the line segment being sought was 'not measurable *in length*' (i.e. that its length was 'not a number'), after they had realised that it was the *diagonal* of the original square. Since,

---

[80] On this topic see the chapter entitled 'Why the geometric formulation?' in *Erwachende Wissenschaft*, pp. 204–6.

however, this line yielded a square of twice the original area, the idea of measuring linearly incommensurable segments by their *squares* must have suggested itself immediately. This is the reason why Plato (*Politicus* 266b3) calls the diagonal of the unit square δυνάμει δίπους ("two feet when measured by the square constructed on it").

We could say that the diagonal of a square was recognized to be linearly incommensurable *at the same time* as it was recognized to be commensurable in square. Hence it is unlikely that this magnitude was ever thought of as "*absolutely* indeterminate from a numerical point of view". The diagonal of a square can be given a numerical value (when measured by its square). This case is similar to the introduction of the new mathematical concept of *dynamis*. The value of the square of a given rectangle whose sides were not 'similar plane numbers', could be given a precise numerical value as an area, even though the sides of the square concerned were incommensurable with the unit of length. *Thus the Pythagorean doctrine 'everything is number' was in no way shaken by the discovery of linear incommensurability.* Although the length, which could not be assigned a numerical value, was described at first as an ἄρρητον, the initial surprise gave way immediately to the realisation that the line whose *length* could not be given a numerical value should be *measured by its square*.

Similarly I disagree with those who maintain that the discovery of irrationality led the Greeks to "turn away from numbers" or "to pass from the domain of numbers to that of geometric magnitudes". It is of course true that some earlier mathematics must have been reformulated after the existence of incommensurability was recognized. (In particular the notion of 'ratio', λόγος, which until then had applied only to numbers, would have had to be redefined so as to include incommensurable magnitudes.) But the interpretation outlined above (the Greeks turned away from numbers, etc.) would only be warranted if one could prove that the *non-geometric* problem of the irrationality of $\sqrt{2}$ occurred to the Greeks before the geometric solution to this same problem made its appearance. As far as I can see, however, such an idea is not supported by any sources. On the contrary, all our Greek sources speak unequivocally in favor of the view that doubling the square and the measurability or non-measurability of the diagonal of a square were authentic geometrical problems right from the start, and that the problem of $\sqrt{2}$ is just a modern equivalent of these problems which has been removed from the original, purely geometric frame-work. In my opinion, therefore, there is no justification (at least in this case) for speaking about a subsequent 'geometrical formulation'.[81]


### 1.14 DOUBLING THE SQUARE AND THE MEAN PROPORTIONAL

My previous conjecture that in all likelihood the problem of *doubling the square* led to the problem of the *diagonal of the square*, enables us to discern some other historically interesting connections. We just have to realize that in the minds of the ancients, the problem of doubling the square was identical to another problem. This can be seen from the following.

In his historical treatment of the problem of doubling the cube, Becker remarks,[82] "In this form, the task seemed intractable to ancient mathematicians. Therefore Hippocrates of Chios transformed the problem of doubling the cube into another problem, namely *to find two mean proportionals x and y between a* (the edge of the cube) *and 2a*."

It can easily be seen that Hippocrates' proposal does lead to a solution of the problem. For if $a : x = x : y = y : 2a$

then
$$ay = x^2 \quad \text{and} \quad xy = 2a^2$$
bence
$$a : x = x^2 : 2a^2$$
i.e.
$$x^3 = 2a^3$$

Now the question is how a Greek mathematician of the 5th century chanced upon the happy idea that the problem of doubling the cube could be reduced to the problem of 'two mean proportionals', even though he was not in a position to carry out his proposal.

My opinion is that Hippocrates just used an *argument by analogy* to obtain his proposal. He wanted to treat an unsolved problem in solid geometry after the manner of a problem in plane geometry,

---

[81] I do not want to go into the problem of the so-called 'geometric algebra' of the Greeks in this book (see however the appendix).

[82] Cf. n. 76 above.

which had been solved a long time before. He may have started out with the idea that 'square' and 'cube' are, in a sense, like figures. Furthermore, a square can be doubled by finding *one* mean proportional between its side ($a$) and twice this side ($2a$), i.e. by finding an $x$ such that $a : x = x : 2a$. Hence one will be able to double a cube by finding *two* mean proportionals between its edge and twice this edge.

This argument of Hippocrates leads me to conjecture that the ancients considered *doubling a square* and *finding a mean proportional between a number and twice that number* to be synonymous. These two problems are not just the same in essence, but are very closely related to each other from the historical point of view as well. I think this conjecture is worth noting for the following reason.

Aristotle's assertion to the effect that the essence of *tetragonismos* (the transformation of a rectangle into a square of the same area) consisted in finding a mean proportional (between two arbitrary line segments) has already been quoted in an earlier chapter. Furthermore, it was indicated there that the problem of *tetragonismos* may itself have led to the new mathematical concept of *dynamis* and hence to the recognition of linear incommensurability. The idea that the problem of '*doubling the square*', or '*finding a mean proportional between a number and twice that number*' may have occasioned the discovery of incommensurability fits in very well with these earlier conjectures. In the last analysis, 'doubling the square' is nothing but a special case of *tetragonismos*. In both cases one has to find a mean proportional between two numbers (or magnitudes). The only difference is that in the first case a mean proportional between a number (magnitude) and twice that number is sought, whereas *tetragonismos* requires one to find a mean proportional between any two numbers.

This insight suggests the conjecture that the discovery of linear incommensurability, as well as the realization that linearly incommensurable straight lines were commensurable in square, was brought about by attempts to find a mean proportional between two straight lines (magnitudes). The problem of the 'mean proportional' was very important in early Greek mathematics. Thus it appears that the problem of irrationality originated in the theory of proportions.

For this reason I will attempt to discuss the early history of the Greek theory of proportions in somewhat greater detail. Such a discussion is contained in Part 2 of this book.

## 2. THE PRE-EUCLIDEAN THEORY
## OF PROPORTIONS

### 2.1 INTRODUCTION

To the extent that it has been clarified by previous research, the early history of the theory of proportions found in Euclid can be summarized roughly as follows:

Euclid begins his discussion of the theory in Book V of the *Elements*. The fundamental definition of this book is *Definition 5*: "Magnitudes are said to be in the same ratio *(ἐν τῷ αὐτῷ λόγῳ)*, the first to the second as the third to the fourth, if any equal multiples of the first and third are alike, greater than, equal to or less than any equal multiples of the second and fourth taken in the corresponding order."

From this definition it follows that the equality $a : b = c : d$ holds, if and only if for any whole numbers $m$ and $n$,

$$ma > nb \text{ implies } mc > nd$$
$$ma = nb \text{ implies } mc = nd$$

and

$$ma < nb \text{ implies } mc < nd.$$

This ingenious definition, whose completeness has often been a source of astonishment,[1] is usually attributed to Eudoxus, a younger contemporary of Plato. According to a scholium[2] which is unfortunately of unknown authorship, the whole of Book V is essentially the work of Eudoxus.[3] At all events, the definition could only have been formulated at a time when incommensurable magnitudes were known to exist and an effort was being made to ensure that the theory of proportions, which up to then had only been applied to *numbers*, could be extended to incommensurable magnitudes.

---

[1] Cf. B. L. van der Waerden, *Science Awakening* (Oxford University Press 1961), pp. 175–6.

[2] Cf. Euclides, *Elements* (ed. J. L. Heiberg, Lipsiae 1888–1916), Vol. V, p. 280.

[3] T. L. Heath, *Euclid's Elements*, Vol. 2, p. 112.

This immediately suggests the question as to what definition was used before the discovery of incommensurability, when the theory of proportions was only applied to numbers. This question could easily be answered by referring to another definition of proportionality which was simpler than the one mentioned above and applicable only to numbers, namely *Elements* Book VII, *Definition 21*: "Numbers stand in the same ratio[4] when the first is the same multiple, the same part, or the same number of parts, of the second as the third is of the fourth."

So these two definitions could be said to mark two different epochs in the historical development of the theory of proportions. The more highly developed theory which could handle incommensurable magnitudes as well as numbers was made possible by Eudoxus' definition (V.5) and this definition must undoubtedly have been of later origin than the other one (VII.21).

This simple chronology may indeed seem illuminating at first glance, but the historical problems which it raises and leaves unsolved cannot be concealed. As an example, let us consider Euclid's proposition VI.13 – the construction of a mean proportional between two arbitrary line segments. A full proof of this proposition can only be given *on the basis of Eudoxus' definition of proportionality*. The other definition fails in this case, because it can only be applied to numbers, and the length of the mean proportional between two line segments whose lengths cannot be written as similar plane numbers (cf. Definition VII. 22) is itself an *incommensurable magnitude*.

Thus one is led to conjecture that the construction of a mean proportional between two arbitrary line segments only became possible after Eudoxus had laid down his definition. This conjecture, however, contradicts the well-documented historical fact that the geometrical construction of a mean proportional was familiar to the Pythagoreans before the time of Eudoxus, and must even have been known to Hippocrates of Chios.[5] The question is how this construction could have been proved at that time. Any answer to such a question must necessarily be conjectural. In the present case, the following three conjectures suggest themselves:

---

[4] On this translation of the expression ἀνάλογον, see pp. 148–57ff.

[5] See the paper by van der Waerden in *Zur Geschichte der griechischen Mathematik*, p. 225, n. 28.

(1) Originally, before the time of Hippocrates of Chios, a full proof of Proposition VI.13 could not be given. At this time the 'intuitive validity' of the construction had to suffice instead of a proof. Even if the construction of the mean proportional were in fact older than the proof, the plausibility of this conjecture is lessened by the fact that the standards of rigour demanded of proofs in 5th century Greek mathematics were already very high.[6] In my opinion it is almost unthinkable that Hippocrates, for example, would have been satisfied with a so-called 'naive concept of proportion'.

(2) Another possibility is to conjecture that Definition V.5 is in reality much older than has been believed. It should not be forgotten that there is no ancient source which actually credits Eudoxus with this definition. It has only been attributed to him because of the above-mentioned scholium which asserts that the substance of Euclid's fifth book is the work of Eudoxus.

(3) Finally, one could imagine that there was a pre-Eudoxian definition of proportionality which Euclid fails to mention. This definition allowed a precise treatment of irrational as well as rational proportions, and made possible a full proof of the geometric construction of the mean proportional.

In the last few decades scholars believed that they had in fact rediscovered just such a definition. Aristotle mentions somewhere (*Topica* VIII.3, p. 158b 29–35) an interesting definition of sameness of ratio (the so-called *anthyphairetic* or *antanairetic* definition) which runs as follows: "*Those magnitudes are in the same ratio whose anthyphairesis is the same.*" The earlier interpretation of this passage was that Aristotle is pointing out the difficulties which, in the absence of a suitable definition, the discovery of incommensurability raised for mathematical proof. The anthyphairetic definition is supposed to have helped in overcoming these difficulties.[7]

I showed some years ago[8] that this interpretation of Aristotle's text *(Topica* VIII.3, p. 158b 29–35) is completely mistaken. Aristotle

[6] Ibid. pp. 215–6.

[7] Cf. O. Becker, 'Eudoxos-Studien I' in *Quellen und Studien zur Geschichte der Mathematik*, etc. B, **2** (1933), p. 311–33.

[8] Á. Szabó, 'Ein Beleg für die voreudoxische Proportionen-Lehre?' *Archiv für Begriffsgeschichte* (Bonn) **9** (1964), pp. 151–71.

is certainly not talking about the difficulties of mathematical proof
in this passage, nor does he intend to ascribe any such historical
significance to the anthyphairetic definition as the modern interpretation attributes to it.

Notwithstanding my objections, the fact remains that Aristotle does
quote this definition. The only question is whether the Greeks themselves *laid down the definition in this form with incommensurable magnitudes in mind* (as Becker and, following him, some other scholars
believed),[9] or whether they *applied it only to commensurable magnitudes*
(as Reidemeister conjectured).[10] As far as I can see, a decision between
these alternatives cannot be made with any certainty on the basis
of the arguments advanced up to now. (Of course I hope that the
discussion contained in the chapters which follow will cast some new
light on this whole question. My only reason for not returning later
to the historical problem which is concerned with the different definitions of proportionality is that it is not central to the subject of
my investigations.)

Thus it can be seen that up to now research into the historical
development of the theory of proportions has been concerned mainly
with the influence which the discovery of incommensurability had
on the further development of the theory. Hence there was less interest
in the early history of proportions. Only passing reference was made
to the fact that the terms which Euclid uses for this theory reveal
a connection with the theory of music.[11]

I would now like to elucidate the development of the pre-Euclidean
theory of proportions by giving a history of these terms. In so doing,
I shall draw extensively on the results of my latest publication on
this subject.[12]

---

[9] Cf. O. Becker, *Das Mathematische Denken der Antike* (Göttingen 1957),
p. 103, n. 25.

[10] K. Reidemeister, *Das Exakte Denken der Griechen*, Hamburg 1949, p. 22.

[11] See, for example, the paper 'Die ἀϱχαί in der griechischen Mathematik',
*Archiv für Begriffsgeschichte* (Bonn) **1** (1955), pp. 13–103.

[12] Á. Szabó, 'Die frühgriechische Proportionen-Lehre im Spiegel ihrer Termino-
logie', *Archive for History of Exact Sciences* **2** (1965), pp. 197–270.

## 2.2 A SURVEY OF THE MOST IMPORTANT TERMS

It would be wrong to believe that a history of the terms which were used in the mathematical theory of proportions will be able to throw light on the very beginnings of this theory. Such a history has much more limited objectives in this case. So as to be clear from the outset about what may realistically be expected from an investigation of this sort, let us recall the kind of things which we were able to learn about the theory of irrationals by investigating the history of a mathematical term.

It was shown in Part 1 of this book that the following facts about a mathematical term like *dynamei symmetros* (measurable with respect to the square constructed on it) can be established with considerable certainty. (The term is introduced in Definition X.2 of the *Elements*.) The expression *dynamis* was taken over into mathematics from the language of finance. Just as when converting one currency into another, 'having the same value' was expressed by the verb *dynasthai* or by the noun *dynamis*, so the 'square having the same area as a given rectangle' was described by these same two words in geometry. Thus the word *dynamis* came to have the special meaning of 'the value of the square of a rectangle' or 'square' in the technical language of geometry. However, since the transformation of such rectangles with sides which were expressed as numbers often resulted in squares whose sides were not measurable in length, it was desired to measure these same sides by their squares and not by the lengths. That is the origin of the expression *dynamei symmetros*. In this case, therefore, the history of a term could be said to have elucidated the genesis of some completely new mathematical concepts, namely 'value of a square', 'commensurable' or 'incommensurable' in length and in square.

Such results are not to be expected for the theory of proportions, because the concept of proportion was not a new creation of Greek mathematics in the same way that such notions as commensurability and value of a square were. By all appearances, the concept of proportion played a significant role in pre-Greek mathematics. For example, it has been remarked that the whole of Egyptian mathematics can be said to have been dominated by the idea of proportion.[13]

---

[13] P.-H. Michel, *De Pythagore à Euclide*, Paris 1950, pp. 365ff.

It would be a hopeless undertaking, in my opinion, to attempt an explanation based on terminological considerations of how the idea of proportion actually came into being, especially since this idea is clearly a very ancient one whose beginnings can be traced back to the most primitive human thought.

So the present investigations will not concern themselves with the earliest beginnings of a mathematical theory of proportions. They are only intended to throw light on the origins of the theory of proportions which is presented in the *Elements* and which undoubtedly originated with the Greeks. It is my view that in all likelihood this pre-Euclidean process can be accurately reconstructed by investigating those expressions which were later adopted permanently into Greek science. Let me therefore start out by listing the most important of those terms whose origins are to be investigated more fully in the sequel.

*

Euclid's customary way of referring to 'ratio' and 'sameness of ratio' is illustrated by the following passage: *Elements* VII.11: "*If a whole (say $A = a + b$) is to another whole (say $B = c + d$) as the subtracted part of the one is to the subtracted part of the other ($A : B = = a : c$)* – ἐὰν ἦ ὡς ὅλος πρὸς ὅλον, οὕτως ἀφαιρεθεὶς πρὸς ἀφαιρεθέντα, *then the remainder of the one is also to the remainder of the other as the one whole is to the other ($b : d = A : B$)* – καὶ ὁ λοιπὸς πρὸς τὸν λοιπὸν ἔσται, ὡς ὅλος πρὸς ὅλον.*"* As we can see, Euclid just uses the preposition 'to' *(πρός)* to express 'ratio' *(ὅλος πρὸς ὅλον* and *λοιπὸς πρὸς λοιπόν)*, whereas he uses a so-called adverbial clause of comparison[14] for *'sameness of ratio' (ὡς ... οὕτως ...)*. This is a very common way of expressing these notions. It is also to be found in an archaic-sounding fragment of Archytas, where he discusses the so-called geometric mean.[15]

| | |
|---|---|
| γεωμετρικὰ δέ, ὅκκα ἔωντι οἷος ὁ πρῶτος (or ὅρος) ποτὶ τὸν δεύτερον, καὶ ὁ δεύτερος ποτὶ τὸν τρίτον. | A geometric mean is present, when the first term is to the second as the second is to the third. |

[14] R. Kühner and B. Gerth, *Ausführliche Grammatik der griechischen Sprache*, vol. 2, p. 490; 4th edn., Hanover 1955.

[15] H. Diels and W. Kranz, *Fragmente der Vorsokratiker*, vol. 1, p. 436 (Archytas B2).

The only grammatical difference between these two quotations is that in the second, sameness of ratio is expressed by a so-called adjectival clause of comparison (οἷος . . . καὶ . . .).

This way of expressing 'ratio' and 'sameness of ratio' is hardly very illuminating from a historical point of view. The only thing which can be said is that this use of the preposition πρός (to describe a ratio) may well have been a specialized one in mathematics. There is probably no other way of explaining how Aristotle was able to change the meaning of this mathematical expression so that it referred to his category of 'relation' (τὸ πρός τι).

More interesting is the fact that the four-place relation 'proportion' (or 'sameness of ratio', as it is usually called in this book), which is expressed by the formula $a : b = c : d$, was denoted by ἀναλογία in the terminology of mathematics. Euclid's only use of this term is in Definition V.8 of the *Elements*: "*The least proportion consists of three terms*" ('Αναλογία ἐν τρισὶν ὅροις ἐλαχίστη ἐστίν). Because this definition is completely redundant as far as Euclid's propositions are concerned, it has often been regarded as a 'later interpolation'.[16] However, this question need not concern us here. The word ἀναλογία (in the sense of 'geometric sameness of ratio') is undoubtedly an old mathematical term. On one occasion Aristotle explains the formula $a : b =$ $= c : d$ as '*geometric analogia*', in the following words:[17] 'As term $a$ is to term $b$, so is term $c$ to term $d$' (ἔσται ἄρα ὡς ὁ α ὅρος πρὸς τὸν β, οὕτως ὁ γ πρὸς τὸν δ). There are also quotations from Euclid which show that the word ἀναλογία is an old mathematical term. Euclid never uses solely the words ὁ αὐτὸς λόγος to mean 'the same ratio' (see for example, Definition V.5); sometimes he employs also the archaic expression ἀνάλογον (for 'in the same ratio'), as for example in Definition VII.21, which was quoted above.

I intend to devote some later chapters to a thorough investigation of the fundamental expressions of the theory of proportions (in particular λόγος, ἀνάλογον and ἀναλογία). However, there is another mathematical expression which is the key to the following historical discussions, and I would like to start out by concentrating on it.

---

[16] T. L. Heath, *Euclid's Elements*, Vol. 2, p. 131.
[17] *Nicomachean Ethics*, 1131b5.

In Greek a 'term' in a proportion (sameness of ratio) is called ὅρος. This word occurs only once in the *Elements*, namely in Definition V.8: "The least proportion consists *of three terms*" (. . . *ἐν τρισὶν ὅροις*). Again, however, there is a great deal of evidence which supports the view that this word (ὅρος in the sense of '*term* of a proportion') was actually a well-known mathematical term. Incidentally the definition states that the least proportion has three terms, i.e. $a : b = b : c$. Aristotle expresses this same fact by saying that every proportion has at least *four* terms. According to him, even the least *analogia* (the so-called *ἀναλογία συνεχής*) actually has *four* terms; it is just that in this case the middle term ($b$) is counted twice. In explaining this,[18] Aristotle uses the word ὅρος to denote *terms* of a proportion, just as Euclid does in the definition mentioned above. There are many other examples which could be given of this same usage of the word ὅρος in mathematics.

It is my opinion that a historical explanation of the origins and development of the pre-Euclidean theory of proportions has to start from the word ὅρος. My reasoning is as follows:

Although the term ὅρος is used only once by Euclid to mean the '*term* of a proportion', he uses this same word on other occasions in two completely different senses:

(1) Ὅροι is most frequently used in the *Elements* to refer to the definitions. This usage undoubtedly comes from the language of philosophy, which I will discuss more fully in Part 3 of this book. The use of ὅρος to mean 'definition' has nothing to do with the theory of proportions.

(2) The other meaning of ὅρος is more important to us in this connection. In many passages Euclid uses the Greek word ὅρος (masculine) in its everyday meaning of 'boundary' or 'endpoint'. This meaning was clearly taken over into geometry. For example, Definition I.13 of the *Elements* states that ὅρος ἐστίν, ὅ τινός ἐστι πέρας, which Heiberg translates as "*Terminus* est, quod alicuius rei extremum est".

Now I conjecture that this latter meaning of ὅρος (terminus or endpoint) was also taken over into the theory of proportions. After all, we still talk about the *terms* of a proportion. But the sense in

---

[18] Ibid. 1131 a31–b5.

which a proportion can be said to have *endpoints* has yet to be explained.

We have to answer such questions as why the Greeks spoke of *endpoints* in connection with proportions and ratios; why they maintained that every 'sameness of ratio' (ἀναλογία) had four endpoints and every 'ratio' (λόγος) had two; and whether this nomenclature is significant. I do not believe that this can be accomplished just by considering the two terms λόγος and ἀναλογία, for they make the name ὅροι seem obscure and meaningless. If, however, we bear in mind that in the mathematical theory of music developed by the Pythagoreans, and also in the *Sectio canonis* which has come down to us as a work of Euclid, the 'ratio of two numbers to each other' (which later received the name λόγος in arithmetic and geometry) was called διάστημα, then we are well on the way to answering these questions.

In the theory of music διάστημα had two meanings. On the one hand it means 'musical interval' (and also the 'ratio between numbers' which expressed this interval), and on the other hand it had its everyday meaning of 'line segment' or 'distance between two points'. (Incidentally it should be remarked that Euclid frequently uses διάστημα in this latter sense. His third postulate, for example, says "It is postulated that a circle can be drawn with any centre and any line segment" – παντὶ διαστήματι.)

The latter meaning of διάστημα ('line segment' or 'distance between two points') immediately makes the choice of the name ὅροι (end points) seem meaningful and evident, because a line segment does in fact have two end points. If we could show some kind of connection between 'ratios' and 'line segments', then these segments or ratios really would have two end points (ὅροι). This is exactly what I hope to do in the following chapters.

I will proceed by showing that:

(1) Musical intervals which were expressed as 'ratios between numbers', were called διάστημα (the distance between two points) in the most ancient Greek (Pythagorean) theory of music. A *diastema* really had two end points (ὅροι) which (as we shall soon see) were assigned numbers.

(2) By all appearances, the 'ratio of two numbers to each other' received the name λόγος somewhat later. A λόγος is also said to have two ὅροι (end points) because originally (in the theory of music) this

word meant something similar to *diastema*. As a matter of fact, λόγος in geometry soon came to mean the same as *diastema* in the theory of music.

### 2.3 CONSONANCES AND INTERVALS

My principal aim now will be to elucidate the musical term διάστημα, because the history of this scientific term gives great deal of information about the whole pre-Euclidean theory of proportions as well as about the origin of the expression 'ὅροι of an *analogia*'. Before starting out, however, it is worthwhile to recall some well-known facts about the theory of music in antiquity.

### (A) *diastema* as consonance

In the theory of music διάστημα meant interval. This fact can be confirmed by consulting any Greek lexicon. For the time being, we shall regard an interval simply as the 'distance between two tones', without worrying about the origin or the true meaning of this word.

In antiquity, theorists of music were interested mainly in the so-called *concordant intervals* or *consonances*.[19] In musical practice the usual name for intervals of this kind was συμφωνία, because they had to do with the *concord (συμφωνεῖν)* of two sounds. In the present work we shall only be concerned with the three most important concordant intervals; these are the *octave*, the *fourth* and the *fifth*. The Greek names of these intervals throw some light on the origin of the expression διάστημα as it was used in the theory of music, so I have listed them below. (My reference is Pape's Dictionary, 1849.)

The *octave* was called διαπασῶν (actually ἡ διὰ πασῶν χορδῶν συμφωνία, the accord which passes through all the (eight) strings). The name of the *fourth* was διατεσσάρων (actually ἡ διὰ τεσσάρων χορδῶν συμφωνία, the accord which passes through four strings) and correspondingly the *fifth* was called διάπεντε, ἡ διὰ πέντε χορδῶν συμφωνία (the accord which passes through five strings).

---

[19] Cf. B. L. van der Waerden, 'Die Harmonielehre der Pythagoreer', *Hermes* **78** (1943), 163–99.

It seems, however, that these names are of relatively late origin. They presuppose that the strings were numbered consecutively in a way which was customary in musical practice and which is still reflected in our names for these intervals (*octave* [8], *fifth* [5] and *fourth* [4]). However, this was not so in the earliest times.

Originally the strings were not numbered, instead each one had its own name. For example, the topmost string (which was the longest, and produced the lowest note) was called ὑπάτη (the highest) and the bottommost string (which was the shortest, and produced the highest note) was called νήτη. The remaining strings lay between these two outermost ones; of these, the only ones whose names need be mentioned here are the μέση and the παραμέση.[20] Hence the concordant intervals could be described simply by the names of the strings concerned (i.e. those which produced the tones in question). So, for example, the *octave* was known as the concord of the *hypate* and *nete*.[21]

We know of some earlier names for these three consonances, which are even more interesting. The *fourth* used to be called συλλαβά; this word actually means 'holding together', i.e. 'holding together the first and last strings of a tetrachord'. (These two strings were *held together* and sounded at the same time, or directly after one another; in this way a fourth was produced.) Similarly Nicomachus reports that amongst the Pythagoreans the *fifth* was called διοξεῖα (also written δι᾽ ὀξειᾶν or δι᾽ ὀξειῶν).[22] This name clearly comes from the fact that to produce a fifth, two tetrachords were joined together and the note from each one which was sounded was usually desribed as the 'sharp' (i.e. the *high*) note ὀξεῖα. Finally the *octave* was just called ἁρμονία, 'joining together' (i.e. the '*joining together of two tetrachords*'),[23] because the two outermost strings of two tetrachords joined together were sounded to produce an octave.

---

[20] Cf. Plato, *The Republic*, Book IV. 443d.

[21] 'Pseudo-Aristoteles, "De Rebus musicis problemata" ' *Musici scriptores graeci*, ed. C. Janus (Lipsiae 1895), problems 23 and 25.

[22] See the entry under διοξεῖα in Pape's Dictionary (1849), also Porphyrios *Kommentar zur Harmonielehre des Ptolemaios*, ed. Düring (Göteborg 1932), p. 96. 21.

[23] See the article by H. Koller in *Museum Helveticum* 16 (1959), pp. 238–48; also B. L. van der Waerden, ibid. 17 (1960), 111ff.

These ancient names for the three most important consonances are all mentioned in a fragment of Philolaus from which I would like to quote a few words here:[24]

"The extent of an *octave* is a *fourth* and a *fifth* (ἁρμονίας δὲ μέγεθός ἐστι συλλαβὰ καὶ δι' ὀξειᾶν). ... For the distance between the topmost and the middle string is a fourth (ἔστι γὰρ ἀπὸ ὑπάτας ἐπὶ μέσσαν συλλαβά) and the distance between the middle and the bottommost is a fifth (ἀπὸ δὲ μέσσας ἐπὶ νεάταν δι' ὀξειᾶν) . . .".

(B) *diastema* as interval

In my opinion one just has to survey the various names and descriptions of strings and consonances which have been listed up to now, to see that without doubt they all stem directly from musical practice. Ultimately it was in *musical practice* that strings first received names and later were assigned numbers; also the consonances were described by the strings which produced them.

This fact needs to be emphasized from the outset, because the other term which interests us most in this connection, the word *diastema* meaning the *interval* itself (i.e. not the concord *(symphonia)* of two sounds, but their distance from each other), derives from musical *theory*, not practice. This matter will be taken up later.

The other fact about the technical terms discussed above which I would like to stress is that they are all completely *concrete*. The consonances received their names from the strings on the musical instrument which produced them. There is clearly no question of any so-called metaphorical transfer of names in these cases. This latter fact needs to be stressed, because the modern literature on the subject is full of assertions which lead one to believe that the musical term *diastema* (interval) is understood as some kind of metaphor. To see how this term is interpreted nowadays, let me quote the following sentence from Burkert's book:[25] "For us, the representation in terms of straight lines is suggested particularly by musical notation and the

---

[24] Diels and Kranz, *Fragmente der Vorsokratiker*, vol. 1, 44 B6.10ff.
[25] W. Burkert, *Weisheit und Wissenschaft*, p. 348.

piano key board, *yet it was also known to the Greeks, as the term 'interval'* διάστημα *indicates."*

Whoever wrote this sentence was clearly unaware of the fact that to the Greeks the musical interval *(diastema)* was something as real and concrete as συλλαβά, the 'holding together of the two outermost strings of the tetrachord' (i.e. the fourth). It is misleading to talk of a *"representation* in terms of straight lines". One should rather point out that the musical interval *(diastema)* was a real and concrete straight line which was measurable in length. It is even more misleading to read in the same author[26] that "The conception (of intervals) as straight lines is linked to the name of Aristoxenus, the theory of musical proportions is associated with the Pythagoreans . . .".

I must confess that I have not succeeded in discovering any kind of reasonable meaning in this assertion. "The conception (of intervals) as straight lines" and "the theory of musical proportions" cannot be contrasted with one another, because the word *diastema* (interval) means both 'straight line' and 'numerical ratio of a musical interval'. Furthermore, it is impossible for this fact to have escaped the notice of anyone who has read with sufficient care even a single sentence from the *Sectio Canonis* (in the Greek original).

It is worthwhile recalling here that the false (metaphorical) interpretation of the meaning of *diastema* (musical interval) was not unknown in antiquity. Let me just indicate here some of these false interpretations, as found in late classical writers.

Cleonides writes that "An interval *(diastema)* is that which is contained *(τὸ περιεχόμενον)* by a low and a high note".[27] Literally almost the same explanation can be found in Gaudentius.[28] Nicomachus describes a musical *diastema* as a 'way' which leads from a low note to a high one, or vice versa *(ὁδὸν ποιὰν ἀπὸ βαρύτητος εἰς ὀξύτητα ἢ ἀνάπαλιν)*.[29] Such learned 'explanations', of course, were possible only because it was known exactly what the word διάστημα meant in ordinary usage, namely 'distance between two points' or 'straight line'. If one reads only the interpretations cited above, then one would in fact be inclined to believe that it is just a metaphor which is being

---

[26] Ibid., p. 349.
[27] *Musici scriptores graeci*, ed. C. Janus (Lipsiae 1895), p. 179.11.
[28] Ibid., p. 329.23ff.
[9] Ibid., p. 243.2.

dealt with here. The two     nes of the consonance are thought of as
two points in space and ju.,ι as between two points in space there lies
a 'straight line', so according to the metaphor, between the two tones
there can be said to lie a *diastema*. (Thus the fact that the *diastema*
between two tones was *also a real straight line in space* is forgotten.)

Probably the first person who wanted to explain 'musical interval'
in a metaphorical sense (and who by all appearances deliberately
wanted to change its meaning), was Aristoxenus. In the 4th century
B.C. he wrote:[30]

> "An interval is that which is bounded by two sounds having different
> tensions (i.e. pitches). Thus according to the basic concept an
> interval manifests itself both as a *difference in tension (διαφορά τις
> εἶναι τάσεων τὸ διάστημα)* and as a *space* capable of taking in those
> tones which are higher than the lower pitch bounding the interval,
> and lower than the higher one *(τόπος δεκτικὸς φϑόγγων)*. A difference
> of pitch, however, consists in being more or less taut."

It is clear that this change in the meaning of *diastema*, from a
completely concrete to a metaphorical one, is very closely tied to the
basic teaching of Aristoxenus, for he was an opponent of the Pythagoreans. Van Jan has summarized his teachings as follows:[31]

> "In his harmonics he did not inquire into the origin of a tone, nor
> did he ask *whether it was a number or a speed*. The ear need only
> listen unaffectedly to the range of tones; it will tell us with certainty
> which tones harmonize with each other. ... His system is based
> on the fourth and the fifth which are easily perceived consonances,
> and without asking *which numerical ratios underlay them*, he was
> able to determine from them the whole and half tones, etc."

As a matter fact, Aristoxenus' own words also attest to his *anti-
Pythagorean* tendencies:[32]

[30] *Die Harmonischen Fragmente des Aristoxenos*, ed. P. Marquard, Berlin
1868, pp. 20ff.

[31] C. Van Jan, 'Aristoxenus' (in *Realencyclopädie* III, pp. 1057–65).

[32] See P. Marquard (ed.), *Die Harmonischen Fragmente des Aristoxenos*,
pp. 46–7 (32, 19ff. in Meibom's numbering).

"We are attempting to draw conclusions which are in agreement with the data, as opposed to the theorists who preceded us. Some of them introduced completely foreign viewpoints into the subject and dismissed sensory experience as imprecise; hence they made up intelligible causes and stated that there were *certain ratios between numbers and speeds (λόγους τέ τινας ἀριθμῶν εἶναι καὶ τάχη πρὸς ἄλληλα)* on which the pitch of a note depended. These were all speculations which are completely foreign to the subject and absolutely contrary to appearances. Others renounced reasons and arguments completely and proclaimed their assertions as though they were oracular sayings; they likewise paid insufficient attention to the data."

Aristoxenus is undoubtedly talking about the Pythagoreans when he says that some theorists wanted to account for the pitch of a note by means of "certain ratios between numbers". Furthermore, he was obliged to reinterpret the meaning of *diastema* in a metaphorical way, both because he wanted to discount the teaching of the Pythagoreans and because he attributed differences in pitch to whether the string was more or less taut. (We will see in the next chapter that it was the Pythagoreans who coined the musical term *diastema* and that for them this word denoted a *straight line* whose *end points* yielded a *numerical ratio*, namely the numerical ratio of the consonance concerned. Aristoxenus had to change this meaning, since he did not want to admit that numerical ratios had anything to do with consonances.) This explains his extraordinary interpretation of a musical interval *(diastema)* as either a 'difference in tension', or something like 'a space capable of taking in those tones which are higher than the lower pitch bounding the interval and lower than the higher one'. It is apparent from this forced explanation that Aristoxenus wanted to deprive the Pythagorean concept *diastema* of its original concrete meaning. As a matter of fact, he was partially successful in this endeavour.

According to the original Pythagorean concept, there were always *two* numbers associated with a musical *diastema*. As Porphyry says in his commentary on Ptolemy's theory of harmony,[33] "Most κανονικοί

[33] Porphyrios, *Kommentar zur Harmonielehre des Ptolemaios*, ed. Düring (Göteborg 1932), p. 92, 22–3: καὶ τῶν κανονικῶν δὲ καὶ πυθαγορείων οἱ πλείους τὰ διαστήματα ἀντὶ τῶν λόγων λέγουσιν.

8 Szabó

and Pythagoreans say *interv*[...]αστήματα) instead of *numerical*
*ratios* (λόγοι)." These words [...]that 'interval' (διάστημα) and
'numerical ratio' (λόγος) are e[...]nt concepts in the terminology
of Pythagorean musical theory[...]conclusion is confirmed not only
by the fact that the term διάστ[...]consistently used to mean λόγος
in the *Sectio Canonis*, but als[...]nother passage from Porphyry.
Although the Pythagoreans a[...]t explicitly mentioned in this
latter passage, it is clear that [...]erence is to them.[34]

> "Some call a *numerical ratio*[...]ι end points, diastema (τὸν λόγον
> καὶ τὴν σχέσιν τῶν πρὸς ([...]ις ὅρων τὸ διάστημα καλοῦσι);
> these could be characterized [...]ις of their end points as λόγοι, as
> well as διαστήματα, namely th[...]h would be *epitritos logos* (4 : 3),
> the fifth would be *hemiolios l*[...]: 2) and so on."

This quotation is interesting, [...]st because it shows that the two
concepts 'musical interval' (di[...]) and 'numerical ratio' (*logos*)
were equivalent as far as the P[...]ιreans were concerned, but also
because it clearly implies that e[...]nts (ὅροι), which must also have
been numbers, could function l[...]ι 'end points of *diastemata*' and
as 'end points of *logoi*'.

It now remains for us to sho[...]the word διάστημα could mean
'the distance between two poin[...]e musical interval between two
tones' and 'numerical ratio' all[...]ι same time.

### 2.4 THE 'DIASTEMA'    EEN TWO NUMBERS

I shall now try to explain th[...]ιsis of two interesting concepts,
διάστημα and ὅροι, which figur[...]e Pythagorean theory of music.
Although as far as I know th[...]no ancient source which deals
directly with the question of [...]ιese concepts came into being,
there is a noteworthy experim[...]ιscribed by more than one late
classical author, which could al[...]be said to provide an answer to
it. Before quoting one of these[...]ιrs in translation, there are two
facts which should be stated:

[34] Ibid., p. 94, 31ff. My translatic[...]his passage is provisional. See also
that part of the text marked by n. [...]w.

(1) As has already been mentioned, the Pythagoreans expressed musical intervals as numerical ratios.

(2) There is a whole series of evidence which shows that they always used the same fixed numbers for this purpose.[35] Furthermore, the numerical ratios which corresponded to the three most important consonances were always 12 : 6 (= 2 : 1, the octave), 12 : 9 (= 4 : 3, the fourth) and 12 : 8 (= 3 : 2, the fifth).

Now Gaudentius tells of the following experiment by means of which Pythagoras is supposed to have discovered the numerical ratios corresponding to the three most important musical intervals.[36]

He stretched a string across a ruler, a so-called *canon*, and divided this (ruler) into twelve parts. First of all he plucked the whole string and also half of it which comprised six units; he found that the tone of the whole string harmonized with that of the half (12 : 6) according to the *octave* ... Then he plucked the whole string once more and also three parts of the whole[37] (4 : 3 = 12 : 9), and he found that these two tones harmonized according to the *fourth*. Finally he plucked the whole string and two parts of the whole (3 : 2 = 12 : 8), and found that this time the two tones harmonized according to the *fifth* etc.

The first thing to be emphasized is that the acoustic experiment described in the above quotation cannot be characterized as 'physically impossible'. It must be admitted that in late antiquity acoustic observations and experiments which are physically impossible were frequently attributed to Pythagoras,[38] but experiments with the 'canon' are to be distinguished from these. Even those modern scholars who are rightly skeptical about such stories agree for the most part that "the Pythagorean theory of music can be verified to some extent"[39]

---

[35] Cf. H. Koller, *Museum Helveticum* 16 (1959), 240–1.

[36] Gaudentius lived in the 4th century A.D. See *Musici scriptores Graeci*, ed. C. Janus (Lipsiae 1895), p. 341. 13ff., for the quotation; cf. also B. L. van der Waerden, *Science Awakening*, p. 95.

[37] The phrase 'three parts of the whole' means *three quarters*: similarly 'two parts of the whole' in the next sentence means *two thirds*.

[38] W. Burkert, *Weisheit und Wissenschaft*, pp. 354ff.

[39] Ibid., p. 353.

8*

by means of the canon. This of course still leaves open such questions as whether it is correct to attribute any such experiments to Pythagoras himself, whether the 'canon' was perhaps just an "artificial piece of apparatus which was devised later" and whether the proportional numbers of the musical consonances were really discovered in the course of experiments of this kind. However, we may postpone answering these questions for the time being, since we are interested in a different aspect of the experiment described above.

The 'canon' mentioned by Gaudentius was a stretched string (monochord) on a measuring rod which was divided into twelve parts. The point of having a duodecimal scale was clearly to ensure that the ratios between the lengths of string which were sounded could be easily determined. So we may assume not only that the measuring rod was divided into twelve parts, but also that each part was numbered. The above passage deals really with three experiments. In each one the whole string (all twelve units of the measuring rod) was sounded first and then a shorter section of the string was plucked so as to produce a tone consonant with that of whole string. To carry out the second step of each experiment, it must have been necessary to prevent a section of the string from vibrating. Although Gaudentius does not say anything about how the string was shortened in practice, we do have some information about this from other sources. A small bridge (ὑπαγωγεύς) was moved under the stretched string.[40] This is the reason why I conjecture that the canon was not only divided into twelve parts, but was also numbered. This would enable one to see at a glance where (i.e. at which number) the ὑπαγωγεύς stood, when the string was plucked a second time, and hence how much of the string was vibrating and how much was kept silent.

As an example, let us look more closely at Gaudentius' account of how Pythagoras is supposed to have discovered the *fifth*. First he plucked the whole string and then he shortened it to 'two parts', i.e. when he plucked the string a second time, a *third* of it was kept silent and *two thirds* was sounded. So when he plucked the string for the first time, the bridge (ὑπαγωγεύς) was positioned at the end of the 'canon' (i.e. at the number 12), and it was at the number 8 when the

---

[40] Cf. Porphyrios, *Kommentar zur Harmonielehre des Ptolemaios*, ed. Düring (Göteborg 1932), p. 66, 24ff.

string was plucked a second time. The two tones which were produced were consonant according to the *fifth* and between them lay a musical interval which could be detected by the ear. This same 'interval', however, was also visible to the eye, for the bridge was first positioned at 12 on the canon, and then at 8. These two numbers are the *end points of the interval* of the fifth, or ὅροι τοῦ διαστήματος[41] as they were called by the Greeks. From this we see immediately why the Pythagoreans maintained that the proportional numbers of the fifth were 12 : 8 or 3 : 2.

This interpretation also helps us to understand why Porphyry says in the passage which has already been quoted above,[42] that the Pythagoreans thought of the *diastema* (musical interval) as a 'numerical ratio' *(λόγος)* and as a 'relationship between end point' *(σχέσις τῶν πρὸς ἀλλήλους ὅρων).*

I believe that the above satisfactorily explains the origin of the musical terms *diastema* and *horoi*, even though neither of them is explicitly mentioned in the text of Gaudentius which we have been discussing. In the acoustic experiments of the Pythagoreans, the word διάστημα meant 'straight line' and referred to that *section of string on the 'canon' which was prevented from vibrating when the second tone of a consonance was produced* (i.e. after the whole string had already been sounded) *and was in this way necessary for the creation of a musical interval.*[43] Hence a word which actually meant 'straight line' came to have the meaning of 'musical interval' as well. Furthermore, since the end points *(ὅροι)* of this straight line were numbers on the 'canon', the word διάστημα was also used to describe the 'relationship between two numbers' which was exhibited by the proportional numbers of the consonances (12 : 6, 12 : 9, 12 : 8 and so on).

I hope that the previous interpretation has thrown some light on the genesis of two important concepts from the theory of music.

---

[41] According to Plato's dialogue *Philebus* (17c–d), anyone who wants to be an expert in musical theory has to know "the *intervals* of high and low notes" *(τὰ διαστήματα . . . τῆς φωνῆς ὀξύτητός τε πέρι καὶ βαρύτητος)* and the "end points of these intervals" *(καὶ τοὺς ὅρους τῶν διαστημάτων).*

[42] See n. 34 above.

[43] This also enables us to make sense of a fragment by Philolaus (Diels and Kranz, *Fragmente der Vorsokratiker*, I. 440 25): τινὲς . . . διάστημα ἐκάλεσαν εἶναι ὑπεροχήν. The ὑπεροχή is the *difference* between the two sections of string on a 'canon' which produce the two notes of a consonance.

These are διάστημα, 'the musical interval between two tones which is expressed as a ratio between two numbers' and ὅροι, 'the end points of a *diastema* which are expressed as numbers'; hence ὅροι also came to mean 'the numbers themselves in the numerical ratio (of a musical consonance)'. According to the explanation given above, these two fundamental concepts from the Pythagorean theory of music only become comprehensible and meaningful when they are thought of in connection with the canon, for they were developed in the course of acoustic experiments with this instrument.

However, there is a widespread view according to which the monochord and canon of the Pythagoreans is supposed to be "an artificial piece of apparatus which was devised later".[44] Indeed it has even been conjectured that there was no measuring scale on the 'canon' until after the time of Aristoxenus, i.e. not before about 300 B.C.[45] The major piece of evidence in favor of this conclusion is the fact that the 'canon' is not mentioned in the *Musical Problems* (a spurious work of Aristotle) nor in the *Sectio Canonis;* similarly it is not mentioned by writers of the 4th and 5th century in general. The explanation given for this is that most probably no such musical measuring instrument existed at that time. I think that this view is mistaken, because the canon can be conclusively proved to have existed at least at the time of Plato. To see this, let us recall the following.

There is an interesting parenthetical remark made by Plato,[46] which runs as follows: ἐν μέσῳ δὲ τοῦ ἕξ πρὸς τὰ δώδεκα συνέβη τό τε ἡμιόλιον καὶ τὸ ἐπίτριτον. A correct translation of these words would be:[47] "*The ratio* $1\frac{1}{2}$ *and the ratio* $1\frac{1}{3}$ *are to be found in between* (the ratio of) *the 6 to the 12.*" Using the terminology of the Pythagoreans, ἡμιόλιον (= $1\frac{1}{2}$ = 3 : 2) and ἐπίτριτον (= $1\frac{1}{3}$ = 4 : 3) are the *proportional numbers of the fifth and fourth*. So Plato is saying that the proportional numbers of the fifth and the fourth (which of course can also be written as 12 : 8 and 12 : 9) lie between 12 and 6 (the proportional numbers of the octave, 12 : 6). [We have already quoted a fragment of Philolaus[48] which

---

[44] Cf. nn. 38 and 39 above.

[45] Cf. B. L. van der Waerden, *Hermes* **78** (1943), 177.

[46] *Epinomis* 991a.

[47] The translation and interpretation of this passage is also discussed by van der Waerden, *Hermes* **78** (1943), 186–7.

[48] Cf. n. 24 above.

states that in musical practice the octave consisted of the combination of a fourth and of a fifth. Plato's remark stresses that the proportional numbers of the fifth and fourth lie right in between the proportional numbers of the octave (12 : 6). This proves amongst other things that attempts had already been made in the theory of music to explain the resolution of the octave into a fifth and a fourth. This point will be taken up later.]

The only questions now are how it ever occurred to anyone to place the proportional numbers of the fifth and fourth (which Plato does not write as 'proportional numbers', but in accordance with an archaic convention as the corresponding fractions $1\frac{1}{2}$ and $1\frac{1}{3}$) right '*in between* (the ratio of) *the 6 to the 12*' and why 12 and 6 were chosen as the proportional numbers of the octave, when this interval can also be described as 2 : 1. I believe that these questions cannot be answered properly unless one bears in mind that the measuring instrument of the Pythagoreans (the canon) which was used to illustrate the proportional numbers of the consonances was *divided into twelve parts*. In other words, Plato's remark proves convincingly that the canon existed at that time. Modern attempts to regard the canon as "an artificial piece of apparatus which was devised later" have not been successful.

## 2.5  A DIGRESSION ON THE THEORY OF MUSIC

The present investigation is chiefly concerned with the early history of the theory of proportions. It touches on particular problems in ancient musical theory only incidentally. It seems to me that in discussing the genesis of the concepts διάστημα and ὅροι, both of which were originally applied only to the theory of music, an essential contribution has also been made to the history of the theory of proportions. As we have seen, both the concept of *diastema* (interval expressed as a numerical ratio) and that of *horoi* (the actual numbers in a numerical ratio) were developed in the course of acoustic experiments with the monochord and canon.

From now on we could concentrate our attention on the theory of proportions itself, for there is no longer any need to concern ourselves with particular historical problems in the theory of music. However, since I believe that the method applied above can also shed new

light on many questions about ancient musical theory, let me digress
here and discuss these questions, even though they are not strictly
relevant to the history of the theory of proportions.

The question of how the Pythagoreans came to express intervals
by means of numerical ratios has frequently been discussed in earlier
research.[49] The answer which was given to this question can be said
to have had two aspects. On the one hand, it was stressed that the
Pythagoreans could not be held to have "obtained the numerical
ratios of the consonances solely by observing the various lengths of
string", whereas on the other hand it was insisted that "the empirical
method used to measure the tones with numbers was a secondary
matter for the Pythagoreans" and that "there could be different opin-
ions about is". This latter assertion is borne out by Theon of Smyrna
who wrote:[50] "Some want to understand the (numerical ratios of the)
consonances in terms of weights, others in terms of sizes or in terms
of movement, and still others in terms of vessels."

It is my opinion that the way in which the question was posed
inevitably led to the view outlined above (and furthermore that this
view is only partially true). The rather abstract question of how the
Pythagoreans came to express intervals by means of numerical
ratios can only be answered by attempting to survey the rich variety
of classical and late classical literature which deals with this question.
These accounts are of uneven worth, some being completely unreliable
and if one looks only at them, then one is forced to accept the above
view. If, however, some facts about the scientific language of the
Greeks are taken into consideration and in addition the previous
question is given a more concrete formulation, then a totally different
conclusion will be reached.

The linguistic facts to which I am referring are the following.
In ancient musical theory, musical tones were usually called τόνοι,
from the verb τείνω (to stretch). Thus a musical tone was above all
the tone of a stringed instrument. Furthermore, the consonances were
called χορδῶν συμφωνία (concord of strings) in Greek. These two facts

---

[49] See E. Frank, *Plato und die sog. Pythagoreer* (Halle 1923), pp. 160–1;
W. Burkert, *Weisheit und Wissenschaft*, pp. 348–64; also van der Waerdens'
important paper (see *Hermes* **75** (1943)).

[50] Theon of Smyrna, ed. E. Hiller (Lipsiae 1878), p. 59.

clearly indicate that Greek musical theory was based predominantly on experiences and experiments with *stringed* instruments (or with just a single string).

If a more concrete form is given to the previous question, i.e. if one asks why the Pythagoreans used the names διάστημα and ὅροι to denote musical intervals expressed as numerical ratios, then it becomes immediately apparent that these proportional numbers must originally have represented *ratios between lengths*. These same two expressions also indicate how important experiments must have been at one time in Pythagorean science. *The concepts 'diastema' and 'horoi', which have been analyzed above, could never have originated without musical experiments using the canon.*

The assertion that the empirical method used to ascertain the numerical ratios of the consonances was a secondary matter for the Pythagoreans (which is correct as it stands), receives a new emphasis in view of the above. Although the concepts *diastema* and *horoi* originated in the course of experiments with the canon, Hippasus was already able to establish the most important consonances by means of bronze *diskoi*.[51] Similar experiments were also carried out with vessels containing different amounts of water and with wind instruments.[52] Clearly the Pythagoreans wanted to show that other kinds of experiments led to the same proportional numbers for the consonances as did those original experiments with the monochord and canon. This explains why these same Pythagoreans could subsequently hold the view that the empirical method used to ascertain the proportional numbers of the consonances was a secondary matter. The actual numerical ratios of the consonances which had been established once and for all were important to the Pythagoreans, not the empirical method used to ascertain them.

I also believe that the above account of the concepts *diastema* and *horoi* enables us to give a better explanation of many things from the patently false tradition which originated in late antiquity. As an example, let us consider the following passage:[53]

---

[51] Diels and Kranz, *Fragmente der Vorsokratiker*, I. 18. 12; cf. also W. Burkert, *Weisheit und Wissenschaft*, pp. 355–6.

[52] Cf. B. L. van der Waerden, *Hermes* **78** (1943), 172.

[53] Ibid., p. 170.

"There is a nice story found in Nichomachus (p. 10 Meibom), Gaudentius (p. 13 Meibom) and Boethius (pp. 10–1 Friedlein) which tells
how the Pythagoreans came to represent intervals by means of
numerical ratios. It cannot possibly be true however. According
to the story, Pythagoras was passing by a smithy and heard the
tones of the falling hammers produce various intervals, in particular
the octave, fourth and fifth. He carefully weighed the hammers
and found that their weights were in the same ratio to each other
as were the numbers 12, 9, 8 and 6. He then went home, suspended
four identical strings and attached weights to them, which were
proportional to the weights of the hammers. Now he established
that the string weighted with 12 units, produced a note which
was an octave higher than the one produced by the string weighted
with six units. Similarly the strings weighted with 9 and 8 units,
produced notes which were a fourth and a fifth lower than the one
produced by the original string. He corroborated these results by
experimenting with stretched strings and other instruments (the
aulos, syrinx, etc.)."

Among the reasons for rejecting this traditional story as false, is the
following:[54]

"If one had actually tried to carry out this experiment and measure
the numerical ratios (of the consonances) by means of suspended
weights, as Pythagoras is said to have done, then the attempt
would inevitably have ended in failure. For the weights corresponding to the octave are in the ratio $1 : \sqrt{2}$, not $1 : 2$."

So we can see how modern research has uncovered some false-
hoods in traditional accounts. In doing so, however, the question is
left untouched of how this experiment, which could only have been
thought out and never carried out, ever occurred to anyone. Yet it
seems to me that as far as this question is concerned, at least a part
of the above account is very informative. When it is alleged that Pythagoras "suspended four identical strings and attached weights to

[54] Ibid., p. 173.

them", it is clear that he wanted to use the weights to measure the *tension* of the strings. An experiment of this kind could only have been devised after it was realized that the pitch of a note depended on the tension of the string in question. This, however, is exactly what Aristoxenus, the opponent of the Pythagoreans (see p. 113) who refused to acknowledge the connection between proportional numbers and consonances, succeeded in doing. Thus I believe that the experiment described above was in part a 'Pythagorean answer' to the teachings of Aristoxenus. The person who invented this impractical experiment and attributed it to Pythagoras, wanted to convince his readers that the traditional proportional numbers of the consonances could also be demonstrated by considering the tension of the string. Of course this cannot be done, yet the so-called experiment is not without importance for the history of science.

In other words, my position is that the late classical tradition, which is concerned with the science of music and which gave rise to the story quoted above, originated in attempts to vindicate the Pythagorean theory and that these attempts met with only partial success. At first, when the expressions *diastema* and *horoi* were coined (with reference to the monochord and canon), one did not need to worry overmuch about the tension or thickness of the single stretched string. At this time no one asked whether the tone itself was a speed, a movement of the air, a vibration or something else. Hence it could be maintained that the consonances depended simply on *numbers*. (These numbers, of course, originally expressed ratios between the lengths of individual sections of string on the monochord.) It was only later, when it became necessary to defend the ancient teaching against its opponents, that these same numbers were also supposed to represent ratios between *weights* (i.e. between various *degrees of tension* of the string) and between the various *thicknesses* of the objects which produced the tones (in this case, bronze *diskoi*). This also explains the invention of partly nonsensical stories like the one mentioned above, which tell of impossible observations and impractical experiments. Furthermore, it explains why Theon of Smyrna stressed that it was not important how the proportional numbers of the consonances were obtained.[55]

---

[55] See n. 50 above.

In the same way, I believe that I can explain why ancient musical theorists were inconsistent in their method of associating a number with the pitch of a note. Originally, when the starting point was experiments with the canon, a *larger number* (a longer section of string) had naturally to be associated with a *deeper note*. But this method had to be turned around completely when it became necessary to defend the ancient teaching against subsequent objections, and take newer speculations into account. As van der Waerden wrote:[56] "It is clear that those who sought to interpret the numerical ratios in terms of weight or speed, were obliged to assign *a larger number to a higher note* ..."

There is yet another interesting transitional state which is worth noting. I am referring to a case in which a newer theory has already been taken into account, although the original method of associating proportional numbers with intervals is still used. Let me quote once more from van der Waerden's work:[57]

"A noticeably vacillating position is taken in the *Section Canonis:* whereas in the introduction it states that two tones are in a numerical ratio to each other because tones are quantities which *are increased when the tone is raised and diminished when it is lowered*, in the main body of the text tones are represented by straight lines and higher tones correspond to shorter lines."

## 2.6 END POINTS AND INTERVALS PICTURED
## AS 'STRAIGHT LINES'

I believe that many details of the pre-Eudoxian theory of proportions, as well as of the Greek mathematical theory of music, can be seen in a new light if one bears in mind the explanation for the origins of the concepts *diastema* and *horoi* which was discussed fully in chapter 2.4. So let me recapitulate here and repeat the etymologies of these two words.

---

[56] *Hermes* **78** (1943), pp. 173 f.
[57] Ibid., p. 173.

In the field of music διάστημα (straight line) originally meant the piece of string on a 'canon' which was prevented from vibrating when the second tone of a consonance was produced (i.e. after the entire string had already been plucked). If this piece of string had not been kept still, it would not have been possible to produce the second note of the desired consonance on the monochord.

The ὅροι were the end points of this piece of string (the *diastema*) and were shown as two numbers on the *canon*.

It would not be superfluous to indicate here the precise way in which the concept of musical interval (*diastema*, in the above sense of the word) developed into that of *the ratio between two numbers*.

According to the previous explanation, the *diastema* was the *actual interval* between two tones. It could be determined acoustically and was also visible as a length of string on the 'canon' (namely the length by which the section of string producing the first note differed from the one producing the second). This is the only interpretation which makes sense of such musical expressions as ὅροι τοῦ διαστήματος (end points of an interval). This length of string on the monochord was specified by its two end points (by two *numbers* on the canon), just as a straight line is given by its two end points in geometry.

These matters, however, can be viewed in another way which differs in part from the above. A musical interval is characterized by the consonance of the two *notes* which bound it. These two *notes* harmonize with each other and form the interval which is said to lie between them. Hence it was often the case in musical *practice* that the consonance of two notes was described by the two strings which produced it. For example, the octave (as has already been mentioned[58]) was called the concord of the ὑπάτη and νήτη. This suggests the idea that in the *theory* of music *diastema* may also have been construed as the concord produced on the monochord by *two different lengths of string*, instead of as a piece of string which did not vibrate. Of course these two lengths of string could still be characterized by the end points of the latter piece. As a matter of fact, this is the viewpoint found in the *Sectio Canonis*. To see this, we need only look at the beginning of the proof of Theorem 1 (in the *Sectio Canonis*): ἔστω διάστημα τὸ ΒΓ . . . . Heiberg translates this as "sit intervallum *BC* . . .". A reading of

----

[58] See n. 21 above.

the Greek words together with their Latin translation would lead one to think at first that the letters $B$ and $C$ (or $B$ and $\Gamma$) refer to the *end points of a straight line* (the *diastema*). This agrees with the explanation offered in chapter 2.4 according to which the *diastema* was the section of string which did not vibrate (the straight line upon which the musical interval depended), and was specified by its two end points. Surprisingly, however, the diagram which accompanies this passage in the *Sectio Canonis* makes it clear that the letters $B$ and $\Gamma$ (two numbers on the canon) represent *two straight lines of different lengths*.

The only way to explain this fact is to say that by the time the propositions in the *Sectio Canonis* received their final form, *diastema* no longer meant simply "the length of string between two numbers on the canon, which did not vibrate (and by which the section of string producing the first note differed from the one producing the second)". Instead it referred to the *two sections of string which produced the notes* in question and, since these could be characterized by the same two numbers which had originally been used to specify the actual interval (the piece of string which did not vibrate), it had come to mean '*the ratio between two numbers*'.

This slight change in outlook (namely the characterization of a musical interval in terms of the two lengths of string which produced the notes bounding it, instead of as a piece of string which did not vibrate) led not only to the creation of a new concept (ratio between two numbers), but also to a curious anomaly in the use of language.

I am referring to the fact that the word διάστημα (in the singular) which originally denoted a *single straight line* in the theory of music, had to be illustrated by *two straight lines of different lengths*. These two lines represented the lengths (expressed as *numbers*) of the sections of string which produced the notes, and hence could also be regarded as representations of the numbers themselves. Diagrams of this sort become even more interesting if one considers that what were earlier called the end *points* of the interval' (ὅροι τοῦ διαστήματος) must have been shown as *straight lines* in them.

In this way it also becomes clear why the two straight lines which in the diagram from the *Sectio Canonis* are supposed to represent the two sections of string *could each be designated by only one letter*. As long as *diastema* meant 'the piece of string on the monochord, which did not vibrate', the obvious way to characterize this straight line on the 'canon', was in terms of its two 'end points' *(horoi)*. One end point was usually the number 12, i.e. the end of the canon (because for the most part a note consonant with the key note of the whole monochord was sought). The other end point was the number at which the ὑπαγωγεύς stood when the second note was produced, i.e. at the end of the shorter section of string. This method of characterizing a *straight line* in terms of its two end points (in terms of two numbers on the canon) was also the usual one in Greek geometry.

However, when the word *diastema* underwent the slight change in meaning described above, it was no longer possible to use this method of characterization. The two sections of string which produced the notes could not be specified by their end points. The reason for this was that the numbering on the canon began with 0 and the Greeks did not have any symbol for this number. So the common end point of these two lines was situated at one end of the monochord and there was no number to describe it. This made it all the easier to characterize each one in terms of its other end point and describe it *by a single number* (or letter). Of course the two numbers (or letters) belonging to the sections of string which produced the notes were identical with the end points *(horoi)* of the piece of string which did not vibrate.

In the *Sectio Canonis* a *distema* is in fact always illustrated by two lines, each one of which is labelled *by a single letter*. So in the Greek theory of music, a *number* was always pictured as a *line* and this line was designated by a single letter, *not by its two end points* (as was the case in geometry).

This method of representing *numbers* by lines was taken over from the theory of music into Euclid's arithmetic, although not in a very consistent manner. Euclid always represented numbers as lines. It is instructive to see the way in which letters were used to designate these lines which were supposed to stand for numbers, in Book VII of the *Elements*, for example:

(1) There is a series of propositions (e.g. VII. 3, 12, 13, 14, 16 and 17) in which lines are described *by just a single letter*. This way of picturing numbers is the same as in the *Sectio Canonis*.

(2) In the same book, however, there are propositions (VII. 1, 2, 7 and 8 amongst others) in which numbers are represented by lines and

these lines are labelled by two letters, one at the beginning and one at the end. This was the usual method in geometry.

(3) There are even some propositions (e.g. VII. 4, 5, 6 and 9) which use both the above methods at the same time. In these propositions, one number is shown as a line labelled by a *single letter*, whereas another is shown as a line labelled by *two letters*.

So we can see that at Euclid's time a decision had not yet been made about the way to designate (number) lines in arithmetic, i.e. about whether to adopt the 'musical' convention (deriving from experiments with the canon) which used a single letter, or the more 'geometric' one which used two.

## 2.7 'DIPLASION', 'HEMIOLION' AND 'EPITRITON'

In the course of our investigations up to now we have learned to distinguish two stages in the development of the concept *diastema* as it applies to the theory of music.

(1) During the first stage *diastema* (musical interval) was the length of string on the monochord which did not vibrate, and by which the section of string producing the first note differed from the one producing the second. It was characterized in terms of its two end points *(horoi)* and these could be read off as two numbers on the 'canon'.

(2) The account of this concept given in the *Sectio Canonis* clearly belongs to a later stage in its development. Here *diastema* does not refer to the piece of string on the monochord which was kept from vibrating, but rather to the two sections of string which produced the notes. At this time the word unequivocally meant the *ratio* between the two numbers which expressed the lengths of these two sections of string.

Now I believe that I can even point out a third stage which preceded the other two. This stage might be said to have prepared the way for the concept *diastema*. Since in my opinion it gives us a great deal of information not only about the theory of music but also about the whole history of Greek mathematics, I would like to discuss it rather more fully. My starting point is the passage from Gaudentius, which has been quoted once:[59] "Pythagoras stretched a string across a ruler, a so-called *canon*, and divided this (ruler) into twelve parts, etc.". These words indicate that the numerical ratios of the three most important consonances were found only *after the 'canon' had been divided into twelve parts*. The canon with its twelve divisions was the instrument with which the Pythagoreans performed experiments and by whose help they were able to ascertain the numerical ratios of the important musical intervals (12 : 6 for the octave, 12 : 9 for the fourth and 12 : 8 for the fifth). In fact, on the basis of Gaudentius' account of these experiments together with our knowledge of the canon, we were able to find out the original (and exact) meanings of the musical terms *diastema* and *horoi*.

It is time, however, to look somewhat more critically at Gaudentius' account. It seems unlikely that the experiment he outlines, was originally carried out by first *dividing the canon into twelve parts*. This would not have been absolutely necessary, so long as one merely wanted to exhibit the proportional numbers of the three most important consonances (the octave, fourth and fifth) separately on the monochord, without regard to the connections between them. To obtain the octave, it is only necessary *to halve* (2 : 1) the stretched string. The fifth is obtained by *dividing in into three parts*. (The whole string and two thirds of it produce notes which are a fifth apart, 3 : 2.) Similarly, it suffices to divide the string *into quarters* to obtain a fourth (4 : 3). Even the words which Gaudentius chose to describe these experiments are such as to enable the attentive reader to establish for himself that the canon must originally have been divided into two, three and four parts, before it was divided into twelve. *Thus it can hardly be the case that the original division of the musical canon was into twelve parts.*

In other words, I construe Gaudentius' account as an outline of how these musical experiments were performed at a relatively late stage. Originally they were carried out simply with a monochord. The canon was only introduced later, when it was discovered that there were good reasons for dividing the measuring rod beneath the stretched string into twelve parts.

---

[59] See n. 36 above.

**9** Szabó

I hope to be able to show that the above is a correct reconstruction
of the process of development, by explaining the meanings of some
terms from the theory of music which clearly came into being *before
the introduction of the canon*. The terms to which I am referring are
the oldest known names for the *octave, fourth* and *fifth*.

In the Pythagorean theory of music[60] the most common name for
the numerical ratio of the *octave* was διπλάσιον διάστημα or, in a fragment
of Philolaus,[61] διπλόον. This expression is translated by the formula
2 : 1 or, bearing in mind the twelve divisions on the canon, by 12 : 6.
However, one should also attempt to translate it literally and then consider it in connection with the monochord. The technical term
διπλάσιον διάστημα means literally 'doubled line'. It is clear that the
octave received this name because after the stretched string had
been plucked once, it was shortened to half its length and the shorter
section of string which resulted produced a note lying an octave
above the one produced by the whole string.

This still leaves the question of what the word *diastema* itself
means in this context (i.e. in the expression διπλάσιον διάστημα 'doubled
line'). It clearly does not have the meaning which I explained earlier,
namely "the piece of string on the monochord which was kept from
vibrating". When used as a name for the numerical ratio of the octave,
the expression διπλάσιον διάστημα can *only be referring to the whole
monochord*. In my opinion the only conceivable way to account for
this name is the following.

The half of the stretched string (i.e. the section of string which
was plucked to produce the second note of an octave)[62] was regarded
as the 'unit'. Relative to this unit, the whole monochord becomes a
'doubled line'. So if the monochord is treated as a 'doubled line', i.e.
if the whole length of string is plucked first and then the unit chosen
for the present case (half of the monochord) is plucked, the consonance
of the octave is obtained.

---

[60] Cf. *Nicomachi Enchiridion*, ed. *C*. Janus 250. 22ff; also Plato's *Timaeus*, 36.

[61] See n. 24 above.

[62] The other possibility is that the *muted section* of string was regarded as the
unit, but this seems unlikely in view of the related terms ἡμιόλιον and ἐπίτριτον διάστημα.

At first glance this explanation is rather surprising and may not seem to make much sense. Nonetheless I must also insist that there is no possibility of explaining the ancient Pythagorean names for the proportional numbers of the *fifth* and *fourth* except in this manner.

On the monochord, a fifth is obtained by first plucking the whole string and then holding a third of it still, so that only two-thirds of the whole monochord are sounded the second time (3 : 2 = 12 : 8). The ancient Pyrhagorean name for this interval was ἡμιόλιον διάστημα[63] i.e. 'line $1\frac{1}{2}$ (units) in length'. This description also refers to the whole monochord, for in this case as well, the section of string which produced the second sound (the length '0–8' on the canon) was the unit. This was regarded as a 'whole' *(ὅλον)* and in relation to it, the complete monochord (i.e. the length '0–12' on the canon) was 'one whole together with half of the whole', namely a 'line $1\frac{1}{2}$ (units) in length' *(ἡμιόλιον διάστημα)*. So if the monochord is treated as a 'line $1\frac{1}{2}$ (units) in length', i.e. if the whole string is plucked first, followed by the unit (two-thirds of the string) relative to which the whole string is a 'line $1\frac{1}{2}$ (units) in length', then one obtains a fifth.

The Pythagorean name for the numerical ratio of the fourth (4 : 3), ἐπίτριτον διάστημα,[64] can be explained in a similar way. On the monochord a fourth is obtained by keeping a quarter of the string from vibrating when it is plucked for the second time. Thus the second note is produced by three quarters of the string (12 : 9). If this length (0–9) is regarded as the unit, then the whole monochord (0–12) becomes a 'line $1\frac{1}{3}$ (units) in length', ἐπίτριτον διάστημα. (The Greek word ἐπίτριτον means 'a third in addition to it', i.e. 'in addition to the unit'.)

Surprising confirmation for the above explanation of the terms ἡμιόλιον and ἐπίτριτον διάστημα is to be found in Theorem 8 of the *Sectio Canonis*. To prove its correctness, however, it would be necessary to reprint here almost the entire Greek text of this theorem together with its accompanying diagrams. (The only thing which should not be forgotten when interpreting this proposition [*Sectio Canonis* 8], is that a *diastema*, the ratio between two numbers, is always illustrated

---

[63] See n. 21 above (problem 41) or Plato's *Timaeus*, 36.

[64] Cf. Plato, *Timaeus*, 36; or Philolaus, fragment 6 (in Diels and Kranz, *Fragmente der Vorsokratiker*).

by *two straight lines* in the *Sectio Canonis*. Hence the whole monochord and the piece of string which produces the second note, the unit, are represented by *two distinct straight lines*.)

The etymological investigation conducted above has led us to two conclusions:

(1) The introduction of the terms *diplasion, hemiolion* and *epitriton diastema* into the theory of music antedates *the division of the canon into twelve parts*, although the proportional numbers of the three most important consonances were known already at the earlier time and were given as *ratios between lengths*. The question of whether these ratios were discovered by the use of an actual monochord or by shortening one of the strings on a multi-stringed instrument is immaterial. In the end the monochord, a one-stringed experimental instrument, was adopted; prior to this, experiments were certainly made to produce any desired note using only one string of a multi-stringed instrument, so that one would be able to manage with just a single string.

(2) Another important fact to which I must call special attention is that by the time the names *diplasion, hemiolion* and *epitriton diastema* were introduced, the technical meaning of the word *diastema* was already very much the same as what we understand by 'musical interval', although in a rather special sense. In these names *diastema* does not mean 'the piece of string which was kept from vibrating' (the visual picture of the actual audible interval); instead it refers to the *whole monochord*, the intention being to measure this as a different *diastema* in the case of each of the three most important consonances. The names can be read as different methods of measuring the monochord as a *diastema*, so as to obtain the three consonances. The monochord becomes a *diplasion diastema* (a doubled line) in the case of the octave, because here the whole string was plucked first, followed by the *unit* (which in this case was a half). In the case of the fifth however, the same monochord is a *hemiolion diastema* (a line $1\frac{1}{2}$ units in length), because this time the whole string is $1\frac{1}{2}$ times as long as the unit which produces the second note. The case of the fourth can be explained similarly. We see that at this time there could not as yet be any talk of *end points on the canon (ὅϱοι)*.

A comparison between the system of measurement described above, and the way in which the numerical ratios of the consonances were specified on the canon, immediately reveals the ingenious simplification which was brought about by dividing the 'canon' into twelve parts. While the names *diplasion, hemiolion* and *epitriton diastema* were current, it was necessary to seek three *different units of length* for measuring the monochord. In addition, this system of measurement led to the use of *fractions* in two cases *(hemiolion* and *epitriton).* However, when a measuring rod which had already been divided into twelve parts was laid under the monochord, the whole process suddenly became much simpler. It was no longer necessary to find a new unit of length for each of the three consonances, nor to express the length of the same monochord in various different units. Instead of this, the length of *the piece of string which was kept from vibrating* could simply be specified by two numbers on the canon. Thus the introduction of a numbered canon seems to have brought with it the following important innovations:

(a) The word *diastema* no longer referred to the whole monochord (as in *diplasion, hemiolion* and *epitriton diastema*), but just to the section of string on the monochord which was kept from vibrating. Only through this did a *diastema* come to be a musical interval in our sense of the words, i.e. an *interval* between two successive sounds.

(b) *"Ὅϱοι, the end points of the piece of string which did not vibrate, expressed as two numbers* did not enter into the theory of music until after the introduction of the numbered canon. (Of course the concept of *logos*, in the sense of the 'ratio between two numbers', also did not exist before the introduction of the canon. On this topic see chapter 2.18, 'The creation of the mathematical concept of λόγος').

When the technical terms we have been discussing were first introduced into music, there were as yet no such things as *horoi* because at that time the whole monochord, not the section of string which did not vibrate, was treated as the *diastema*.

(c) After the introduction of a numbered canon, the fractions which had been used earlier in the theory of music ($1\frac{1}{2}$ and $1\frac{1}{3}$) became redundant. The fifth and fourth were from now on described not as lines $1\frac{1}{2}$ or $1\frac{1}{3}$ units in length, but simply in terms of the numbers which gave the end points of a certain section of string on the canon (namely that section which did not vibrate when the second note of the consonance in question was produced). Hence they could be specified *in terms of whole numbers.* Of course the old names (for the fractions)

were still retained, but these were no longer understood as referring
to fractions, instead they denoted ratios between whole numbers.
Hence the ancient term ἡμιόλιον διάστημα, for example, which originally
meant just a *'line* $1\frac{1}{2}$ (units) *in length'*, came in classical times to
mean 3 : 2 as well as 12 : 8 or 9 : 6 (all proportional numbers of the
fifth).

## 2.8 THE EUCLIDEAN ALGORITHM

The etymology of the musical terms *diplasion, hemiolion* and
*epitriton diastema* has led us back to the ancient period, during which
experiments to ascertain the numerical ratios of the three most important consonances were made with only a stretched string and
without a canon. It is wortwhile to take a closer look at the method
which was employed in these experiments, since it also played an
important role in Greek arithmetic.

The problem was to find a note which, together with the tone produced by the whole stretched string, would make a fifth or a fourth.
It was already known at this time (from experience with wind-
instruments) that the object which produced the sounds (in this case
a stretched string) would have to be diminished in order to obtain
the desired consonance; the only question was, by how much. Of
course the answer to this question could only be obtained after much
trial and error. When it was finally discovered how  long a piece of
string had to be kept from vibrating so that the remainder would
produce a note which was a fourth or a fifth above the one produced
by the whole string, the section of string which yielded the second
note of the consonance was regarded as the *unit*. The last step was
to express the length of the piece of string which was kept from
vibrating as a fraction of the unit which had been chosen. In the case
of the fourth it was a third and in that of the fifth, a half.

The squence of steps which made up this method may be summarized
as follows:

(a) On discovering the shorter section of string which, together
with the whole string, yielded the desired consonance, one was actually faced with *two lines of different lengths*. The one was the whole
monochord and the other was the section which produced the second
note.

(b) The shorter line was considered to be the unit in this case, and was *subtracted* from the longer one. This left the piece of string which had been kept from vibrating as a *remainder*.

(c) To ascertain the length of the remainder, one tried to see *how many times it could be subtracted from the shorter of the two lines*. This could be done twice in the case of the fifth and three times in the case of the fourth, because in the former case the piece of string which did not vibrate was *half* as long as the one which produced the second note, whereas in the latter, it was only a *third* as long.

Thus the shorter line was first subtracted from the longer one and then the remainder was subtracted from the shorter, until in the end nothing was left over.

The method described above is well known to readers of Euclid. It is the so-called 'Euclidean algorithm', or method of *successive subtraction*, as it is termed in the more recent literature on the history of Greek mathematics.[65] In the *Elements* there are two contexts in which it occurs. The first is in Book VII where it is used to find the greatest common divisor of two numbers. The method is applied to answer the question of whether or not two numbers are relatively prime. As *Proposition VII. 1* states:

"Given two unequal numbers, if one subtracts successively [i.e. the process is repeated with the subtrahends and remainders] the smaller from the greater and no remainder measures the preceding number exactly, until the unit is reached, then the original numbers are relatively prime."[66]

On the other hand, in Book X successive subtraction is used to answer the question of whether two quantities are commensurable. If this method is applied to two incommensurable magnitudes, then the process of successive subtraction will not terminate (cf. *Elements* X. 2).

As far as I know, the origin of this method has never been elucidated. The only thing which was known about it (from Zeuthen's[67] work),

---

[65] Cf. B. L. van der Waerden, *Erwachende Wissenschaft*, pp. 208 and 236.
[66] Heath's translation.
[67] *Det Kgl. Danske Vidensk. Selsk. Skrifter, nat. og math.* (Copenhagen), vol. 8, Raekke I, 1917, pp. 199–381.

was that it was known already to Aristotle.[68] As Junge once wrote[69]: "The method may well be very ancient; one is inevitably led to it, if one wants to ascertain the ratio between the lengths of two lines."

The explanation presented above of the musical terms *hemiolion* and *epitriton diastema* suggests that the Euclidean algorithm (successive subtraction) may well have originated in the Pythagorean theory of music. More precisely, this method was developed in the course of experiments with the monochord and was used originally to ascertain the *ratio* between the lengths of two sections on the monochord. In other words, successive subtraction was first developed in the musical theory of proportions. This conjecture fits in very well with the other known facts about the method. Let me recall some of these here.

Euclid described the operation of successive subtraction with the verb ἀνϑυφαιρεῖν.[70] The noun derived from this is ἀνϑυφαίρεσις.[71] Aristotle mentions this word once[72] and at the same time alludes to an ancient definition of *sameness of ratio* which Euclid did not include in the *Elements*. This definition is quoted in the commentary of Alexander of Aphrodisias as well as in the Suda (see under ἀνάλογον) and runs as follows: ἀνάλογον ἔχει μεγέϑη πρὸς ἄλληλα, ὧν ἡ αὐτὴ ἀνϑυφαίρεσις, 'Magnitudes having the same *anthyphairesis* are in the same ratio'.

In my opinion the true meaning of this ancient definition can be understood correctly only if it is borne in mind that the method of *successive subtraction (ἀνϑυφαίρεσις)* has already been used to calculate the *ratio* between two lengths of string on the monochord. Since the ratio between two line segments was established by means of successive subtraction, it seemed natural to define ratios as being *the same*, if in each case the method of *successive subtraction* (applied to the two quantities concerned) *yielded the same result*.

[In this way the Greeks may also have had their original interest in the concept of 'sameness of ratio' aroused by considerations hav-

---

[68] Aristotle, *Topica* 3, p. 158b 29–35; cf. also nn. 7, 8 and 9 above.

[69] G. Junge, 'Von Hippasus bis Philolaus, "Classica et Mediaevalia" ', *Revue Danoise de Philologie et d'Histoire* **19** (1958), pp. 41–72.

[70] *Elements* VII. 1.

[71] Alexander Aphrodisiensis (Vol. 2 of *Commentaria in Aristotelem Graeca*, ed. M. Wallies, Berlin 1892), p. 545.

[72] See n. 68 above.

ing to do with the theory of music. It must have been immediately evident to them that in establishing the proportional numbers of the fifth, for example, it was all the same whether a monochord with just three divisions, or a canon with *twelve*, was used. The numerical ratios 3 : 2 and 12 : 8 (or even 9 : 6 for that matter) were equal to each other. Furthermore, in all these cases successive subtraction applied to the two quantities concerned yielded the same result.]

## 2.9 THE CANON

In chapter 2.7 [points (a), (b) and (c) on p. 133] I summarized the principal ways in which experiments on the monochord were facilitated by the introduction of a numbered measuring rod. Now, however, I would like to talk about some other advantages it had.

It was well known from musical practice that if a pair of tetrachords were joined together, the two outermost strings (the ὑπάτη and the νήτη) produced an octave.[73] Furthermore, it was recognized that the octave was composed of two smaller musical intervals. As Philolaus wrote in the fragment quoted previously:[74] "The extent of an octave is a *fourth* and a *fifth* ... For the distance between the topmost *(ὑπάτη)* and the middle string *(μέσσα)* is a fourth and the distance between the middle and the bottommost *(ἐπὶ νεάταν)* is a fifth."

Now the question is whether these same facts can also be illustrated on a monochord, i.e. whether it can be shown on a monochord that the *diastemata* of the fourth and fifth together make up the *diastema* of the octave.

Of course it was impossible to give a satisfactory answer to this question as long as three different units (depending on whether an octave, fourth or fifth was being considered) were used to measure the length of the monochord. When a measuring rod divided into four equal parts was placed beneath the monochord, however, it immediately became much easier to provide an answer. The *diastema* (the piece of string which was kept from vibrating) of the octave

---

[73] See n. 21 above.
[74] See n. 24 above.

could now be characterized in terms of the two end points of the string by the numbers 4 and 2 on the measuring rod (4 : 2). So it was easy to show that this *diastema* was in fact composed of two smaller *diastemata*, namely the fourth (4 : 3) and the fifth (3 : 2). An octave is a *ratio composed* of a fourth and a fifth,[75] more precisely, on the monochord the *length* of an octave (4 : 2) is composed of the *length* of a fourth (4 : 3) plus the *length* of a fifth (3 : 2).

Thus experiments with the monochord and numbered measuring rod revealed the same facts as had been learned from musical practice. Such experiments with a canon, which perhaps was originally divided into only four parts, may also have helped to give τετρακτύς the almost mythical meaning which it had for the Pythagoreans.[76]

It should not be forgotten, however, that there are no sources which mention a canon divided into four parts. Although the invention of the canon is traditionally attributed to Pythagoras (and whatever one may think of the half legendary figure of Pythagoras, this dating is not to be doubted), the sources for this tradition either do not discuss the way in which the canon was divided, or they only mention that it was *divided into twelve parts*. Hence the existence of a canon with four division can only be conjectured. Nevertheless it is quite a plausible conjecture, if one bears in mind the proportional numbers of the consonances (4 : 3, 3 : 2 and, putting these two together, 4 : 2) and the Pythagorean *tetraktys*.

On closer inspection the disadvantages of a canon with four divisions (if an instrument of this kind was ever really used by the Pythagoreans) as opposed to one with *twelve*, soon become apparent. As an example, let us consider the following case.

Of course in musical practice it was immaterial whether, given a note, one which differed from it by a fourth was found first, followed by another differing by a fifth from the latter, or whether the order was reversed. In either case the combination of the two intervals resulted in a tone which differed from the original one by an octave. It is impossible, however, to illustrate this fact (that the order in which the two intervals are found may be reversed) by means of a *canon with four divisions*. In this case it is only possible to combine

---

[75] B. L. van der Waerden, *Erwachende Wissenschaft*, p. 183.
[76] Ibid., p. 157.

the two *diastemata* if the fourth precedes the fifth (4 : 3 and 3 : 2 together make 4 : 2). On the other hand, a canon with twelve divisions enables one to choose 12 : 9 (a fourth) followed by 9 : 6 (a fifth), or alternatively 12 : 8 (a fifth) followed by 8 : 6 (a fourth). In either case an octave (12 : 6) results from combining the two *diastemata*.

The reason for dividing the measuring rod into *twelve* parts can now be appreciated as well. We need only bear in mind that originally the *length* of the monochord was measured in *three different units* corresponding to the three most important consonances. In the case of an octave, the monochord was divided into *two* parts; in the case of a fourth, it was divided into *four*, and in the case of a fifth, it was divided into *three*. To illustrate these three consonances on a single monochord all at the same time, it was first necessary to find the *least common multiple* of 2, 3 and 4, which is exactly 12.

Apart from allowing a fourth and a fifth to be combined in any order, a canon with twelve divisions also makes it possible to illustrate another interesting connection between the numerical ratios of the consonances.

I am referring to the fact that in ancient sources the interval of the fifth is frequently described as being *greater* than that of the fourth. (Let me remark here in passing that this assertion by itself already reveals the way in which it was customary to measure intervals on the canon. The longer section of string, or the whole monochord, must always have been plucked before the shorter one. Hence the pairs of numbers obtained were 3 : 2 and 4 : 3. The way in which the ancients thought precluded the possibility of reversing the order of either pair (i.e. of associating 2 : 3 with the fifth, or 3 : 4 with the fourth).[77] If such a change had been permitted, the assertion that the fifth was greater than the fourth would certainly not have been correct. (Although $3/2 > 4/3$, $2/3 < 3/4$.) Now this can be shown very easily on a canon with *twelve divisions*. In this case the *diastema* of the fifth was the length between the numbers 12 and 8, whereas the *diastema* of the fourth was represented by the length between the numbers 12 and 9. Hence it was immediately apparent *by how much*

[77] Cf. the German translation of Philolaus' words found in Diels and Kranz, *Fragmente der Vorsokratiker*, vol. 1, p. 409.

*the fifth was greater than the fourth, namely by the length between the numbers 9 and 8 on the canon.* As Philolaus wrote:[78] " . . . the fifth exceeds the fourth by an ἐπόγδοον (9 : 8)".

## 2.10 ARITHMETICAL OPERATIONS ON THE CANON

It is worthwhile to take a closer look at Philolaus' assertion that " . . .the fifth exceeds the fourth by an *epogdoon*". Of course ancient literature on the theory of music is full of such statements. For example, there is a passage from Ptolemy's *Theory of Harmony* which discusses the subject matter of Philolaus' remark in somewhat more detail.[79] This passage is of interest to us chiefly because it talks about the proportional numbers *hemiolios logos* (3 : 2) and *epitritos logos* (4 : 3), instead of *fifths* and *fourths* and goes on to say that the 'difference' (ὑπεροχή) between these two ratios is an *epogdoos logos* (9 : 8).

The method one would use to obtain this difference (between the fractions 3/2 and 4/3) is obviously division; for 3/2 : 4/3 = 9/8. Yet the Greek expression for 'difference' used here (ὑπεροχή) customarily refers to a *difference obtained by subtraction*, not to a 'quotient resulting from division'. There is, for example, a passage in Porphyry's commentary on Ptolemy's *Theory of Harmony* in which it is explicitly stated that although the pairs 6, 3 and 2, 1 have the same *logos* (6 : 3 = 2 : 1), their *differences* are not equal (αἱ δ'ὑπεροχαὶ ἄνισοι).[80] In the one case the *difference* is 3 (6–3 = 3) and in the other it is 1 (2–1 = 1). So the word ὑπεροχή, just like its opposite, ἔλλειψις, is undoubtedly a technical term which had to do with *subtraction;* both words denote the *difference.*

The question now is why the *quotient* of the ratios 3 : 2 and 4 : 3 (i.e. 9 : 8) was called a *difference* in the theory of proportions, even though the smaller of these two ratios (4 : 3) did not represent a subtrahend but a divisor. I believe that this question can be easily

---

[78] See the previous note. This fragment of Philolaus is further evidence for the fact that the canon with its twelve divisions was already in existence at his time.

[79] *Die Harmonielehre des Klaudios Ptolemaios*, ed. I. Düring, Berlin (1930). The whole passage is translated in van der Waerden, *Hermes* **78** (1943), 166ff.

[80] Düring's edition, p. 92.1ff.

answered on the basis of some facts about the canon which I explained above. The proportional numbers of the fifth (12 : 8, *hemiolios logos*) and fourth (12 : 9, *epitritos logos*) both referred to *lines* on the canon, the former to the piece of string between the numbers 12 and 8, and the latter to the piece between 12 and 9. In order to find out how much smaller the ratio of the fourth (12 : 9) was than that of the fifth (12 : 8) was on the same canon, it was necessary to subtract the smaller line from the greater. This left a 'difference' (ὑπεροχή), namely the *line* between the numbers 9 and 8 on the canon.

Thus, if in the Greek theory of proportions the quotient of two ratios is described as a *difference*, the historical explanation for this remarkable fact is that the operation which today is understood as the division of one fraction by another, was originally conceived as an operation on a canon with twelve divisions and was simply the *subtraction of a shorter line from a longer one.*

Of course it is not only the term ὑπεροχή which testifies to the fact that ascertaining the 'difference' between the ratios 3 : 2 and 4 : 3 was originally not a matter of division for the Greeks, but (since this operation was carried out directly on a 'canon') of subtraction. Indeed, it even seems that the operation which we would denote by 3/2 : 4/3 could only be described in terms of *subtraction* in Greek. This explains Proposition 8 of the *Sectio Canonis:* ἐὰν ἀπὸ ἡμιολίου διαστήματος ἐπίτριτον διάστημα ἀφαιρεθῇ, τὸ λοιπὸν καταλείπεται ἐπόγδοον (i.e. 3/2 : 4/3 = 9/8) which Heiberg translates as: "Si ab intervallo sesquealtero intervallum sesquetertium *aufertur*, quod *relinquitur* sesqueoctavum est." The verb forms ἀφαιρεθῇ and καταλείπεται are characteristic of *subtraction*, not division. (Compare this with *Elements* Book IX, Proposition 27: ἐὰν ἀπὸ περισσοῦ ἀριθμοῦ ἄρτιος ἀφαιρεθῇ, ὁ λοιπὸς περισσὸς ἔσται; "Si a numero impari par *subtrahitur, reliquus* impar erit.")

Those terms from the Greek theory of proportions which describe the *multiplication* of one ratio by another as *addition* can be explained in the same way (as originating from operations on a canon). I have already mentioned that the octave was considered to be a *compound interval*, διάστημα συντεθέν. If it is borne in mind that the piece of string which had to be kept still in order to produce the second note of an octave (i.e. the *length* between the numbers 4 and 2 on a canon with four divisions) was actually compounded of the *length* of a fourth (the piece lying between 4 and 3) and that of a fifth (the piece lying between 3 and 2), then this use of a term for 'addition' immediately becomes comprehensible. The verb συντιθέναι is nothing but the technical expression for *adding*.[81] Obviously the two ratios 4 : 3 and 3 : 2 (or even the fractions 4/3 and 3/2) considered by themselves must be *multiplied* together, *not added*, in order to produce the ratio of an octave (4 : 2 or 2 : 1). The unexpected use of a term for addition, instead of one for multiplication (πολλαπλασιάζειν), is explained by reference to the original operation on a canon.

The addition and subtraction of lengths on a canon had far-reaching effects not only on musical terminology, but also on the terms used in the Greek theory of proportions. Even Euclid in the *Elements* frequently employs the expression συγκείμενος λόγος, where the verb συγκεῖσθαι does not have its usual meaning of addition but refers instead to multiplication. Let me give some examples here to illustrate this:

*Elements* VI.23: "Equiangular parallelograms have to one another the *ratio compounded* (λόγον ἔχει τὸν συγκείμενον) of the ratios of their sides."

*Elements* VIII.5: "Plane numbers have to one another the *ratio compounded* (λόγον ἔχουσιν τὸν συγκείμενον) of the ratios of their sides."

Another example is furnished by Definition 5 of Book VI, which is probably an interpolation: "A ratio is said to be *compounded* of ratios (λόγος ἐκ λόγων συγκεῖσθαι λέγεται) if it is formed by *multiplying* together the magnitudes of ratios."[82]

The expression συγκείμενος λόγος *(compound ratio)* derives originally from an operation on the canon, as does the related technical term διπλασίων λόγος *(doubled ratio)* which appears in Definition V.9 of the *Elements*. Concerning the latter, van der Waerden has written:[83] "Doubling a ratio clearly corresponds to multiplying a fraction by itself. For if $a : b = b : c$, then $a/c = (a/b) \cdot (b/c) = (a/b)^2$. The Greeks

---

[81] Euclid usually employs the verbs συντιθέναι and συγκεῖσθαι to mean add; see, for example, Elements IX.21 and 22.

[82] It is worthwhile to recall here Thaers' remark (to be found in his translation *Die Elemente von Euclid*, part 1, n. 60): "Definition VI. 5 is probably spurious, but it may well be part of a theory of proportions which antedates Book V."

[83] *Math. Ann.* **120** (1947/9), 133–4.

were certainly no less able to multiply fractions then we are."Moreover, Tannery had already traced the expression 'doubled ratio' back to the theory of music.[84] The only thing which I have to add is that although it is of course appropriate to refer to the theory of music in this context, it is not by itself sufficient. The technical terms we have been discussing (including the expression 'doubled ratio') remain unexplained unless mention is also made of the arithmetical operations which were carried out on the canon.[85]

There are two other points worth making in connection with these operations:

(1) Frank was already aware of the fact that the Greeks used the words 'addition' and 'subtraction' in the musical theory of proportions to mean 'multiplication' and 'division' respectively.[86] One can hardly blame him for being puzzled by this. Yet instead of making a serious attempt to explain this surprising fact, he hit upon the frivolous idea that 'compounding ratios' was too complex an operation for the Greeks to have been able to perform it prior to the 4th century. Van der Waerden's response to this was:[87] "Compounding ratios in this way is such an elementary and obvious operation that I do not understand why Frank insists that the Greeks did not have a clear idea about if before the time of Archytas and Eudoxus." Van der Waerden was completely correct in rejecting Frank's frivolous view; on the other hand, however, he failed to understand what Frank found 'problematic'. (He did not see that Frank was unable to resolve the confusion which *seemed* to exist in the Greek theory of proportions between addition and *multiplication* on the one side, and subtraction and *division* on the other. So neither the philologist nor the mathematician succeeded in explaining the fact which had puzzled Frank.)

(2) The other fact which has to be emphasized here is the following: In the Greek theory of proportions the words for addition and subtraction are indeed used to mean multiplication and division respectively. This usage comes from the time when ratios were still calculated directly by *lengths* on a canon. If, however, one considers the theory

[84] Van der Waerden; see the previous note.
[85] Burkert, *Weisheit und Wissenschaft*, p. 348, n. 3, uses the word 'Strecken-vorstellung' in this connection.
[86] *Plato und die sog. Pythagoreer*, pp. 158–61, especially p. 158, n. 1.
[87] *Hermes* **78** (1943), p. 179.

of proportions as presented in Euclid, one sees that almost the same expressions for addition and subtraction occur in its historically later parts, this time with the meaning of *real addition* and *subtraction*. In Definitions V.14 and 15 of the *Elements*, for example, the words σύνθεσις λόγου or διαίρεσις λόγου, and ὑπεροχή refer to actual *addition* and *subtraction* of the terms of λόγοι, not to the addition (i.e. multiplication) or subtraction (i.e. division) of two λόγοι. Of course this later usage has nothing to do with operations on the canon.

### 2.11 THE TECHNICAL TERM FOR 'RATIO' IN GEOMETRY

In the last chapter we discussed some arithmetical operations on the 'canon' ('addition' and 'subtraction' of *diastemata*) which had a lasting effect on the terminology used subsequently in the theory of proportions. Such important expressions as *'compound ratio'*, *'doubled ratio'*, *'difference between two ratios'* etc., which in reality refer to *multiplication* and *division*, would have to be regarded simply as linguistic enigmas if each of them could not be traced back to a particular operation on the canon. We shall have to return to the subject of arithmetical operations on the canon in a later chapter, but for the time being let us interrupt our discussion of the musical theory of proportions and devote the next few chapters (2.12–2.16) to the consideration of some other geometrical terms.

It is astonishing that the musical term διάστημα (the ratio between two numbers) is never used in geometry. The word διάστημα in geometry means 'line' and never anything else. (For example, Euclid's third postulate, which has already been quoted once, states that a circle can be drawn with any center and any *line segment – παντὶ διαστήματι*.) In geometry the concept of *'ratio between two numbers or quantities'* was not expressed by διάστημα, but by the word λόγος.

The conjecture that these two terms *(λόγος and διάστημα)* must have something to do with each other is confirmed by the following two facts:

(1) As was mentioned above,[88] λόγος and διάστημα were *equivalent concepts* in the Pythagorean theory of music.

[88] See p. 113–4; also nn. 33 and 34.

(2) Just as a musical *diastema* had two 'end points' *(ὅροι)*, so in geometry one referred to the *ὅροι* of a *λόγος* (as is shown by Definition V.8 of the *Elements*, for example).

These two facts, however, do not tell us anything about the origin of the concept *λόγος (a : b)*. In particular, we do not know whether this term originated within the theory of music, perhaps as a synonym for a musical *διάστημα*, or whether it existed earlier in geometry and was only subsequently equated with a musical *διάστημα*.

To be able to answer these questions, it will be necessary to conduct a more thorough investigation of the geometric term *λόγος* and its etymology. For the time being, however, our point of departure will not be the term *λόγος* itself, but the Greek name for 'proportion', i.e. the word *ἀναλογία*. I have already mentioned (in chapter 2.2) that the four-place relation of *proportion* (or sameness of ratio) was called *ἀναλογία* in Greek geometry. Our first task will be to investigate how the word *ἀναλογία* acquired this remarkable meaning.

### 2.12 *'Αναλογία* AS 'GEOMETRIC PROPORTION'

This investigation gains some added interest from the fact that the word is still in current use, although its meaning has changed somewhat. Furthermore, we frequently speak of an 'analogy', by which we mean something other than a 'geometric proportion'. Let us now take a look at the history of this Greek word.

Modern variants of the word *ἀναλογία* are to be found in all European languages. Moreover, they all have roughly the same meaning. 'Analogy' means similarity, conformity, relationship, or the extension of a rule to similar cases. These meanings already make it clear that the modern word derives from the Greek *grammatical term* (by way of the Latin word *analogia* of course). Greek grammarians have used this word with its present day meaning since Hellenistic times.

It is less well known that this same word was not originally a grammatical or linguistic term, but a *mathematical* one. The root of the word *ἀναλογία* is obviously *λόγος*, which in mathematics meant the 'ratio' between two numbers or quantities *(a : b)*. *'Αναλογία* itself described a 'pair of ratios'. Following Cicero,[89] it was translated into Latin by *proportio (a : b = c : d)*. The Greek grammarians of Hellenistic times undoubtedly borrowed their term ἀναλογία from the language of mathematics. So in the last analysis we are indebted to Greek mathematics for our word 'analogy'.

Our principal concern now is to find out the original and precise meaning of the mathematical term ἀναλογία. Before attempting to do so, however, there are two important facts which have to be stated first.

(1) The expression ἀναλογία *did not figure in the everyday language of the Greeks*, i.e. it was not a part of the common vocabulary of Greek as it was spoken in classical times. There are, it is true, some passages from Plato[90] which almost lead one to believe that at his time the word ἀναλογία had already been adopted into everyday speech, but in reality all the evidence (including these same passages) points to the fact that ἀναλογία was actually an artificial coinage, a *mot savant*. Moreover, it originally had no meaning outside mathematics, having been coined by mathematicians to express a mathematical notion. Subsequently the word was taken over from mathematics into educated speech and later still it was transformed into a technical term of another branch of learning, namely grammar. The fact that it became part of the common vocabulary of Greek shows the enormous influence which the mathematics of the 6th century exerted on the whole of Greek thought. As we shall soon see, the term ἀναλογία was coined in the 6th century and by Plato's time it was already quite a common word in the speech of the educated.

(2) It would appear that the compilers of those Greek lexicons, which are most widely used nowadays, did not know the exact and true meaning of the word ἀναλογία. Although they record the meaning of this word in a manner which initially satisfies the uncritical reader, they reveal on other occasions that they have not really understood it. Let me illustrate this point by quoting some sample passages from these dictionaries.

[89] *Timaeus seu de Universo*, 4 §12: "Id optime assequitur, quae Graece ἀναλογία Latine (audendum est enim, quoniam haec primum a nobis novantur) *comparatio proportiove* dici potest."

[90] See, for example, Plato's *Politicus* 257b. or Aristotle, *Nicomachean Ethics*, 5. 3.

In Passow's dictionary (1841) ἀναλογία is explained as follows: "entsprechendes oder richtiges Verhältnis, Proportion, Analogie etc." [corresponding or correct ratio, proportion, analogy]. Similarly, Pape's dictionary (1849) gives: "das richtige Verhältnis, Proportion, Übereinstimmung etc." [the correct ratio, proportion, conformity]. Finally, Liddel and Scott (1948) list under ἀναλογία: "mathematical proportion, analogy etc." (and then go on to quote the most important authorities).

Naturally there is for the most part no difficulty in applying the meanings given above to the various ancient texts in which the word occurs. Nonetheless, as soon as one attempts to find out the derivation of the expression ἀναλογία, using these dictionaries as references, it becomes apparent that the true meaning of the word has not been understood. In the end, the word ἀναλογία is connected with ἀνάλογος or ἀνάλογον in the same way that φιλολογία is connected with φιλόλογος. So let us look at the way in which these dictionaries explain ἀνάλογος or ἀνάλογον.

Passow gives the following meanings (which are almost literally the same as those found in Pape) under ἀνάλογος: "dem λόγος entsprechend, verhältnismässig, übereinstimmend, gemäss, einer bestehen den Regel entsprechend" [in accordance with the λόγος, proportional, conformable, in conformity with, in accordance with an existing rule]. Liddel and Scott have: "according to a due λόγος, proportionate, conformable".

The critical reader can see immediately from these entries that the authors did not think it necessary to explain the exact meaning of the preposition ἀνά in the words ἀναλογία and ἀνάλογος; indeed they may not even have understood it. This suspicion is confirmed particularly by the otherwise excellent lexicon of Liddel and Scott. They mention that the adverb ἀνάλογον is sometimes written as ἀνὰ λόγον; but instead of discussing the meaning of this important phrase, they refer the reader back to λόγος. If, however, one tries to look up the exact meaning of ἀνὰ λόγον under λόγος, there is a disappointment in store, because it is translated there simply by 'analogically'. There could be no more striking documentation of the fact that the true meaning of the expression ἀνὰ λόγον has not been understood at all.[91]

---

[91] The same holds for Diels and Kranz, *Fragmente der Vorsokratiker, I*, p. 436, where the phrase ἀνὰ λόγον in the second Archytas-fragment is translated by the German word 'analog'. I have no idea what the translator had in mind when he wrote this.

10*

An exact understanding of the mathematical term ἀναλογία can only be gained by starting out from the following, rather obvious, fact. Although the noun ἀναλογία seems to be a direct derivative of the adjective ἀνάλογος, this latter is not the root word whose explanation will lead us to the solution of our problem (the meaning of ἀναλογία). It would appear that the adjective ἀνάλογος itself was derived from the adverbial expression ἀνὰ λόγον, i.e. originally there was only the phrase ἀνὰ λόγον. It is written as two words not only in some parts of Plato (e.g. in the *Phaedo*, 110a: ἀνὰ τὸν αὐτὸν λόγον), but also in a fragment of Archytas which "represents perhaps the only surviving, authentic mathematical text from the time before Autolycus and Euclid".[92] From this phrase there came on the one hand the familiar expression ἀνάλογον (see for example *Elements*, Definitions V.6, VII.21, etc.), and on the other hand the adjective ἀνάλογος as well as the noun ἀναλογία. So let us now try to find out the meaning of ἀνάλογον or ἀνὰ λόγον.

## 2.13 Ἀνάλογον

The first thing which has to be emphasized is that our knowledge of the expression ἀνάλογον comes exclusively from the language of mathematics.[93] It is true that on one occasion Plato (*Phaedo*, 110d) uses ἀνὰ λόγον and ἀνὰ τὸν αὐτὸν λόγον in contexts which are not obviously mathematical, none the less a closer inspection of the passages concerned soon reveals that Plato is in fact using a mathematical term. In any event, the meaning of this expression can only be explained by reference to *the language of mathematics*.

[92] See O. Töplitz, *Quellen und Studien zur Geschichte der Mathematik*, etc. B, **2** (1932), 288, n. 5. See also M. Timpanaro-Cardini, *Pitagorici, Testimonianze e Frammenti*, fasc. 2 (Florence 1962, p. 372).

[93] Burkert (*Weisheit und Wissenschaft*, pp. 414ff.), who believes that the mathematical term λόγος was derived from the language of finance, writes misleadingly: "When several quantities 'correspond to the same amount' they were said to be ἀνὰ λόγον". [The quantities were coins of different denomination or currencies.] I fail to see how he can make such an assertion without citing any evidence for it. As far as I know, there is no ancient text where the phrase is used with this meaning. In fact there is no evidence that ἀνὰ λόγον was used at all *outside mathematics*.

In the *Elements* there are two different usages of the word ἀνάλογον which have to be clearly distinguished. Let me give an example of each of these here. (In both cases the Greek text is quoted together with Thaer's German translation.)[94]

An example of the first usage is Definition V.6:

τὰ τὸν αὐτὸν ἔχοντα λόγον μεγέθη ἀνάλογον καλείσθω.

Die dasselbe Verhältnis habenden Grössen sollen *in Proportion stehend* heissen.

[Magnitudes having the same ratio are said to be in proportion.]

The other usage can be illustrated by the statement of Proposition VI.12:

τριῶν δοθεισῶν εὐθειῶν τετάρτην ἀνάλογον προσευρεῖν.

Zu drei gegebenen Strecken die vierte *Proportionale* zu finden.

[To three given straight lines to find a fourth *proportional*.]

As we can see, in our first example the word ἀνάλογον refers to all those magnitudes which constitute the proportion, and is translated by Thaer as "*in Proportion stehend*", whereas in our second example the word refers to only *one magnitude (τετάρτη ἀνάλογον)* and hence is translated as "*Proportionale*". The latter example would suggest that each one of the magnitudes which goes to make up the proportion is itself an ἀνάλογον. However, we are left with the question of what exactly an ἀνάλογον is, and why the word does not decline. The connection between these two usages is by no means settled at the moment.

It turns out that the usage which occurs in the second quotation is the later of the two (i.e. it belongs to a later stage of linguistic development). Hence we shall return to it in a subsequent chapter and deal with the other one first.

Our starting point will be Definition V.6; moreover, we will leave the word ἀνάλογον untranslated for the time being. Thus our provisional

---

[94] C. Thaer, *Die Elemente von Euklid*, Leipzig 1933-7.

interpretation of the definition runs as follows: "*Magnitudes having the same ratio (λόγος) are said to be ἀνάλογον.*"

It is necessary to know what a mathematical 'ratio' is, in order to understand this definition. For the moment, we will accept the Euclidean conception of λόγος, according to which the λόγος between two numbers or magnitudes *(a* and *b)* is just *a : b* (as we would write it nowadays). I should perhaps mention right away that the Euclidean conception of mathematical ratio *(λόγος)* is by no means the original one. We shall see later that in pre-Euclidean times λόγος was a somewhat more comprehensive notion; but this fact need not concern us just now. The Euclidean concept of λόγος is sufficient for our present purpose of discovering the meaning of ἀνάλογον. If we know what the mathematical ratio *(λόγος)* of two magnitudes is, then the interpretation of Definition V.6 poses no problem. The definition clearly states that if the ratio between two magnitudes *a* and *b* is *a : b*, and if the ratio between two other magnitudes *c* and *d* is the same as that between *a* and *b* (in our notation, if *a : b = c : d)*, then these *four* magnitudes *(a, b, c* and *d)* collectively are said to be ἀνάλογον.

This interpretation also accords with another definition in the *Elements*, which is arithmetical in nature. I am referring to Definition VII.21:

ἀριθμοὶ ἀνάλογόν εἰσιν, ὅταν ὁ πρῶτος τοῦ δευτέρου καὶ ὁ τρίτος τοῦ τετάρτου ἰσάκις ᾖ πολλαπλάσιος ἢ τὸ αὐτὸ μέρος ἢ τὰ αὐτὰ μέρη ὦσιν.

*Analogon* sind die Zahlen[95] wenn die erste von der zweiten Gleichvielfaches oder derselbe Teil oder dieselbe Menge von Teilen ist, wie die dritte von der vierten.
[Numbers are *analogon* when the first is the same multiple, the same part, or the same number of parts, of the second as the third is of the fourth.]

Both these definitions show that the word ἀνάλογον describes at least four numbers or magnitudes, two of which have the same ratio

---

[95] Heath translates ἀνάλογον in this context by *proportional*, while Thaer uses *in proportion*.

to each other as the other two. We cannot apply either of the preceding definitions and speak of ἀνάλογον unless there are at least four numbers or magnitudes present.[96] Of course this interpretation does not contradict Definition V.8 of the *Elements*, which expressly states that "the least *analogia* consists of *three* terms". In a case of this kind (the ἀναλογία συνεχής, as it is called) the middle term is taken twice *(a : b = b : c)*, so that all in all there are four terms, not three. Aristotle has explained this point in greater detail.[97]

The other important fact which follows from the preceding definitions (V.6 and VII. 21) is that the word ἀνάλογον describes *sameness of ratio (λόγος)* between two pairs of numbers or magnitudes (i.e. it describes the relation which holds between two pairs of numbers or magnitudes each having the same ratio – τὸν αὐτὸν λόγον). Its meaning would be expressed in English by the phrase 'in the same ratio'. Aristotle[96] also attributes this meaning to the word ἀναλογία. He writes: "*Analogia* is *sameness* of ratios, i.e. λόγοι."

## 2.14 THE PREPOSITION ἀνά

Our next task is a purely linguistic one, namely to explain how the word ἀνάλογον is able to express the notion of *'being in the same ratio'*. The word λόγος clearly means ratio in this context, so our task reduces to explaining how *equality* between ratios can be expressed by prefixing the preposition ἀνά.

I have already mentioned that the adverb ἀνάλογον was at one time written as two words *(ἀνὰ λόγον)*. It takes this form in the second fragment of Archytas. We now want to find out exactly what the preposition ἀνά means in the expression ἀνάλογος and ἀναλογία. First, however, let me just say that as far as I know this question has never before been clearly formulated. Nonetheless, it has received an answer which is both erroneous and widely accepted. In dictionaries (see chapter 2.12) the word ἀνάλογος is explained as follows: "dem λόγος entsprechend, gemäss, verhältnismässig" or "according to a due λόγος." These explanations offer a translation of the preposition ἀνά as part

---

[96] Aristotle, *Nicomachean Ethics*, 1131a31: ἡ γὰρ ἀναλογία ἰσότης ἐστὶ λόγων, καὶ ἐν τέτταρσιν ἐλαχίστοις.

[97] Ibid.

of the compound ἀνάλογος, yet it would seem to be a mistaken one.

I cannot believe that the original use of the preposition ἀνά was to express 'entsprechend', 'gemäss' or 'according to', for quite different prepositions were usually employed in such cases. As an example, let us consider the preposition κατά. Xenophon (*Cyropaedia*, 8.6, 11) writes: κατὰ λόγον τῆς δυνάμεως (in proportion to their power). Similarly, in the works of other authors we encounter such phrases as κατὰ τὸν αὐτὸν λόγον or τρόπον (Herodotus, I. 182), κατὰ τοῦτον τὸν λόγον (Plato, *Protagoras*, 324c), κατὰ λόγον τὸν εἰκότα (Plato, *Timaeus*, 30b), κατὰ τὸν λόγον τὸν αὐτόν (Aristotle, *Nicomachean Ethics*, 1131b30) and κατὰ τὴν ἀναλογίαν (Plato, *Politicus*, 257b). In all these instances just that function of the preposition ἀνά which one thought to find in the compounds ἀνάλογον, and ἀναλογία, is performed by the other preposition κατά. Even Heraclitus, when he wants to say 'according to the same nature (or proportion)', does not use ἀνά; instead he writes: εἰς τὸν αὐτὸν λόγον.[98]

These examples suggest the following two conjectures:

(a) It is unlikely that ἀνά (in the expression ἀνὰ λόγον) had originally the same meaning as κατά and εἰς in the above examples. It is especially unlikely that ἀνά and κατά should have been synonymous. These two prepositions, at least originally, had directly opposite meanings. 'Ἀνά meant 'upwards' or 'up', whereas κατά meant 'downwards' or 'down'.

(b) As we shall soon see, a great deal of historical information can be obtained once we have established exactly what the preposition ἀνά means in the phrase ἀνὰ λόγον. Nonetheless, I want to emphasize here that the explanation of ἀναλογία which I am going to propose holds good only for the meaning of this mathematical term *at a very early period*. It would be wrong to think that this word had precisely the same meaning in later texts, such as those of Plato. The quotation from the *Politicus* given above shows that Plato uses the word ἀναλογία in a sense which differs from the ones actually listed in the dictionary. The phrase κατὰ τὴν ἀναλογίαν indicates that by Plato's time the original meaning of ἀνά in the word ἀναλογία must already have been forgotten. (This is the only way to explain the fact that a new adverbial phrase could be formed from ἀναλογία by prefixing the preposition κατά, even

[98] See fragment 31 in Diels and Kranz, *Fragmente der Vorsokratiker*.

though ἀναλογία itself was just *an adverb which had been converted into a noun.*) The same is also true of ἀνάλογον. In educated speech this word undoubtedly came (even in *classical times*) to mean roughly the same as our adjective 'analogously'. Proclus,[99] for example, writes: κατὰ τὸ ἀνάλογον, which Schönberger[100] quite correctly translates as "in *analoger* Weise" [in an *analogous* way]. This, however, is not the meaning which interests us here. Our present concern is with the original meaning of this mathematical term, i.e. with the meaning which was probably current in the 6th century B.C. and traces of which are to be found in many of Euclid's propositions (since the propositions concerned date from a much earlier, pre-Platonic period).

In my opinion it is not difficult to ascertain what ἀνά means in the expression ἀνὰ λόγον, and hence to discover the original and exact meaning of the mathematical term ἀναλογία. Amongst the various meanings of the preposition ἀνά, all the major Greek lexicons list a so-called *distributive* one which occurs especially in numerical contexts. For example, in Xenophon we read: ἀνὰ πέντε παρασάγγας τῆς ἡμέρας (*at the rate of five* parasangs a day — *Anabasis* 4, 6, 4) and: ἀνὰ ἑκατὸν ἄνδρας ("*a hundred* men *apiece*" — ibid., 3. 4. 21). Similarly Diophantus (*Arithmetica* IV 20) writes: ἀνὰ δύο *(two at a time)*.

The examples which the dictionaries quote to illustrate this particular meaning of ἀνά all have to do with numbers. We can correct the impression which this leaves, however, by adding the remark that ἀνά was used with this distributive meaning not only in numerical contexts, but also occasionally before nouns having some connection with the concept of 'number' or with that of 'division'. An example of this usage, the expression ἀνὰ μέρος,[101] is quoted in Kühner's Grammar[102] where it is translated as "wechselweise" (by turns). 'Ανά is being used distributively in this phrase, since the meaning 'by turns' clearly derives from the original literal meaning of ἀνὰ μέρος, which was 'a piece at a time' or '(taken) in parts'.

---

[99] *Procli Diadochi in primum Euclidis Elementorum Librum commentarii,* ed. G. Friedlein (Lipsiae 1873), 117. 2.

[100] Proklus Diadochus, *Kommentar zum ersten Buch von Euklids Elementen,* translated by P. L. Schönberger, ed. M. Steck, Halle 1945.

[101] Euripides, *The Phoenician Women,* 478.

[102] R. Kühner—B. Gerth, *Ausführliche Grammatik der griechischen Sprache. Satzlehre I. Teil,* 4th edition, Hannover 1955, p. 474.

Now I believe that the phrase ἀνὰ λόγον has to be understood in the same way; for just as ἀνὰ μέρος meant '(taken) *in parts*', so ἀνὰ λόγον meant '(taken) *in logoi*'. There are two very important facts which support this explanation.

(1) The dictionaries tell us that the preposition ἀνά had its 'distributive meaning' especially in *numerical* contexts. Bearing in mind the expression ἀνὰ μέρος, this remark has to be supplemented by the observation that ἀνά could also be used distributively before nouns which were related to the concept of 'division'. Μέρος (part) and 'division' (*distribuere* in Latin) are very closely related concepts. On the other hand, the phrase ἀνὰ λόγον shows that ἀνά could be used in its distributive sense before nouns which (unlike πέντε, ἑκατόν and δύο in the above examples) were not themselves numerals, but which at one time could not be separated from the concept of 'number'. *Logos* was originally just the 'ratio between two *numbers*'.

At the very least, these considerations make it plausible that the preposition in the phrase ἀνὰ λόγον is being used *distributively*.

(2) A second important argument in favor of this view is provided by the way in which the expression ἀνὰ λόγον was used in the language of mathematics. Two pairs of numbers were said to be proportional whenever the members of one pair were in the same ratio as those of the other. Hence the terms of the relation (of proportionality) were first *distributed* for the sake of contrast, and then compared. There is no doubt that the exact meaning of the expression ἀνὰ λόγον was '(taken) in *logoi*'.

## 2.15 THE ELLIPTIC EXPRESSION ἀνὰ λόγον

The above considerations are simple, yet they establish conclusively that the preposition ἀνά has a distributive meaning in the phrase ἀνὰ λόγον, and take us a step further in our investigation. Aristotle emphasizes that "*analogia* is *sameness* of *logoi*". So, in order to establish that four numbers or quantities were ἀνὰ λόγον, one had to show that they were '*equal* (when taken) in *logoi*'. This implies that ἀνὰ λόγον ([taken] in *logoi*) actually meant '*equal* (when taken) in *logoi*', or in other words that ἀνὰ λόγον was a so-called *defective* or *elliptic* expression. The phrase used initially in Greek mathematics must have been ἀνὰ λόγον ἴσοι or ἴσα which was subsequently abbreviated to the standard expression ἀνὰ λόγον (or ἀνάλογον, as it was sometimes written).

It must be admitted that the more complete version of this phrase does not occur in any of the ancient texts which have been handed down to us. As far as I know, they contain only the elliptic form ἀνὰ λόγον (or ἀνάλογον). Nonetheless, there is no doubt that we have succeeded in correctly reconstructing the complete expression. Evidence for this is furnished by the two definitions from Euclid (*Elements* VII.21 and V.6) which have already been quoted above. To interpret and translate these satisfactorily, they have to be supplemented (or at least it has to be kept in mind that they need to be supplemented) in the following way.

*Definition VII. 21:*

ἀριθμοὶ ἀνάλογον ⟨ἴσοι⟩ εἰσίν, ὅταν ὁ πρῶτος τοῦ δευτέρου καὶ ὁ τρίτος τοῦ τετάρτου ἰσάκις ᾖ πολλαπλάσιος ἢ τὸ αὐτὸ μέρος ἢ τὰ αὐτὰ μέρη ὦσιν.

Numbers are *equal* (when taken) in *logoi*, if the first is the same multiple, the same part, or the same number of parts of the second, as the third is of the fourth.

*Definition V. 6:*

τὰ τὸν αὐτὸν ἔχοντα λόγον μεγέθη ἀνάλογον ⟨ἴσα⟩ καλείσθω.

Magnitudes which have the same logos are said to be *equal* [when taken] in *logoi*.

Of course there are many similar passages from Euclid[103] which need to be complemented in exactly the same way as the above. Let me mention a few of these here.

In the *Elements* Book VIII, Definition 22, the Pythagorean concept of 'similar plane' or 'solid numbers' is described as follows: "similar plane or solid numbers are those whose sides (i.e. factors) are *equal* (when taken) *in logoi*" (οἱ ἀνάλογον ⟨ἴσας⟩ ἔχοντες τὰς πλευράς).

---

[103] For example, Definition VI. 1 and Proposition VI. 6.

Proposition VI.5 of the *Elements* is concerned with triangles whose sides are 'proportional', i.e. '*equal* [when taken] *in logoi*' *(ἐὰν τρίγωνα τὰς πλευρὰς ἀνάλογον ⟨ἴσας⟩ ἔχῃ)*. Furthermore, in such propositions as *Elements* VIII. 1, 3, 6 and 7, the Greek phrase ἐὰν ὦσιν ὁποσοιοῦν ἀριθμοὶ ἑξῆς ἀνάλογον . . . which Thaer translates as "Hat man beliebig viele Zahlen in geometrische Reihe . . ." (given arbitrarily many numbers which form *a geometrical progression* . . .), is rendered equally well by the more literal translation: "given arbitrarily many numbers which taken *in succession, are equal in logoi* . . .".

It is easy to understand how the elliptic expression ἀνὰ λόγον came into being. In comparing *logoi*, one was usually concerned with whether or not they were equal. Hence the phrase '[taken] in *logoi*' was used for the most part in conjunction with the word 'equal'. This meant that after a while the word ἴσοι or ἴσα, which expressed equality, could be omitted since it was implied, so to speak, by the rest of the phrase. Thus the truncated phrase ' . . . [taken] in *logoi*' came to have the meaning '*equal* [when taken] in *logoi*'. The fact that ἀνὰ λόγον soon came to be written as one word is independent evidence for the thesis that this elliptic expression developed into a new word with its own meaning. The new word of course was the adverb ἀνάλογον ('equal [when taken] in *logoi*'). It was only after this adverb had been accepted in mathematical language, that it was possible to derive the noun ἀναλογία from it. 'Αναλογία was the mathematical property possessed by numbers or quantities which were ἀνάλογον. In other words, ἀναλογία was 'sameness of ratio'.

We cannot be certain about when mathematicians coined the technical term ἀνάλογον ([taken] in *logoi*) nor about when this adverb gave rise to the noun ἀναλογία (sameness of ratio), for no mathematical texts have survived from this period. However, these developments certainly took place before Plato's time. From his time on, the archaic expression ἀνὰ λόγον (equal [when taken] in *logoi*) was hardly used any more, except by Euclid. On the other hand, even Plato no longer used ἀναλογία with just its mathematical meaning. Although it was originally a purely mathematical term, it seems already by his time to have been current in educated speech. The impression left by Aristotle's writings is of ἀναλογία as a philosophical term whose mathematical origins seem to have been almost completely forgotten.

There are, however, some indications that the archaic word ἀνάλογον survived as a mathematical term and that its meaning subsequently underwent some changes. These will be discussed in the next chapter.

### 2.16 THE SUBSEQUENT HISTORY OF ἀνάλογον AS A MATHEMATICAL TERM

In the preceding chapter we showed how the phrase ἀνὰ λόγον (an abbreviation for ἀνὰ λόγον ἴσοι) came to have the meaning 'equal [when taken] in logoi' (or, put more simply, 'in the same ratio'), and we established that it probably acquired this meaning during the archaic period. The text of Euclid's *Elements* is full of examples which could be used to illustrate this particular meaning of the phrase. However, there are two remarks which need to be made in this connection; the first is that in classical times Greek mathematicians no longer used the word ἀνάλογον when they wanted to say 'in the same ratio', and the second is that by this time the word itself (in its role as a mathematical term) had undergone a slight change of meaning. Let us begin by discussing the second point.

It was empasized in the last chapter that the elliptic expression ἀνάλογον *(equal* [when taken] in *logoi)* referred originally to *a group of at least four* numbers or quantities, which had the property that it could be split up into pairs whose members stood in the same *logos*. Euclid, however, also uses this word in a different context. We have already given an example of this second usage, namely, *Elements* VI. 12: "To three given straight lines, to find *a fourth proportional" (τετάρτην ἀνάλογον).*

From the Greek words in brackets it is apparent that the above was written at a time when the word ἀνάλογον was viewed in a somewhat different way. No one took account any longer of the fact that it was used (as in Definitions VII. 20 and V. 6, for example) to denote at least four numbers (or quantities) which were ἀνὰ λόγον ἴσοι (or ἴσα): they reasoned instead that if four numbers (or quantities) were ἀνάλογον, then each one of them had to be ἀνάλογον (to the others). This usage must have suggested the idea that ἀνάλογον (a word which did not decline) was an unusual kind of *neuter noun*. So, in addition to its original meaning, ἀνάλογον came to denote what we would call a 'proportional'. For this reason it became possible to speak of a τρίτη ἀνάλογον or a μέση ἀνάλογον,[104] and even of μέσος ἀνάλογον ἀριθμός and μέσοι ἀνάλογον ἀριθμοί.[105] This latter meaning (which originated in the pre-Platonic period) is of course different from the archaic one and postdates it. Furthermore, it seems to have prepared the way for the creation of the hybrid adjective ἀνάλογος. (It should be mentioned, however, that this adjective never figured in mathematical language.)

As we can see, in the course of time the word ἀνάλογον came to have two meanings in mathematics. It retained its original meaning of 'equal [when taken] in logoi' and also came to mean a 'proportional'. This ambiguity seems to have been partly responsible for the fact that it later became more usual to employ a different expression for the notion of being 'in the same ratio'. I am referring to ὁ αὐτὸς λόγος ('the same logos' or 'the same ratio'), which was the phrase used most frequently in Greek mathematics to express sameness of ratio. Actually Euclid and the mathematicians who followed him employed this expression in two slightly different ways. They said that pairs of numbers (or quantitites) 'had the same ratio' (τὸν αὐτὸν λόγον ἔχουσι — see, for example, Elements VII. 17) and also that they 'stood in the same ratio' (ἐν τῷ αὐτῷ λόγῳ εἰσίν — see Elements VII. 14).

[Hence, if the phrase ὁ αὐτὸς λόγος is included, mathematicians had at their disposal three ways of expressing the concept of being 'in the same ratio'. These were:

(1) The archaic expression ἀνάλογον — 'equal [when taken] in logoi'

(2) The adverbial clause of comparison ὡς ... οὕτως — as used, for example, in Elements VII. 11 (cf. p. 104)

(3) The phrase ὁ αὐτὸς λόγος.]

Now the term used most frequently for 'in the same ratio' in Elements V (a book which is supposed to have been mostly the work of Eudoxus) seems to be ὁ αὐτὸς λόγος (the same ratio) rather than the archaic ἀνάλογον (equal in logoi). It is true that the word ἀνάλογον also occurs in this book, but it appears only in proposition which can be classified as the work of earlier mathematicians or as new formulations of earlier results. This can be seen from the following.

[104] See Propositions VI. 11 and 13.
[105] Elements, VIII. 11 and 12.

In the fundamental, so-called Eudoxian definition (V. 5, quoted above on p. 99), 'sameness of ratio' is defined as ὁ αὐτὸς λόγος. However, this definition is followed immediately by Definition V. 6: 'Let magnitudes which have the same ratio (τὸν αὐτὸν ἔχοντα λόγον) be called ἀνάλογον (equal in logoi).'

The latter is clearly not a proper mathematical definition. It serves only to introduce an alternative way of describing the concept (ὁ αὐτὸς λόγος) defined previously. Nonetheless, these two definitions, when read in the order in which they occur, lead one to expect that both ὁ αὐτὸς λόγος and ἀνάλογον will be used to refer to 'sameness of ratio' in Book V. Let us see how often each of these phrases is used.

Altogether, the notion of being 'in the same ratio' occurs in *seventeen* of the propositions in Book V. In *twelve* of these (V. 4, 7, 9, 11, 13, 14, 15, 20, 21, 22, 23, and 24) it is referred to by the phrase ὁ αὐτὸς λόγος. On the other hand, ἀνάλογον is used in only *five* of these propositions (V. 12, 16, 17, 18 and 25). This statistic is by itself sufficient to suggest the conjecture that the propositions in Book V were written down at a time when the usual term for 'in the same ratio' was no longer ἀνάλογον, but rather ὁ αὐτὸς λόγος. In at least two cases, however, we can go further than this and explain why the archaic expression ἀνάλογον was used with this meaning in Book V. Before discussing these in greater detail, I have to mention an interesting peculiarity of Book V, to which insufficient attention has been paid up to now.

Proposition V. 19 of the *Elements* states that: "If, as a whole is to a whole, so is a part subtracted to a part subtracted, the remainder will also be to the remainder as whole to whole."

If the above is read in the original Greek, then one is astonished to find that it is word for word the same as a theorem to be found in Book VII, namely, Proposition VII. 11. The two versions differ only in that the adjectives ὅλος, ἀφαιρεθείς and λοιπός, whose form is neuter in Book V, are put into the masculine in Book VII. This difference, of course, arises simply from the fact that 'number' in Greek is a masculine noun (ὁ ἀριθμός), whereas 'magnitude' is neuter (τὸ μέγεθος).[106] The proposition in Book V applies to arbitrary 'magnitudes'; in Book VII, however, it is formulated only for 'numbers'. The difference of gender also reveals which of the two versions is the earlier one. There

---

[106] Heath (*Euclid's Elements*, Vol. 2, p. 312) makes the same point.

is no doubt that the proposition could originally be formulated only for numbers, as it is in Book VII. (Quite apart from this, there are other considerations which have led scholars to conjecture that the propositions in Book VII antedate those in Book V.[107]) Only after the theory of proportions had been extended to cover incommensurable magnitudes as well, could this same proposition (with just a slight grammatical change in its formulation) be incorporated into Book V. Of course, when this took place a new proof was also required. The old one, which held for the case of 'numbers', was not valid for arbitrary 'magnitudes'. Thus from a mathematical point of view Proposition V. 19 is completely different from VII. 11, even though it appears to be simply a minor linguistic variant of the latter.

In my opinion, what we have just learned from comparing these two propositions (V. 19 and VII. 11) also provides us with an explanation of why the archaic term ἀνάλογον is used for the notion of being 'in the same ratio' in some of the propositions in Book V. Let us look more closely at the following two propositions:

V. 12: "If any number of *magnitudes* be proportional (ἀνάλογον), as one of the antecedents is to one of the consequents, so will all the antecedents be to all the consequents."

V. 16: "If four *magnitudes* be proportional (ἀνάλογον), they will also be proportional alternately (καὶ ἐναλλὰξ ἀνάλογον ἔσται)."

[I.e., if $a : b = c : d$, then $a : c = b : d$.]

If we confine our attention to the statement of these two propositions, then they are seen to reappear in Book VII formulated almost word for word in the same way. Correspondingly to V.12, there is:

VII. 12: "If there be as many *numbers* as we please in proportion, (ἀνάλογον), then, as one of the antecedents is to one of the consequents, so are all the antecedents to all the consequents." And corresponding to V. 16 we have:

VII. 13: "If four *numbers* be proportional (ἀνάλογον), they will also be proportional alternately (καὶ ἐναλλὰξ ἀνάλογον ἔσονται)."

As we can see, Propositions 12 and 13 in Book VII apply only to *numbers*, whereas the corresponding ones (12 and 16) in Book V are formulated for arbitrary *magnitudes*. No doubt these propositions

[107] B. L. van der Waerden, *Erwachende Wissenschaft*, pp. 182ff.

could originally be stated and proved only for numbers, and were subsequently generalized so as to apply to magnitudes as well. When they were modified, however, their wording was left unchanged, except that 'number' (ἀριθμός) was replaced throughout by the newly defined concept of 'magnitude' (μέγεθος).

It is immediately apparent from this why the archaic term ἀνάλογον is used in V. 12 and V. 16 to express 'in the same ratio'. It is a relic of the earlier versions (VII.12 and 13) of these propositions. Furthermore, it is precisely because of the use of this earlier terminology that the so-called Eudoxian definition of proportionality (V.5), which employs the phrase ὁ αὐτὸς λόγος, had to be supplemented by another (improper) one (V.6: "Let magnitudes which have the same ratio be called ἀνάλογον").

It is not only in Book V, however, that ὁ αὐτὸς λόγος occurs more frequently than the archaic ἀνάλογον. This is also true of Book VII, even though the propositions in this latter book are probably of an earlier date than those in the former. It is interesting that ἀνάλογον (with the meaning 'in the same ratio') appears in *two* of the definitions (21 and 22) in Book VII, whereas ὁ αὐτὸς λόγος, a phrase of more recent origin, does not occur in any of them. (This fact is a further indication of the great age of these definitions.) Yet there are altogether only *three* propositions (VII. 12, 13 and 19) in this book which mention ἀνάλογον (equal [when taken] in *logoi*), as opposed to *six* (VII. 14, 17, 18, 20, 21, and 22) in which ὁ αὐτὸς λόγος occurs. Thus in the entire text of Euclid, the expression ἀνάλογον (in the sense of 'in the same ratio') appears less frequently than the phrase ὁ αὐτὸς λόγος which has the same meaning.

### 2.17 CUTS OF THE CANON AND MUSICAL MEANS

In the last five chapters (2.12–2.16) our historical investigation of the mathematical terms ἀνάλογον and ἀναλογία has been confined to linguistic matters. Because of this, we have not yet considered how it came about that mathematicians formed λόγοι (or ratios) out of pairs of numbers, nor why they compared numbers '[taken] in *logoi*'. In order to formulate these questions in a more concrete manner, we must resume our discussion of musical subjects.

Although we have already devoted one chapter to an account of the calculations (addition and subtraction of lengths) which were performed on the canon, the Greek name for these *operations* has not yet been mentioned. The Pythagoreans called them *cuts of the canon*. There is, for example, a passage in Ptolemy's *Theory of Harmony*[108] where he sketches the procedure followed by certain of those who performed experiments in the field of music. It runs as follows:

"They do not obtain the divisions (of the string — τὰς κατατομάς) by calculation, but by stretching the string and then moving the little bridge until each of the notes being sought after is heard. They mark the required cut *(σημειοῦνται τὴν τομὴν)* at those places, and do not concern themselves about the ratios."

We see from this account that the word τομή or κατατομή *(cut)* is used to describe the result of trying to find out how long a section of string on the monochord should be kept from vibrating in order to produce a consonance. (Such attempts have been discussed more than once in previous chapters.) Clearly the Pythagoreans differed from the 'Experimentalists' in that the former took numerical ratios into account as well, whereas the latter completely ignored them.

Figuratively speaking, the whole string (the monochord) was '*cut*' at the place on the 'canon' where the little bridge (the ὑπαγωγεύς) eventually stood, i.e. at the point separating the length of string which was plucked to produce the second note from the section which was kept from vibrating. The words τομή and κατατομή are synonymous with each other. They are technical terms from the theory of music and were quite commonly used as such. It is also worth noting that the earliest surviving work of the Pythagoreans on the theory of music (the text of which bears Euclid's name) is called κατατομὴ κανόνος or *Sectio Canonis*.

The *cuts of the canon* are especially important because it is clear that these operations must also have led to the so-called '*musical means*'. The later Pythagoreans dealt with many different kinds of mean; of these, however, there are only three which interest us. It

---

[108] *Die Harmonielehre des Klaudios Ptolemaios*, ed. Düring (Berlin 1930) 66.28 ff

can be shown that these three were known in the fifth century B.C. and probably even earlier. They are the *arithmetic, geometric* and *harmonic* (or *subcontrary*) *means*. Before we look more closely at the text of the earliest account of these which has been handed down to us, it is worthwhile to recall what the significance of the arithmetic and harmonic means was in the study of music. It has been written that:[109]

> The arithmetic and harmonic means effect the division of an interval into unequal subintervals. The subintervals are the same in both cases except that they occur in the reverse order . . . . The classic example, mentioned by most authors, is the division of the octave. The numbers 12 and 6 stand in the ratio 2 : 1. Their arithmetic mean is 9, and their harmonic mean is 8. The former divides the octave into a fourth and a fifth; the latter divides it into a fifth and a fourth.

The interesting connection between arithmetic and harmonic means can be better understood if the canon with its twelve divisions is kept in mind when reading the above. An *arithmetic mean* is obtained by cutting the canon at the number 9, because this divides the *diastema* of the octave, the section of the string between the numbers 12 and 6, into two shorter sections, each having the same length. This cut is situated exactly at the midpoint between 12 and 6, and hence is at the same distance from each of the two endpoints of the original *diastema*. Yet the two equal lengths (12–9 and 9–6) induced by the cut represent two distinct musical intervals (numerical ratios). Furthermore, since $\frac{12}{9} < \frac{9}{6}$, the larger numbers form the smaller interval (12 : 9, a fourth) and the smaller numbers form the larger one (9 : 6, a fifth). The *harmonic mean*, on the other hand, which is the subcontrary of the arithmetic, is obtained by making the cut at the number 8. The fact that $\frac{12}{8} > \frac{8}{6}$ means that in this case the longer of the resulting sections (that between the numbers 12 and 8) corresponds to the greater ratio

---

[109] The following, which is not a literal translation, is taken from van der Waerden's paper, *Hermes* **78** (1943), 182.

11*

(a fifth), while the shorter one (that between 8 and 6) corresponds to the smaller (a fourth).

The connection between the three means is the subject of Archytas' 2nd fragment, which is quoted here mainly because of the interesting terminology used in it.[110]

μέσαι δέ ἐντι τρῖς τᾷ μουσικᾷ, μία μὲν ἀριθμητικά, δευτέρα δὲ ἁ γεω-μετρικά, τρίτα δ'ὑπεναντία, ἂν καλέ-οντι ἁρμονικάν.

ἀριθμητικὰ μέν, ὅκκα ἔωντι τρεῖς ὅροι κατὰ τὰν τοίαν ὑπεροχὰν ἀνὰ λόγον· ᾧ πρᾶτος δευτέρου ὑπερέχει, τούτῳ δεύτερος τρίτου ὑπερέχει. καὶ ἐν ταύτᾳ τᾷ ἀναλογίᾳ συμπίπτει ἦμεν τὸ τῶν μειζόνων ὅρων διά-στημα μεῖον, τὸ δὲ τῶν μειόνων μεῖ-ζον. ἁ γεωμετρικὰ δέ, ὅκκα ἔωντι οἷος ὁ πρᾶτος ποτὶ τὸν δεύτερον, καὶ ὁ δεύτερος ποτὶ τὸν τρίτον. τού-των δ'οἱ μείζονες ἴσον ποιοῦνται τὸ διάστημα καὶ οἱ μείους.

ἁ δ'ὑπεναντία, ἂν καλοῦμεν ἁρμο-νικάν, ὅκκα ἔωντι <τοῖοι· ᾧ> ὁ πρᾶ-τος ὅρος ὑπερέχει τοῦ δευτέρου αὐταύτου μέρει, τούτῳ ὁ μέσος τοῦ τρίτου ὑπερέχει τοῦ τρίτου μέρει. γίνεται δ'ἐν ταύτᾳ τᾷ ἀναλογίᾳ τὸ τῶν μειζόνων ὅρων διάστημα μεῖζον, τὸ δὲ τῶν μειόνων μεῖον.

In music there are three means: the first is the arithmetic, the second is the geometric and the third is the subcontrary, which is sometimes called the harmonic.

The arithmetic mean is present if the three numbers differ, [when taken] in *logoi*, as follows: the second number exceeds the third by as much as the first exceeds the second $(12-9 = 9-6)$. Furthermore this equality [when taken] in *logoi* means that the ratio [interval] of the larger numbers is the smaller one, and that the ratio of the smaller numbers is the larger $\left[\dfrac{12}{9} < \dfrac{9}{6}\right]$. The geometric mean is present if the first number is to the second, as the second is to the third. In this case both the larger and the smaller numbers have the same ratio $[a : b = b : c,$ where $a > b > c]$.

The subcontrary or harmonic mean is present if the numbers are such that the amount by which the first exceeds the second,

---

[110] Diels and Kranz, *Fragmente der Vorsokratiker*, I⁸, p. 436. Reidemeister ("Das Exakte Denken der Griechen". Hamburg 1949, p. 27) claims that this fragment is not authentic. In my opinion, however, his argument is not convinc-ing.

expressed as a fraction of the first, is the same as the amount by which the second exceeds the third, when expressed as a fraction of the third.[111] In this case of equality [when taken] in *logoi*, the ratio [interval] of the larger numbers is the greater and that of the smaller numbers is the lesser $\left[\dfrac{12}{8} > \dfrac{8}{6}\right]$.

As we can see, the terminology employed in this fragment is drawn from the Pythagorean theory of music. There is no need to enter into any further discussion of the terms ὅροι and διάστημα here, but the use of ἀνὰ λόγον and ἀναλογία is worthy of special notice. Even though these words can be translated as '(taken) in *logoi*' and 'equality [when taken] in *logoi*', they are undoubtedly being used in the fragment quoted above with a meaning which differs from all the other ones which we have considered up to now. In this fragment, the pairs of numbers which go to make up the arithmetic mean (12, 9 and 9, 6), are compared ἀνάλογον, and the arithmetic mean itself $(12-9 = 9-6)$ is said to be ἀναλογία. Hence this usage suggests that λόγος refers to the relationship between two numbers which is exemplified by *12 − 9* and *9 − 6*.

[Nowadays we would regard '12 − 9' and '9 − 6' as terms which denote numbers (resulting from the application of an operation to other numbers) rather than as expressing a relationship between *two* numbers. This remark applies to the word 'ratio' (and the notation '*a : b*') as well; although it is used properly to refer to a relation, *the common practice now is to speak of the ratio between two numbers as if it were itself a number* (such notation as '*a : b = c : d*' is evidence of this). These

---

[111] I.e., in the case of a harmonic mean the relationship between the three numbers *(a, b and c)* is expressed by the formula $\dfrac{a}{a-b} = \dfrac{c}{b-c}$. So, in particular, 8 is the harmonic mean between 12 and 6 since $\dfrac{12}{12-8} = 3 = \dfrac{6}{8-6}$.

usages, of course, are a reflection of modern ideas about mathematics (in particular about the closure of the real numbers under the operations of subtraction and division) *which it would be wrong to attribute to the Greeks.* Hence it is necessary to abandon the modern terminology when discussing these concepts. The relation referred to in the preceding paragraph would be denoted in English by the now obsolete phrase *'arithmetical ratio'.*][111a]

Thus the conclusions which I draw from Archytas' second fragment are as follows. In the theory of music, the word λόγος originally meant any kind of *'relation between two numbers'*, and not just the *'ratio between two numbers'* as it did in geometry. Consequently the word ἀναλογία meant any kind of *'equality between pairs of numbers'.*[112] It was only later that λόγος took on the narrower meaning of that relation between two numbers which we would refer to as the *'ratio'* *(a : b)*. Furthermore, this change resulted from the use of the word in geometry, not in the theory of music. The corresponding remark holds for ἀναλογία, namely that in mathematics it came to have the specialized meaning of *'sameness of ratio'* *(a : b = c : d)* instead of referring in general to pairs of numbers whose members bore the same relation to each other.

Apart from the fragment of Archytas which we have just been considering, I know of no other ancient text which uses the mathematical term λόγος in this old sense (i.e. with the meaning 'any kind of relation between two numbers'). In Euclid, λόγος is always used to mean 'ratio' (*a : b*, as we would write it). Nonetheless, Aristotle himself corroborates my conclusion that ἀναλογία originally had a wider meaning in mathematics. On one occasion he writes that the proportion *a : b = c : d* is what "mathematicians refer to as the geometric *analogia*"[113], while he remarks on another occasion, about the notion of μέσον, that "if ten is many and two is few, then six occupies the mean . . ., for it is as much greater than the one as it is smaller than the other, and this is the mean according to the arithmetic analogia" *(τοῦτο δὲ μέσον ἐστὶν κατὰ τὴν ἀριθμητικὴν ἀναλογίαν).*[114]

[111a] Translator's note.
[112] In fact Archytas also refers to the harmonic mean as *analogia* in the fragment quoted above.
[113] *Nicomachean Ethics*, 1131b, 12–3.
[114] Ibid., 1106a33.

So the original meaning of ἀναλογία must in fact have been 'sameness of any kind of relation between two numbers'.[115]


### 2.18 THE CREATION OF THE MATHEMATICAL CONCEPT OF λόγος

We have come across two different meanings of the mathematical term λόγος which have to be clearly distinguished from each other:

(1) In Euclid, and in the mathematical literature generally, λόγος denotes the relation which we would call *ratio (a : b)*.

(2) In the fragment from Archytas discussed in the previous chapter the word λόγος had another meaning. Archytas, when talking about the arithmetic mean, compares pairs of numbers ἀνὰ λόγον. Furthermore, he refers to the arithmetic mean itself $(12 - 9 = 9 - 6)$, and also to the harmonic mean $(12 \ldots 8 \ldots 6)$, as ἀναλογία. Hence he must be using the word λόγος to denote *any kind of relation between two numbers*.

My conjecture is that the first of these meanings (ratio) was obtained by a process of specialization from the second, and is based upon the following reasoning:

*(a)* In the terminology of the theory of music, the word for 'ratio' was διάστημα. The way in which it acquired this meaning can be described roughly as follows. Initially it meant a 'straight line' or 'length'. It then came to mean that 'section of string which was kept from vibrating' (to produce the second note of a consonance on the monochord), and finally it was used to describe the 'ratio between (the) two numbers' (associated with the end points of this piece of string). A *diastema* had of course two ὅροι (end points or numbers). When the various calculations (i.e. *cuts*) were first made on the canon, and it was desired to compare different line segments contained in an interval, the term λόγος was introduced. It was used to denote any pair of numbers which formed the end points of a *diastema*. It goes without saying that according to this ancient musical usage

---

[115] This is why the later Pythagoreans were able to refer to the remaining mean (μεσότητες), which will not be discussed here, as ἀναλογία. On this subject, see P.-H. Michel, *De Pythagore à Euclide*, Paris 1950, pp. 365–411, also the paper of mine listed in n. 12 above, p. 102.

(the traces of which are to be found in our fragment of Archytas) the words λόγος and διάστημα are *not* synonymous.

*(b)* It was only when the theory of proportions, which had initially been confined to music, began to be used in arithmetic and geometry as well that the meaning of λόγος changed from any 'relation between two numbers' to 'ratio'. The decision to assign this meaning to λόγος may have been prompted at least in part by the fact that  the ambiguity of the term διάστημα (straight line and ratio) must have been especially confusing in geometry. It was advisable, therefore, to introduce an unambiguous name for 'ratio' and reserve διάστημα (at least in geometry) for 'straight line'. After the new meaning of λόγος (as a mathematical term) had been settled once and for all, a corresponding change naturally took place in the meaning of ἀναλογία, which henceforth became 'sameness of ratio' or 'proportion'.

### 2.19 A DIGRESSION ON THE HISTORY OF THE WORD λόγος

The next question which presents itself is how the word λόγος acquired its meaning of 'relation between two numbers' and, in particular, 'ratio between two numbers'. This meaning was peculiar to music and mathematics. In everyday speech λόγος was used only with such meanings as 'word', 'speech', 'conversation', 'narrative', 'account' and 'reason'. I believe that this question can be answered only if the following is kept in mind:

Since κατάλογος meant an 'enumeration', in classical times the word λόγος must have had amongst its other everyday meanings one like 'number', 'quantity', a 'series or totality of certain things belonging together'. In Herodotus (III.120), for example, one finds the words: Σὺ γὰρ ἐν ἀνδρῶν λόγῳ – "Are you in the *ranks* (or amongst the *number*) of men?" [i.e. 'Are you worthy of being regarded as a man?'] Similarly he writes on another occasion (III.125): ἐν ἀνδραπόδων λόγῳ ποιεύμενος – "he added them to the *totality* (or to the *number*) of his slaves". In this usage λόγος seems to be a concept similar to ἀριθμός. The phrase οὔτ' ἐν λόγῳ, οὔτ' ἐν ἀριθμῷ,[116] which is quoted in Pape's lexicon, supports this conclusion. Rather than citing a host

---

[116] See the entry under λόγος.

of other examples, let me just quote a passage from Aristotle (*Nicomachean Ethics*, 1131b20), which runs as follows: ἐν ἀγαθοῦ γὰρ λόγῳ γίγνεται τὸ ἔλαττον κακὸν πρὸς τὸ μεῖζον κακόν. – "For a lesser evil compared to a greater one, counts as a good (or is counted in the *ranks* of goodness)."

It seems to me that the distinctive meaning of λόγος as a musical term is derived from the usage of the word which is discussed in the preceding paragraph. Since one of the everyday meanings of λόγος was 'a *series* of things' or 'a *combination* of several objects or numbers', it could be given the technical meaning of 'a combination of *only two numbers* which were related to one another, i.e. a *'relation between two numbers'*. (Of course, this would have required the conscious creation of a new concept.)

In addition, there is an interesting ambiguity about the word λόγος as it was used by the Pythagoreans. In music and mathematics they used it to denote a 'relation between two numbers' $(a-b)$ or $(a:b)$, yet at the same time they followed the general philosophical practice of using it to mean 'understanding', 'thought', or 'reason'. There are innumerable passages from the literature on the theory of music which could be quoted to show that on some occasions it is not at all easy to distinguish sharply between these two meanings (λόγος as 'numerical ratio' and λόγος as 'understanding' or 'reason'). This applies for the most part to passages which are not difficult to understand, and whose meaning is clear and unambiguous. Nonetheless it is difficult, when one reads them, to know whether λόγος should be translated as 'reason', 'understanding' 'reflection' or 'numerical ratio'. Of the many examples which illustrate this ambiguity, I want to mention only two here.

We have already discussed a passage from Ptolemy[117] in which he describes the procedures employed by certain of those who conducted musical experiments as follows: "They do not obtain the divisions of the string by *calculation*, but by stretching the string and then moving the little bridge until each of the notes being sought after is heard, etc."

The meaning of this passage is not problematic at all. If, however, it is read in the original Greek (οὐδ᾽ ὅλως ἔτι τῷ λόγῳ ποιοῦνται τὰς

---

[117] See n. 108.

κατατομάς), then it is not clear whether λόγος is supposed to mean 'reflection', 'understanding' or 'numerical ratio', since any of these senses would fit the context. (The word is translated by 'calculation' in our English version of the passage.)

There is a similar passage among the writings of Aristoxenus, the opponent of the Pythagoreans.[118] In it he maintains that only our senses (in particular, hearing) should be called upon in deciding about musical tones, since they alone are competent in this matter (ἡ αἴσθησίς ἐστιν ἡ τούτου κυρία). He then goes on to add that 'our senses are not in need of any help from logos' (οὐδὲ λόγου δεῖται). As the context shows, this is another case in which λόγος can be taken to mean 'numerical ratio' as well as 'rational thought'. In the last analysis, Aristoxenus is objecting not only to the fact that the Pythagoreans called upon reason (λόγος) instead of relying exclusively on sense perception, but also to the fact that they wanted to make musical intervals, which were supposed only to be heard, dependant upon numerical ratios (λόγοι).

This ambiguity in the meaning of the word λόγος (relation between two numbers, and rational thought or reason) was naturally of considerable importance to those Pythagoreans who wanted to view the 'rationale' of the universe in terms of number.


## 2.20 THE APPLICATION OF THE THEORY OF PROPORTIONS TO ARITHMETIC AND GEOMETRY

The evidence presented up to now is sufficient, I hope, to show that all the important terms of the theory of proportions have their origins in the theory of music. Furthermore, we have seen that this holds not only for the terms themselves (such as διάστημα or λόγος and ὅροι), but also for the concepts to which they refer. Since Greek mathematics without these terms and concepts would be inconceivable, one's first inclination is to think that the original theory of proportions was concerned exclusively with music and that it was subsequently applied to arithmetic and geometry. This conjecture is borne out particularly by the existence of certain geometrical terms ('doubled

---

[118] Harm. Stoich, 53.13ff.

ratio', 'compound ratio', 'difference between two ratios', etc.) which can only be explained, at least from a linguistic point of view, by reference to the calculations performed in the canon (cf. chapter 2.10).

However, the application of these concepts in arithmetic and geometry was far from being a mechanical procedure. It seems rather that nothing more than a start was made in the theory of music, and that the development of what we would understand by the theory of proportions actually took place within arithmetic and geometry after the basic concepts had been taken over from musical theory. Indeed, their application to a new field even brought about some changes in the concepts themselves. To a certain extent, this should already be evident from our discussion up to now.

Although those Pythagoreans who concerned themselves with the theory of music came later on to regard διάστημα and λόγος as equivalent concepts, the fragment from Archytas makes it clear that λόγος did not originally mean 'geometric ratio', and hence that ἀναλογία did not mean 'geometric proportion' but merely 'equality [when taken] in *logoi*' (where *logos* was some unspecified relation). It seems that on the whole the *comparison* of quantities [taken] in *logoi* was much less important in the theory of music than it was in arithmetic and geometry. It is true that pairs of numbers (the constituents of an arithmetic mean) are compared ἀνὰ λόγον by Archytas, nonetheless the notion of 'equality [when taken] in *logoi*' was not of any great importance (from the point of view of further developments) until it became featured in the geometrical arithmetic of the Pythagoreans. To see this, one should look at the Pythagorean definition of similar plane numbers *(Elements* VII, Definition 21).

As was mentioned in Part 1, the Pythagoreans thought that any number, when decomposed into two factors, could be regarded as a *plane number.* (For example, 3 was represented by a rectangle with sides 1 and 3.) The definition (VII. 21) states: "Similar plane . . . numbers *(ὅμοιοι ἐπίπεδοι ἀριθμοί)* are those which have their sides proportional *(οἱ ἀνάλογον ⟨ἴσας⟩ ἔχοντες τὰς πλευράς)*."

So 3 and 12, for example, would be similar plane numbers according to this definition, since they can be represented by a pair of *similar rectangles* with sides (factors) 1, 3 and 2, 6 respectively. It is obvious that the *logoi* formed from the corresponding sides of these rectangles are equal (1 : 2 = 3 : 6). Furthermore, each of them is composed of a pair of line segments, just as a *diastema* (an interval expressed as a numerical ratio) in the theory of music was produced by two *sections of string* which differed in length. The concept of 'equality [when taken] in *logoi*', however, was much more important in geometrical arithmetic (i.e. in the arithmetic of plane and solid numbers) than it was in the theory of music. In the latter its most important use was in the formulation of such results as that the proportional numbers of a *fifth* were always identical, knowledge of which did not lead to any further discoveries.

On the other hand when applied to the corresponding sides (factors) of plane numbers, this same notion revealed something of great importance. The relation of *similarity* between rectangles could be characterized in terms of it. Of course even before the introduction of the concept ἀναλογία, mathematicians must have had an intuitive idea of what it meant for two rectangles to be *similar*. However, it was only after they realised that 'geometrical similarity' (between rectangles in the first instance, and then between rectilinear figures in general) was precisely the 'equality [when taken] in *logoi*' *(ἀναλογία)* of corresponding sides, that this concept become a truly scientific one. So we see that a concept which our etymological investigations have led us to conclude originated in the field of music contributed to a better understanding of *geometrical similarity* when it was applied in geometrical arithmetic. *Similarity* was seen to be the '*analogia* of sides'. This explains how the word ἀναλογία came to be used by the educated in non-mathematical contexts to mean 'similarity' *(ὁμοιότης)*. (As is well known, Aristotle frequently uses the phrase τὸ ἀνάλογον with just this meaning.)[119]

It is worth emphasizing at this point that there is another indubitable evidence to show that the scientific concept of geometrical similarity must have been an invention of the Greeks. The following definition makes this plain:

*Elements* VI.1: "Similar rectilineal figures are such as have their angles severally equal *(ὅμοια σχήματα εὐθύγραμμά ἐστιν, ὅσα τάς τε*

---

[119] *Metaphysics*, N. 6.1093b17ff. For ἀναλογία in the sense of ὁμοιότης, see *Philodemus: On Methods of Inference, A Study of Ancient Empiricism*, P. H. Lacy and E. A. Lacy, Philadelphia 1941. Similarly Alexander Aphrodisiensis, (vol. 2 of *Commentaria in Aristotelem Graeca* ed. M. Wallies, Berlin 1892) p. 545, 1ff, explains ἀνάλογον as ὅμοιον. On this latter, see my paper listed in n. 8 above.

γωνίας. ἴσας ἔχει κατὰ μίαν) and the sides about equal angles proportional (καὶ τὰς περὶ τὰς ἴσας γωνίας πλευρὰς ἀνάλογον scil. ἴσας ἔχει)."

Thus similarity is defined as having *equal angles*, and sides *in the
same ratio* (ἀνάλογον). This presupposes both the concept of 'angle',
which was developed during the early Ionian period,[120] and that of
ἀνάλογον, which came from the Pythagorean theory of music.

The difference between the above definition of similarity for rectilinear figures and the definition of similar plane numbers is immediately
apparent. Plane numbers were always thought of as *rectangles* in
geometrical arithmetic. (The Pythagoreans also considered other kinds
of plane numbers, such as *triangular* and *pentagonal* ones, but these
are never discussed in Euclid's account of arithmetic.) In the case
of *similar rectangles*, however, there was no need to insist on their
angles being equal, since a rectangle by its very nature has only right
angles. Furthermore, the sides of these rectangles always represented
*integers*. The musical theory of proportions dealt only with integers
and never with *incommensurable magnitudes*; so we have an additional
explanation for the ease with which concepts from this theory could
be applied in geometrical arithmetic.

It was a different story, however, when mathematicians turned to
the similarity of rectilinear figures in general. This problem does not
always yield to the simple arithmetical techniques which work so
well in the case of rectangles. Of course they were no less certain in
their own minds that the sides of similar rectilinear figures stood in
the same ratio that they were about the sides (factors) of similar
plane numbers. Yet it was often impossible to prove this rigorously
(in the case of triangles, for example) so long as the theory of proportions (as in arithmetic and the theory of music) applied *only to numbers*
and not to incommensurable magnitudes as well. It was convenient,
therefore, to base the similarity of rectilinear figures in the first place
on the *equality of their angles*. The fact that corresponding sides stood
in the same ratio came out of this almost as a consequence.

My conjecture is that the concepts of the musical theory of proportions were applied first of all in arithmetic. Their application was
facilitated by the fact that numbers in arithmetic, just as in the theory

---

[120] O. Becker, *Grundlagen der Mathematik in geschichtlicher Entwicklung*,
Freiburg–Munich 1954, p. 27.

of music, were represented by *straight lines*. (Hence the product of
two numbers was a *rectangle*, and product of three factors was a
*solid.)* Furthermore, the application of this theory to geometrical
arithmetic contributed towards an understanding of the problem of
geometrical similarity, and this problem in turn soon led to the problem
of linear incommensurability. I even believe that the theory of music
prepared the way for the discovery of linear incommensurability.
This is the point I wish to take up in the next chapter.


### 2.21 THE MEAN PROPORTIONAL IN MUSIC, ARITHMETIC AND GEOMETRY

In a previous chapter (see p. 164) we discussed a fragment from
Archytas dealing with the *three means in music*, and pointed out the
significance of two of these (the arithmetic and the harmonic) for
the theory of music. [As has been remarked, these two means
effect the division of a musical interval into *unequal subintervals*,
the subintervals being the same in both cases except that they occur
in the reverse order. The classic example is the division of an octave
(12 : 6) into a fourth and a fifth (12 : 9 and 9 : 6) by the arithmetic
mean, and into a fifth and a fourth (12 : 8 and 8 : 6) by the harmonic
one.] The so-called *geometric mean* lies in between these two, but the
role which it played is less clear.

According to Archytas, the geometric mean is that 'end point'
*(ὅρος)* between two other 'end points' ("numbers") which divides
the original interval (say $a : c$) into two *equal subintervals* (i.e. in such a
way that $a : b = b : c$). In the case of an octave, the geometric mean
would be the number $x$ satisfying the equation *$12 : x = x : 6$*. Even
if fractions are allowed as well as whole numbers, there is obviously
no such 'number'. (Certainly there is none in the sense of Euclid's
definition of number.) *An octave cannot be divided into two equal
subintervals by a number.* This was not just a practical difficulty for
the Pythagoreans, but a theoretical one as well. According to their
theory, any musical interval could be expressed as a ratio of whole
numbers.

It is plain that the above applies to fourths *(4 : 3)* and fifths *(3 : 2)*
as well; as with the octave, there is no number which divides them

into equal subintervals, nor is there a (numerical) geometric mean between their proportional numbers.

This leads us to the question of why Archytas counts the geometric mean as a musical one, even though no such mean can exist in music. As far as I can see, no light is shed on this question by the context in which his words appear. Hence any answer to it can only be conjectural. My own view is that an explanation is provided by the following.

It is clear that the *geometric mean* did not receive its name until after mathematicians learned how to construct it in a *geometrical* way. Originally it was a source of considerable puzzlement to those interested in the theory of music. The only thing it has in common with the genuine 'musical means' (i.e. with the arithmetical and harmonic ones) is that it too emerged first in the theory of music.

We have in fact definite proof that the Pythagoreans attempted to divide musical intervals into *equal* subintervals. In other words, they tried to find the geometric means of these intervals and came to the conclusion that this could not be done in an arithmetical way. Our evidence for this is Proposition 3 of the *Sectio Canonis*, which states:[121]

"No mean proportional number can ever be found between two numbers in a *ratio superparticularis*."

To understand this proposition, one needs to know that the Greeks meant by a *'ratio superparticularis'* (ἐπιμόριον διάστημα) what we would denote by '$(n + 1) : n$'. This kind of ratio is exemplified not only by fourths *(4 : 3)* and fifths *(3 : 2)*, but also by octaves *(2 : 1)*.

The proposition states, therefore, that there is no (numerical) geometric mean between two numbers in a *ratio superparticularis*. (It should be kept in mind that all the important musical intervals are in fact examples of this kind of ratio.) Furthermore, the Greeks were well aware of the fact that this result was always true, no matter which multiples of the *ratio superparticularis* were used to test it out.

[121] Most scholars since Tannery's time have ascribed this proposition to Archytas (see, for example, van der Waerden, *Hermes* **78** (1943), 169). Our source, Boethius (*De Musica* III. 11), objects to "*a proof* of Archytas" when talking about the proposition in question. I do not think that this by itself is a sufficient reason to attribute the entire proposition to Archytas. In my opinion, *Sectio Canonis* 3 was formulated long before his time.

This much is plain from the proof of the proposition in the *Sectio Canonis* and from the other propositions which are assumed in the proof.[122]

Their lack of success with this problem in the theory of music prompted the Pythagoreans to pose it in a more general form. They asked under what conditions it was possible to find between two *numbers* a third having the property that it formed the same ratio with both of them; i.e. they wanted to know when a mean proportional number existed between two numbers. In Book VIII of Euclid's *Elements* there is a whole series of propositions which deal with just this question. Let us recall some of them here:

VIII.11: "Between two square numbers there is *one* mean proportional number, etc."

VIII.12: "Between two cube numbers there are *two* mean proportional numbers, etc."

VIII.18: "Between two *similar plane numbers* there is *one* mean proportional number, etc."

VIII.20: "If one mean proportional number falls between two numbers, the numbers will be *similar plane numbers*."

There is no tradition to confirm that these arithmetical propositions do in fact antedate the discovery of linear incommensurability. However, since it is possible to prove them without such knowledge,[123] it does not seem unreasonable to regard them as having paved the way for the discovery of how to construct a mean proportional to *any* two straight lines in a geometric way (cf. *Elements* VI.13).

Thus an insoluble problem in the theory of music (to divide the interval associated with a consonance into two *equal* subintervals or, in other words, to find a *numerical geometric mean* between the two

[122] As far as these other propositions are concerned, I cannot but agree with van der Waerden's analysis (see the previous note). However, I would like to dissociate myself from the view that they are all due to Archytas himself.

[123] Incidentally, I agree with van der Waerden's remark (see *Zur Geschichte der griechischen Mathematik*, p. 224) to the effect that *the problems in Book VIII are very closely connected with the theory of irrationals.* These problems actually prepared the way for the theory of irrationals. This is one more reason why Book VIII could not have been written during the time of Archytas. The theory of irrationals, as we can see from the concept *dynamis,* dates back much earlier than this.

numbers associated with a consonance) may have led, by way of arithmetic, first to the problem of geometric similarity (for rectilinear figures) and then to the geometrical construction of a mean proportional to any two straight lines. In a sense, however, the latter assumes a knowledge of linear incommensurability. For the mean proportional between two straight lines whose lengths are not similar plane numbers is itself a straight line incommensurable in length (cf. Part 1 of the present work).

### 2.22 THE CONSTRUCTION OF THE MEAN PROPORTIONAL

The construction of a mean proportional to two given line segments is discussed by Euclid in Proposition VI.13 of the *Elements*. We know that Hippocrates of Chios, the famous 5th century geometer, must already have been familiar with it.[124] The ancients, however, have not left us any information about the age of the proposition (VI.13), or about the steps which led up to the discovery of this construction. The following attempt to reconstruct these facts could therefore be dismissed right from the start as mere conjecture. Yet it seems to me that it is still worth making, because the whole point of studying the history of mathematics is to try to understand how the development of science has led from one problem to another.

The first thing to remember about Proposition VI.13 is that *similar right-angled triangles* are used in its proof. The problem of the *mean proportional* is of course inseparable from the problem of *similarity* for geometrical figures. It was a more thorough investigation of geometric similarity which provided the solution to the former problem. Furthermore, their experience with arithmetic must have suggested to the Greeks that this was the right way to approach it. For they already knew that a mean proportional number existed between two similar plane numbers, and were also able to calculate it (see pp. 51–3).

It seems likely that they were familiar with a special case of the construction before being able to carry it out in its full generality (as is done in *Elements* VI.13). One case in particular strikes me as especially simple and obvious.

[124] See n. 37 to Part 1 above.

Recall that at the end of Part 1 of this book, we were led to conjecture that *the problem of doubling a square was originally equivalent to the problem of finding a mean proportional between one number and a second which was twice as large.* In fact, when the Greeks discovered that the diagonal of a square was the line segment needed to construct a second square having twice the area of the original one, they must at the same time have realized that *it represented the mean proportional between one side of the square and a line which was twice as long as this side.* The first fact would have been visually evident to them (cf. p. 93), whereas they may have become convinced of the second in the following way.

*(a)* It must have been self-evident to anyone that the two squares had to be similar. (This would have been so, whether or not they also took into account the fact that if a number was assigned, as a length, to the sides of the original square, the same could not be done for the sides of the second one.) Furthermore, it must have been equally obvious that this was also true of the right-angled triangles obtained by halving the two squares. (To begin with, of course, the evidence for this was only visual and intuitive.)

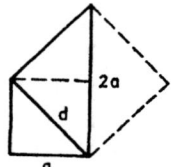

Fig. 5

*(b)* It was now an easy matter to form one *logos* from the corresponding sides of the two squares and another from their diagonals, just as was done in the case of similar plane numbers. Hence the first *logos* was composed of the shorter sides of the two triangles *(a : d)*, whereas the second one was made up of their *hypotenuses (d : 2a).* These *logoi* were evidently equal *(a : d = d : 2a)*, just like the ones formed from the corresponding sides of "similar plane numbers".

In this way the diagonal of the original square was clearly shown to be the required *mean proportional*. [These considerations also provided a solution to the geometrical equivalent of an insoluble problem from the theory of music. I am of course referring to the problem of *halving the octave* (2 : 1), i.e. of finding the mean proportional between two lengths of string which produce an octave.]

The argument sketched above is not valid unless the concepts of λόγος and ἀναλογία are taken as applying to *magnitudes* as well as to *numbers*. Thus they have to be understood in the sense of the so-called Eudoxian definition which is found at the beginning of Book V of the *Elements*. It is an open question whether the definitions of these concepts were extended at the same time it was realized that the diagonal of a square was also a mean proportional, or whether this did not take place until later.

In any event, the way in which mathematicians discovered that the diagonal of a square was the mean proportional between one of its sides and a line twice that length also taught them a great deal about how to construct a mean proportional to two arbitrary line segments.

(1) The first thing which must have been clear to them was that the simplest way to obtain a mean proportional was by using *similar right-angled triangles*.

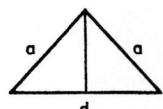 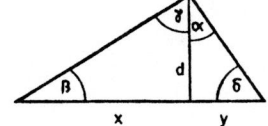

Fig. 6

(2) They must also have realized that the straight line in question was the required mean proportional precisely because it played *two different roles* as far as the two triangles were concerned. (In Fig. 5 the line segment *d* is adjacent to the right angle in the larger triangle, whilst forming the *hypotenuse* of the smaller one.)

The question now was how to utilize this information; how, for example, were they to construct a pair of similar right-angled triangles.

If they started out with a right-angled isosceles triangles (one half of a square), then this latter question was easy to answer. Two similar (in fact, two *congruent*), right-angled isosceles triangles can be got from such a triangle by dropping a perpendicular from the right angle to the hypotenuse. It must have been immediately apparent that the

12*

triangles obtained in this way were similar; but unfortunately they were not suited to the purpose at hand. However, the special case in which similarity coincides with congruence may have been of some use in pointing out the way to the more general one.

If (as in Fig. 6) a perpendicular is dropped from the right angle of a non-isosceles right triangle, two right-angled triangles are again obtained. It could not have taken the Greeks long to realize that these too were similar, since their corresponding angles were equal $(\alpha = \beta$ and $\gamma = \delta)$. The perpendicular $d$ (see Fig. 6) plays a *double role* of the sort discussed in (2) above. It is the longer of the two adjacent sides in the smaller triangle, and the shorter of the two in the larger one. They could now obtain a mean proportional $(x : d = d : y)$ between the lengths $x$ and $y$ simply by forming *logoi* from corresponding adjacent sides (cf. the porism to *Elements* VI.8).

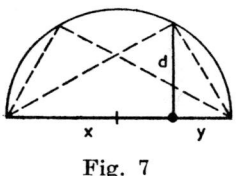

Fig. 7

The construction of the mean proportional in *Elements* VI.13 is actually nothing other than the procedure sketched above in *reverse*. In the preceding paragraph a right-angled triangle was given, and a perpendicular was drawn from the right angle to the hypotenuse. This meant that the perpendicular itself was the mean proportional between the two segments into which it divided the hypotenuse. Suppose that we are given instead two line segments $(x$ and $y)$ and want to find the mean proportional between them. Our first step would be to construct a right triangle of the sort given in Fig. 7.

The two line segments should together $(x + y)$ form its hypotenuse. (This much is obvious from the preceding discussion.) If we regard the length $x + y$ as the *diameter of a circle*, infinitely many right-angled triangles can be constructed upon it, for there is a theorem of Thales[125]

---

[125] See B. L. van der Waerden, *Erwachende Wissenschaft*, p. 145.

which states that the angle in a semicircle is always a right one. What is needed, however, is a right triangle in which the perpendicular from the right angle to the hypotenuse divides the latter into two segments, one of length $x$ and the other of length $y$. We have to draw (in the upper semicircle, say) a line $d$ perpendicular to the hypotenuse, which intersects it at the point where $x$ and $y$ meet. The point at which $d$ intersects the circumference will then be the third vertex of the required triangle, and $d$ itself will be the mean proportional between $x$ and $y$. We see, therefore, that this construction can be carried out using only some very elementary facts about similarity together with a theorem of Thales pertaining to the angle in a semicircle.

As in the previous case, a definition of *'being in the same ratio'* (ἀναλογία or ὁ αὐτὸς λόγος) which applies to *general quantities* is indispensable for a rigorous proof of the above. In my opinion, however, we do not have sufficient knowledge at present to decide whether such a definition was available to the inventors of this construction, or whether they were content to base it on intuitive evidence.

## 2.23 CONCLUSION

At the end of Part 1 of this book we came to the conclusion that the existence of linear incommensurability may have been discovered in the course of attempts to find a *mean proportional* between two numbers (expressed as straight lines) or quantities. From a historical point of view, the Greeks originally thought of the problem of irrationality as belonging to *the theory of proportions*. I hope that Part 2 has shown how the theory of proportions, whose initial development took place in the Pythagorean theory of music, may in fact have led by way of arithmetic to the problem of *geometrical similarity* and thence to problem of *linear incommensurability*.

There is one thing, however, which should not be forgotten. We have up to now confined our investigation to the way in which Greek mathematicians may have come close to discovering that linearly incommensurable straight lines existed. We have not as yet shown how they reached the stage of being able to prove the existence of linear incommensurability in a rigorous manner. Although we can suppose that they knew how to double the area of a square by means of its diagonal, and even that they were convinced by the *intuitive evidence* in favor of the diagonal being for this reason the mean proportional between one side and a line twice its length, more is needed to establish that the side and diagonal of a square are incommensurable in length.

A similar remark applies to the construction of the mean proportional to two arbitrary line segments *(Elements* VI.13), as well as to the creation of the concept *dynamis*. For example, the Greeks may well have known that a rectangle 3 units wide and 5 units long could be transformed into a square of the same area by constructing the mean proportional to two of its sides. Furthermore, their knowledge of arithmetic may well have told them that a mean proportional number existed only between two similar plane numbers (cf. *Elements* VIII.18 and 20). All this, however, would have led them to the realization that the sides of a *dynamis* having an area of 15 were linearly incommensurable (with the basic unit of length), only after the existence of such segments was no longer in doubt. Similarly the definition of proportionality could not have been extended from numbers to general quantities until after the existence of incommensurability was discovered. Furthermore, incommensurability itself is *not something which can be known with certainty in an empirical way. Knowledge of it can only be obtained by a systematic process of reflection.*

For this reason Part 3 of this book is devoted to the problem of how the Greeks transformed mathematics from a practical and empirical body of knowledge into a theoretical and deductive science. After all, the existence of incommensurability can be proved rigorously only by the methods of the latter, not by those of the former. I do not want to conclude Part 2, however, without briefly expressing my opinion about the import of some of our conclusions for future research.

In the last few decades there have appeared several important works on the history of science which concern themselves with pre-Greek mathematics and with that of the Babylonians and Egyptians in particular.[126] This research has led not only to a better understanding of the place of mathematics in pre-Greek culture, but also to a feeling that our earlier picture of the Greek world and of the beginnings of

---

[126] The most important results of these investigations are collected together in the first three chapters of van der Waerden's book, *Science Awakening.*

science needs to be thoroughly revised in the light of this new historical knowledge. As long as nothing was known about Egyptian and Babylonian mathematics, the Greeks could rightly or wrongly be regarded as the originators and founders of science. The situation changed suddenly, however, when it turned out that many important mathematical discoveries which had previously been attributed to the 5th and 6th century Greeks, had been made centuries earlier by the Babylonians and Egyptians. The Greeks seemed to have been robbed of much of their old importance in the history of science by this new knowledge of pre-Greek cultures. Otto Neugebauer, a distinguished authority on the history of this earlier mathematics, felt called upon to write that because we now know something not only about the two and a half thousand years which have passed since the classical age, but also about the two and a half thousand or so which preceded it, the Greeks could no longer be regarded as the prime originators and founders of science.[127] It seemed that their historical position was no longer at the beginning of science, but somewhere in the middle.[128] Attempts were even made to prove that the Greeks did little more than carry on the science of their predecessors, and it was said that[128] "The Babylonian tradition provided the materials out of which the Greeks, or more precisely the Pythagoreans, built their mathematics."

It is clear that students in the future will continue to ask what the Greeks might have learned from their predecessors. In my opinion, however, their research will be given a new turn by the material presented in the first two parts of this book. Although they will still ask to what extent Greek science is just a continuation of its pre-Greek counterpart and to what extent it is a completely original creation, it seems to me that this question has become much more clearly defined than it was before. This is because in the preceding chapters we have come across a whole series of important mathematical con-

---

[127] O. Neugebauer, Studien zur Geschichte der antiker Algebra III, *Quellen und Studien zur Geschichte der Math.* etc. B, **3** (1936) 245–59. He goes on to say: "The substance of elementary geometry, the elementary theory of proportions and the theory of equations is to be found in the Babylonian mathematics upon which the Greeks built their own. In every case the connection between the two is obvious and undeniable."

[128] B. L. van der Waerden, *Math. Ann.* **120** (1947/9) 123.

[129] B. L. van der Waerden, *Erwachende Wissenschaft*, p. 204.

cepts *(diastema, horoi, logos, analogon, analogia, 'compound ratio'* and related notions, 'a cut of the canon', 'mean proportional', 'geometrical similarity defined as equality of angles and *analogia* of sides', *dynamis, symmetros, mekei asymmetros, dynamei symmetros,* etc.) which, at least on the face of it, seem as though they must be *completely original and authentically Greek inventions.* So the question is transformed into one about the extent to which these same concepts or perhaps similar ones figure in pre-Greek mathematics.

I hardly think that a clear and satisfactory answer can be given to this question on the basis of our *present* knowledge of pre-Greek science. Furthermore, the time for making conjectures about what the Greeks might have borrowed from other cultures will come after we have acquired a better understanding of the Greeks and their science.

# 3. THE CONSTRUCTION OF MATHEMATICS
# WITHIN A DEDUCTIVE FRAMEWORK

## 3.1 'PROOF' IN GREEK MATHEMATICS

At the end of Part 2 we discussed the connections which various scholars have tried to establish between Greek mathematics and pre-Hellenic science. As we saw, some of them even maintained that Greek science was built upon the achievements of the Babylonians,[1] and that it could be regarded as a direct continuation of its more ancient, Oriental counterpart.

It is probably no accident that those scholars who tried to establish connections between the Greek world and the ancient Orient did not find it necessary to point out the very substantial differences between these two cultures. Anyone wanting to learn about such differences would have had to consult the works of another group of scholars. Members of this latter group, while not denying that the Greeks were able to learn a great deal from their predecessors (or indeed that they actually did so, which seems very likely to have been the case), did not forget the essentially new features of Greek (as opposed to pre-Hellenic) science. To illustrate this, let me quote the following from an important article by von Fritz:[2]

"Research undertaken during the last few decades has shown that Babylonian mathematics of the pre-Hellenic period was much more highly developed than had been thought previously. The Babylonians were able to find quite good approximations to the solutions of relatively complicated mathematical problems about a thousand years before the beginnings of Greek mathematics. Furthermore it has been shown that the Greeks took over and learnt a great deal from the Babylonians."

[1] See n. 127 to Part 2 above.

[2] K. von Fritz, 'Die ἀρχαί in der griechischen Mathematik'. *Archiv für Begriffsgeschichte* (Bonn), Vol. 1 (1955) 13–103.

The above might lead one to think that the deductive features of Greek mathematics also had their roots in pre-Hellenic science. Yet von Fritz emphasizes:[3] "There is absolutely no evidence to suggest that the Babylonians, not to mention the Egyptians, ever tried to deduce mathematical theorems rigorously from first principles."

Furthermore, the fact that no traces of deductive mathematics have been found in any of the pre-Hellenic cultures cannot be attributed simply to our lack of historical knowledge. For the previous quotation continues: "Everything which we know about ancient Oriental mathematics speaks against the idea that the gaps in our historical knowledge are responsible for our failure to detect the existence of deductive mathematics in the pre-Hellenic period."

The most substantial difference between Greek and Oriental sciences is that the former is an ingenious system of knowledge built up according to the method of logical deduction, whereas the latter is nothing more than a collection of instructions and rules of thumb, often accompanied by *examples*, having to do with how some particular mathematical tasks are to be carried out. Whether such fundamental scientific concepts as 'theorem', 'proof', 'deduction', 'definition', 'postulate', 'axiom', etc. were known to Egyptian and Babylonian mathematicians of the pre-Hellenic period, remains an open question. We do, however, know the following:[4]

"It is not even certain that the Babylonians knew how to formulate general theorems. It is true that in the sacred geometry of ancient India, known as the *rope rules*, certain geometric facts are stated in the form of *sutras* or aphorisms, but these are never proved. *Proofs* do not appear in any of the ancient Oriental texts known to us. At best one sometimes comes across some numerical verifications."

The impartial reader must by now have reached the conclusion that the most striking feature of Greek mathematics, which distinguishes it from its Oriental counterpart, is the presence of *proofs*. Greek science is concerned not only with stating propositions, but with providing genuine *proofs* for them as well. The role played by proofs is no less

[3] Ibid., pp. 13–4.
[4] O. Becker, 'Frühgriechische Mathematik und Musiklehre', *Archiv für Musikwissenschaft* (Trossingen), Vol. **14** (1957), 157.

important in Greek than it is in contemporary mathematics; in fact, it is the same. Furthermore, Euclid's proofs are for the most part models of their kind and have set the standards of mathematical rigour for generations. Notwithstanding the fact that modern mathematicians sometimes find them wanting in one or another respect, the proofs in the *Elements* are by and large exemplary.

The question which now raises itself is what first gave Greek mathematicians the idea that mathematical propositions need to be proved, and how did they become so skilled at proving them that their proofs are still admired today. — A further problem is whether mathematicians were originally satisfied with more primitive and less complete proofs than the ones found in Euclid, and if so, how did techniques of proof become more highly developed later on.

It is disappointing that so little attention has been paid to these questions in earlier research. Although it must be admitted that our historical sources do not seem to provide any answer to them, we still have ancient mathematical texts to consult. Some of these contain technical terms which cannot be much older than the texts themselves. I hope therefore that, by applying the methods used in the first two parts of this book, I will be able to give an analysis of the mathematical terms found in Euclid (and of the concepts to which they refer), which will shed some new light on the above questions.

Let us begin our investigation by considering the mathematical term for *proof* and *to prove*, namely the Greek word δείκνυμι. It is used very frequently by Euclid, especially at the end of mathematical demonstrations. His treatment of theorems always conforms to the following pattern.

First he states the proposition itself, and then a particular instance of it which is taken to be typical. Next follows a proof of the proposition for this instance. The original proposition (which is now repeated word for word) is then inferred from the proof, and the discussion is concluded with the phrase: ὅπερ ἔδει δεῖξαι — "(being) what it was required to prove".

The fact that this phrase is never omitted,[5] shows clearly that the actual *demonstration* is the most essential part of the whole discussion.

---

[5] It should be mentioned that the so-called *problems* do not always conclude with these words. As Proclus remarks in his commentary (p. 81, 5–22 in Fried-

The verb δεῖξαι sums up and places emphasis upon the demonstration. Of course the meaning and etymology of the Greek word δείκνυμι have never been problematic. Furthermore, its use as a mathematical term has been so taken for granted that the question of how it came to be one has not been given much attention.[6] There are, in fact, at least two interesting phases in the development of this expression as a mathematical term. Before discussing these, however, let us consider the non-mathematical usage of this word.

In most dictionaries, the meanings of the word δείκνυμι are split up into three goups as follows: (1) *to show, to point out;* (2) *to make known (especially by words), to explain;* and (3) *to show, to prove* — [or: (1) *faire voir, montrer;* (2) *faire connaître par la parole, expliquer;* and (3) *démontrer, prouver*]. On the basis of this listing, it would seem that the first meaning ('to show', 'to point out' in the most literal sense of the words) is the original one, since it is the most concrete of the three. It is clearly the one which Plato had in the mind when he wrote:[7] τὸ δεῖξαι λέγω εἰς τὴν τῶν ὀφθαλμῶν αἴσθησιν καταστῆσαι (by 'showing', I understand 'to put before the eyes'). Yet we also find δείκνυμι used with the meaning 'to make known by words' in texts as early as the *Odyssey* (12.25). There are in fact some parts of the *Iliad* (e.g., 19. 322) and *Odyssey* (10. 33) where it is not certain, whether the word is being used in the sense of 'to point out with a finger' or 'to point out with words'. The Latin verb *dico*, which comes from the same root as δείκνυμι is also worth noting in this connection. Although it is related to *digitus* (a 'finger' or 'pointer'), it means 'to say' or 'to tell' (i.e., 'to point out with words'). All things considered, it seems most likely that right from

---

lein's edition): " . . . there is a distinction between a problem and a theorem. That Euclid's *Elements* contains both problems and theorems will be evident from the individual propositions and from his practice of placing at the end of his demonstrations sometimes 'This is what was to be done' *(ὅπερ ἔδει ποιῆσαι)* and at other times 'This is what was to be proved' *(ὅπερ ἔδει δεῖξαι)*. The latter phrase is the mark of a theorem . . .". (The translation is by Glenn R. Morrow, *Proclus – A Commentary on the First Book of Euclid's Elements*, Princeton 1970.) Nonetheless the term δεῖξαι does occur in the proof of some problems. It is used in such phrases as προκείσθω ἡμῖν δεῖξαι, ὅτι κτλ. (Our task is to *show* that . . .).

[6] Cf. Á. Szabó, '*Deiknymi*, als mathematischer Terminus für *beweisen*', *Maia* N. S. **10** (1958), pp. 106–31.

[7] *Cratylus* 430 e.

the start the Greek verb δείκνυμι meant 'to point out' in both a figurative and a literal sense, and hence also 'to explain'.

We now want to find out how this word came to be used as a mathematical term, and what the earliest Greek mathematicians meant by it. These questions would probably have to remain unanswered, if we had nothing but the word and its etymology to go on.

It is, however, interesting to note that some scholars, without making any reference to the verb δείκνυμι, have stressed the importance of *visual evidence* in early Greek mathematics.[8] This suggests the idea that the earliest 'proofs' may have involved some kind of 'pointing out' or *'making visible'* of the facts; in other words, δείκνυμι may have become a technical term for 'proof' in mathematics because 'to prove' meant originally 'to make the truth (or falsity) of a mathematical statement visible in some way'.

This idea cannot be supported by any examples from Euclid, for the 'proofs' in the *Elements* are already in a polished form and no longer have anything to do with making facts visible. In Part 1 of this book, however, we quoted an example of mathematics teaching in antiquity[9] which does seem to support our conjecture. The example comes from Plato's *Meno* (82b–85e) and is of interest to us here because it touches upon the problem of mathematical proof.

I have in mind the passage where Socrates asks a slave how the area of a square with sides two units long can be doubled without altering its shape. He then draws a diagram [see (1) in Fig. 8] *showing* the square which is to be doubled. After the slave answers that the square will perhaps be doubled by doubling the length of its sides, Socrates draws a second diagram (2) to *show* that a square whose sides are twice as long

---

[8] K. von Fritz, op. cit. (n. 2 above), p. 94, states that the abundant use which was made of *the method of superposition* in pre-Euclidean mathematics indicates that *visual evidence* must initially have played a not inconsiderable role. (On this method see *Common Notion 4* in the *Elements* and Heath's note on it (Vol. I, pp. 224–5 in his edition, especially his remarks about the verb ἐφαρμόζειν). He goes on to point out that Euclid made considerable efforts to avoid this method and that, being unable to eliminate it completely, he attempted to give an axiomatic foundation for that part of it which could not be dispensed with and to conceal as far as possible its empirical character. Following Reidemeister, von Fritz concludes that the distinguishing feature of Greek mathematics is the turn which it took away from the *visual* and towards the *abstract*.

[9] See n. 77 to Part 1.

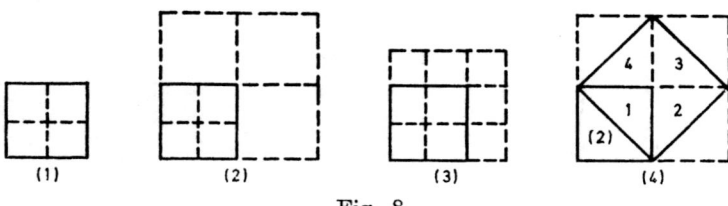

Fig. 8

as those of the original one, has an area four times its size. He then goes on to draw a third diagram (3) *showing* that a square whose sides are three units long, has an area of nine square units, and hence is too large, cannot be the double of the original one. Finally in his fourth diagram (4) he *shows* that the square on the diagonal has exactly twice the original area.

I think that this passage from Plato provides an excellent illustration of the way in which statements were verified at an early stage in the development of Greek mathematics, i.e., it tells us how propositions were 'proved' in an archaic sense of this word. The proposition which has been illustrated and 'proved' above, can be formulated as follows: "A square constructed on the diagonal of another, has twice the other's area."

It was this proposition which Socrates put in the form of a question to an uneducated slave. Diagrams (2) and (3) refute the slave's first two answers and make their defects visible. Diagram (4), on the other hand, not only shows the correct answer (i.e. not only formulates the proposition in question), but is at the same time a visible 'proof' of its correctness. Although this passage contains only one occurrence of δείχνυμι (83e: καὶ εἰ μὴ βούλει ἀριθμεῖν, ἀλλὰ δεῖξον . . . ), in my opinion it gives us a good indication of why this verb subsequently came to mean 'prove' in mathematics.

It would be wrong, of course, to claim that there was ever a time when mathematical *proofs* were nothing more than *making the facts visible*. Even after a fact has been shown or made visible, as in the above example, something more is needed for it to have been *proved*. A *proof* is obtained by reflecting upon what has been seen, and by drawing the correct conclusion from it. It is reflection which transforms what we see into *visible, empirical evidence*. By making a fact visible, we provide only the core of a proof, even if 'proof' is understood in its most ancient sense.

Looking now at the proofs in the *Elements*, it would appear that many of them have evolved from demonstrations similar in form to the one in the *Meno*. They are frequently based on simple visual arguments which can be reconstructed from Euclid's own words. I have as a matter of fact argued elsewhere[10] that the proof of Proposition I. 47 in Euclid's *Elements* (usually called Pythagoras' theorem) is of this kind.

On the face of it, however, Euclid's demonstrations are not concerned with making anything visible. His words suggest that he is less interested in making his propositions visually evident than he is in convincing the reader of their truth by a train of abstract arguments. Contrary to earlier opinion, Euclidean mathematics is not really 'visualizable'. *Reidemeister* was absolutely correct when he wrote:[11] "*The view that Greek mathematics is distinguished by its concreteness or visualizability, is nothing more than a widespread prejudice . . . . The truth is rather that Greek (Pythagorean) mathematics turned away from the concrete and towards the abstract.*" (It is clear that this statement was made with Euclid very much in mind.)

I even think that the above is too weak a claim. In my opinion, visual and empirical methods must at some stage have been deliberately excised from mathematics. Euclid's own text contains evidence of this. One need only read it with a little more care than is usual to see that he goes out of his way to avoid visual arguments. This tendency of his is exemplified by the proofs of the following two propositions:

*Elements* IX. 21: "If as many even numbers as we please be added together, the whole is even."

IX. 22: "If as many odd numbers as we please be added together and their multitude be even, the whole will be even."

Both of these propositions must strike anyone who has had even the slightest acquaintance with arithmetic, as too obvious to require proof. This makes the manner in which they are proved, all the more interesting. In both cases the first step is to state a typical instance of the proposition in question. So, for example, IX 21 begins: "Let as many even numbers as we please, *AB, BC, CD, DE* be added together; I say that the whole *AE* is even."

---

[10] See n. 6 above.

[11] K. Reidemeister, Das exacte Denken der Greichen, Hamburg 1949, p. 51.

It is clear that Euclid thought of these numbers as line segments, since he denoted each of them by a pair of letters *(AB, BC,* etc.). Similarly, the fact that he called their sum *AE* indicates that he must have imagined it as being obtained by laying these segments down one after another in the way pictured below.

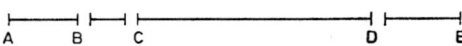

The proof now continues: "For since each of the numbers *AB, BC, CD, DE* is even, it has a half part; so that the whole *AE* also has a half part. But *an even number is that which is divisible into two equal parts.* Therefore *AE* is even. Q. E. D." (The words in italics are exactly the definition of 'even number' given in Book VII.)

It should be apparent that this proof has nothing to do with the kind of *visualizability* which we discussed earlier. It can hardly be claimed that the line segments *AB, BC,* etc., are accurate visual representations of even numbers which *show* us that their sum, the segment *AE,* has to represent another even number. It is nonsense to suggest that a segment of any kind can be the visual representation of an *even* number. After all, in the proof of the next proposition (IX. 22) the same segments (denoted by the same letters) are supposed to represent *odd* numbers. Furthermore, to make the confusion complete, the sum of these odd numbers is an *even* one which, just as in the proof of IX. 21, is designated by *AE.* There is in fact no way to distinguish between odd and even numbers by using line segments, for such segments can always be cut in half. By definition, however, odd numbers cannot be halved, although even numbers can.

There is no doubt that Euclid's proofs of the above propositions are not visualizable. Although he routinely observes the convention of illustrating numbers by line segments, his use of the word δεῖξαι (to show, to point out) is merely figurative. He talks of *pointing* things *out,* but his arguments are essentially abstract and their steps cannot be *visualized.*

It is perhaps easier to see how far Euclid falls short of making the truth of these simple propositions visible (i. e., of 'proving' them, in the most rudimentary sense of the word), if his method of demonstrating them is compared with the original one. Becker has conclusively established that these same propositions were originally *shown* (or 'made visible') by the method of γηφοφορία.[12] This involved representing numbers by groups of pebbles *(γῆφοι)* in such a way that even numbers contained as many black as white pebbles, while odd numbers were obtained from even ones by adding or subtracting a pebble of either colour. So Proposition IX. 21, for example, could be shown by γηφοφορία in the following way:

Let $AB = 4$, $BC = 6$, $CD = 10$ and $DE = 2$, then $AE = 22$.

(1)  oo●●  ooo●●●  ooooo●●●●●  o●

(2)  oo●●oooo●●●ooooo●●●●●o●

(3)  ooooooooooo
     ●●●●●●●●●●

The given even numbers are first arranged in a row (1) and then added by combining them together (2). If their sum is rearranged (3), we can see immediately that it contains as many black pebbles as white ones. Hence it has been shown to be even.

This is a rather primitive technique of proof, but it does make the essential content of the theorem visible. We need only glance at the rows of pebbles to convince ourselves that the proposition is true. They enable us to *see* whether or not a number is even. Furthermore, we can actually see how adding or taking away an odd pebble converts an odd number into an even one. In the proof of Proposition IX. 22 Euclid says: "For since each (of the numbers) *AB*, *BC*, *CD*, *DE* is odd, *if an unit be subtracted from each, each of the remainders will be even.*" Yet, although he says this, he does not show the reader anything. It is not possible for him to illustrate the subtraction of a unit from a line segment which is supposed to stand for an arbitrary odd number. In order to do so, he would need to introduce a shorter segment to represent the unit. The other line segments would then be specific multiples of this unit and hence could no longer represent arbitrary (even or odd) numbers. This, however, is exactly what Euclid wanted to avoid. It is true

---

[12] O. Becker, 'Die Lehre vom Geraden und Ungeraden in neunten Buch der Euklidischen Elemente'. (See Part 1, n. 75 above.)

13 Szabó

that his first step in the proof of IX. 21 is to state a typical instance of this proposition for the numbers $AB$, $BC$, $CD$ and $AE$, but these line segments are not intended to stand for fixed even numbers. The instance is typical in the sense that specific even numbers can be substituted for $AB$, $BC$, etc. The proof which he gives, is valid for any such substitutions. This is in contrast to the method of ψηφοφορία, which only enables one to prove particular cases of the proposition.

It seems that Euclid abandoned visual demonstrations because he wanted his proof to be valid for all possible cases. He turned to logical arguments in his search for greater *generality*. This is why his proof proceeds by recalling that an even number, by definition, is one which can be divided into two equal parts. From this it obviously follows that the sum of even numbers is even; if the halves of such numbers are added separately, their sum can be obtained directly as the sum of two equal parts.

The idea that Euclid was unable to illustrate even and odd numbers in an accurate manner because he was striving for generality brings to mind an interesting passage from Plato. I am referring to the one in which Socrates explains that, as far as arithmetic is concerned, *numbers cannot have visible or tangible bodies*,[13] they are merely *ideal elements which cannot be grasped except by means of pure thought.*[14]

So we see that Greek mathematics, as exemplified by Euclid's proofs with their avoidance and suppression of the visual, also endeavoured to treat its subject matter as something which belonged exclusively to the domain of *pure thought*. This trend in the development of science was responsible for the most remarkable Euclidean proofs. Let me give an example of one of these here.

Proposition VII. 31 of the *Elements* establishes an important property of *composite numbers*. Before examining it more closely, however, let us recall three definitions which are indispensable for its understanding. These are as follows.

*Definition* VII. 2: "A *number* is a (finite) multitude composed of units."

---

[13] Plato, *The Republic*, 525 οὐδαμῇ ἀποδεχόμενον ἐάν τις αὐτῇ ὁρατὰ ἢ ἁπτὰ σώματα ἔχοντας ἀριθμοὺς προτεινόμενος διαλέγηται.

[14] Ibid., 526: ὧν διανοηθῆναι μόνον ἐγχωρεῖ, ἄλλως δ' οὐδαμῶς μεταχειρίζεσθαι δυνατόν.

*Definition* VII. 11: "A *prime number* is that which is measured by an unit alone."

*Definition* VII. 13: "A *composite number* (i.e., one which is *not* prime) is that which is measured by some number."

The theorem (VII. 31) states that *"any composite number is measured by some prime number"*. Euclid's proof of it runs as follows: Let *a* be any composite number. We have to show that some prime number divides *a* (this is what is meant by saying that *a* 'is measured by some prime number'). Since *a* is composite, there must be at least one number, say *b*, which divides it (by Definition VII. 13). Now *b* is either prime or composite, for Definitions VII. 11 and 13 exclude any other possibility. In the former case the theorem is proved, since *b* divides *a* and *b* is prime. In the latter case there must be some number, say *c*, which divides *b* (again by Definition VII. 13). Now *c* obviously divides *a* as well, so if *c* is prime the theorem is proved. If, on the other hand, *c* is composite, there must be some number which divides it, etc., etc. — The proof emphasizes that this process must eventually terminate; otherwise we would obtain an infinite decreasing sequence of (composite) divisors of *a*. (According to the Greeks, a number was always greater than any of its divisors.) The existence of such a sequence for any number *a*, however, would contradict Definition VII. 2. Hence *a* must be divided by some prime number and the theorem is proved.

Nothing is 'shown' or 'made visible' in the proof outlined above; in this respect it fundamentally differs from the argument found in the *Meno*. It is true that Euclid uses line segments to illustrate a typical instance of his proposition (for he represents an arbitrary composite number by one such segment and its divisor by another, shorter one), but this indicates nothing more than his observance of a traditional and outmoded convention. His chain of arguments simply cannot be visualized. If we want to understand it, there is no point in *looking* at the numbers (as represented by line segments); we need instead to keep in mind their *definitions*, since the proof depends solely upon these. Euclid does not even give a name to the prime number which is being sought. The procedure described in his proof may terminate with *b*, *c*, *d* or some other number. We can only be certain that it will eventually terminate, since no number can have an infinite decreasing sequence of divisors.

13*

I believe that the main conclusions of this chapter can be summed up as follows:

(1) A striking difference between Greek mathematics and ancient Oriental science is that propositions are *proved* in the former, whereas there are no proofs to be found in the mathematical texts of pre-Hellenic cultures. ("It is not even certain that the Babylonians knew how to formulate general theorems.")[15]

(2) In the earliest mathematical proofs of the Greeks the truth of a statement was *shown* or *made visible*. This assertion is supported by three pieces of evidence; the first is the original meaning of the verb δείχνυμι (to prove); the second is the example from the *Meno* which was discussed above, and the third is the fact that we can frequently reconstruct the original visual arguments underlying the proofs in the *Elements* on the basis of Euclid's own words (especially in the case of his geometric theorems).

It is possible that those features of Greek mathematics which are mentioned in (1) and (2) above, developed out of pre-Hellenic science. The practical instructions (or 'rules of thumb') for solving mathematical problems, which were known to the peoples of the ancient Orient, may well have been transformed at a later stage into genuine propositions having wider applicability. It must be admitted that no one has yet discovered any such propositions in the mathematical texts of the Egyptians or Babylonians, but this fact hardly justifies the claim that there are fundamental differences between Greek and Oriental mathematics. Many of Euclid's propositions could be regarded simply as more refined versions of ancient calculation rules, while the original proofs of the Greeks (i.e., their 'visual arguments') do not differ in principle from the kind of 'numerical verification' found in Babylonian mathematics.[16] Indeed, it could even be argued that 'numerical verification' prepared the way for the primitive methods of proof which were employed by the earliest Greek mathematicians.

In short, there still remains the possibility that Greek mathematics was nothing more than a direct continuation of pre-Hellenic science. (1) and (2) above do not touch upon the most significant difference between these two traditions. It is therefore necessary to add a third point.

[15] Cf. n. 4 above.
[16] See n. 4 above.

(3) At some time during the pre-Euclidean period, Greek mathematics underwent a remarkable transformation. Visual arguments were no longer accepted as proofs; instead the Greeks sought to 'show' the correctness of their mathematical statements in an entirely different manner. I think that this 'transformation' can best be described as *anti-empirical* and *anti-visual*. It is exemplified in the proofs of Propositions IX. 21 and 22. The simplest way for Euclid to have demonstrated the correctness of these propositions, would have been to illustrate them with the help of pebbles. However, it was quite impossible for him to verify arithmetical propositions empirically, since he used line segments to represent all sorts of numbers. In different propositions the same line segments stand for odd, even, prime and composite numbers; sometimes they are even supposed to illustrate the *unit*. There is a close connection between this 'transformation' of mathematics and the idea expressed by Plato when he wrote[17] that *the subject matter of arithmetic lay within the domain of pure thought*. Other passages from Plato indicate that the Greeks had a similar view of geometry. I do not want to quote any of these here, but I do want to point out that Euclid clearly tried to avoid *visualizability* even where geometry was concerned. Of course this was a more difficult task than avoiding it in the case of arithmetic. Nonetheless, he did his best not to emphasize the visual properties of geometrical figures. (This topic will be dealt with more fully in Chapters 3.27–3.29 below.)

It seems that new kinds of proof appeared at the same time as Greek mathematics was becoming anti-empirical and anti-visual. We have yet to examine how such proofs, which aimed at providing something more than visual evidence, came into being. Before doing so, however, I would like to point out a noteworthy feature of the particular proof discussed above.

Euclid proves his theorem (*Elements* VII. 31: "Any composite number is measured by some prime number") by arguing that its negation leads to a contradiction. If the theorem is false, there is some number which has an *infinite* decreasing sequence of divisors; but this cannot be, since a number is by definition a *(finite)* multitude composed of units. Anyone who believes that Greek mathematics is nothing more than a further development of pre-Hellenic science, must find Euclid's

---

[17] See nn. 13 and 14 above.

use of *proof by contradiction* as inexplicable as his tendency to avoid empirical and visual arguments. I do not think that either of these two phenomena have their roots in ancient Oriental culture. So the next few chapters will be devoted to an investigation of how they arose as part of the development of Greek mathematics.

However, the proof of Proposition VII. 31 has another feature, which I would like to discuss here because it is of some historical interest.

As was already noted, the proof states that "if the investigation be continued in this way, some prime number will be found which will measure *a*. For, if it is not found, an infinite series of numbers will measure the number *a*, each of which is less than the other: *which is impossible in numbers*" (ὅπερ ἐστὶν ἀδύνατον ἐν ἀριθμοῖς). The words in italics (which I have also quoted in the original Greek) obviously refer to the definition of 'number': ἀριθμὸς τὸ ἐκ μονάδων συγκείμενον πλῆθος. In English this definition reads: "A number is a *(finite)* multitude composed of units." (The Greek text contains no word for 'finite', so this is not a literal translation. Nonetheless, it is true to the spirit of Euclid, since he deals only with finite sets in his arithmetic.) I have italicized this part of the quotation for the following reason: Although there is nothing surprising about Euclid's claim that a process of the sort described cannot go on to infinity, it does strike me as strange that he qualified it by adding the phrase '*in numbers*'. He seems almost to be implying that the remark could be false in some other domain.

These considerations bring to mind Zeno of Elea. It was he who argued[18] that a body in motion had first to cover half the distance to its destination; before doing so, however, it would need to cover half of this half, and so on to infinity. Zeno was trying to make the point that a body in motion travels a path made up of *infinitely many distances*, each one shorter than the last. (For this reason he thought that *no* motion was possible – or rather, 'conceivable' – and that the very concept of motion was paradoxical.) The argument can also be interpreted as asserting that any straight line, *AB*, can be split up into infinitely many segments, each one shorter than the last; but this is just another way of saying that AB has *an infinite decreasing sequence of divisors*. It would appear, therefore, that the proof of VII. 31 bears traces of

---

[18] Aristotle, *Physics* 29.239b9ff, and 2.33a21.

Zeno's influence. Its author went out of his way to emphasize that the proof was only valid for numbers, and his motive for doing so was probably that he wanted to exclude the case considered by Zeno.

## 3.2 THE PROOF OF INCOMMENSURABILITY

Our task now is to investigate how it came about that Greek mathematics was transformed into an *anti-empirical* and *anti-visual* science which gave birth to a *new notion of proof*. I hope to show that this development was connected with the discovery of *linear incommensurability*. So let me begin by summarizing our earlier discussion about the origins of this discovery.

The Greeks were led eventually to a knowledge of incommensurability by their attempts to divide the most important musical intervals (the octave, fourth and fifth) into equal subintervals. This problem, which arose first in *the musical theory of proportions*, involved finding the mean proportional (or geometric mean) between the lengths of string (expressed as proportional numbers) which were required to produce the intervals in question. As a result of these investigations, the Greeks must have come to realize that no mean proportional number existed between two numbers in a *ratio superparticularis* (cf. pp. 174 ff.). Nonetheless, the application of new concepts from the musical theory of proportions to geometrical arithmetic (cf. pp. 177 ff.) promoted further research in two different directions. On the one hand, a necessary and sufficient condition for two numbers to have a mean proportional one between them was discovered with the help of plane numbers; on the other hand, the notion of geometrical similarity (for rectilinear figures) was given a precise definition in terms of ἀναλογία (equality of ratios between corresponding sides). Once the Greeks had started to work with similar rectilinear figures, it could not have taken them long to find out how to construct a mean proportional to two arbitrary segments (Proposition VI. 13), even if they were not yet aware that this construction almost always yielded *an incommensurable segment*. Similarly, they may have known that the area of a square could be doubled by means of its diagonal, and even that the diagonal was therefore the mean proportional between the side, *a*, of the square and

a line of length $2a$, without realizing that the side and diagonal were
incommensurable in length.

In order to see how the linear incommensurability of two line segments can be established, let us recall Definition X. 1 of the *Elements*
which states: "Those magnitudes are said to be *commensurable (σύμμετρα μεγέθη)* which are measured by the same measure, and those
*incommensurable (ἀσύμμετρα δέ . . . )* which cannot have any common
measure."

So we see that the *existence* or *non-existence* of a common measure
determines whether or not two line segments are commensurable. It is,
however, by no means obvious that this question can be decided by
*empirical methods*. Let us look more closely at how the Greeks used to
establish the existence of a common measure for two line segments.
We have already seen one example of the method which they employed
in the theory of music.

In Part 2 of this book we discussed in some detail the origin of the
term ἐπίτριτον διάστημα (*a line* $1\frac{1}{3}$ *units in length*) which was used to
denote a *fourth* (4:3). This interval was produced on a monochord by
plucking first the whole string, $AB$, and then a shorter length, $AC$.
Empirical methods were used to determine how much of the string was
to be kept still $CB$, when the second note was sounded. Now the length
$AC$ was regarded as the 'unit' in this experiment. The name ἐπίτριτον,
*a third in addition to it* (i. e., the unit) comes from the fact that $AC$
can in a sense be subtracted from the whole monochord $(AB)$ leaving
a remainder $(CB)$ which goes into it exactly three times. Of course the
length $AC$ need not have been taken as the unit. The whole operation
would actually have been simplified, if $CB$ had been chosen instead.
The whole monochord $(AB)$ would then have been 4 units long, and

the section of string producing the second note would have been made up of 3 units ($1\frac{1}{3} = 4{:}3$).

The above is a reconstruction of how successive subtraction (*ἀνθυφαίρεσις*) was applied in the theory of music. (See pp. 134 ff. for further discussion of this method.) This same method, of course, can also be used *to find the greatest common measure of two line segments*. In view of the fact that musical intervals can always be expressed as ratios between *whole numbers*, the application of *anthyphairesis* to ancient musical theory is best described as *quasi-geometric*. Although it seems to be concerned with *line segments* (and in particular sections of string), these always correspond to *whole numbers on the canon;* hence they are not *arbitrary geometrical quantities*. As described above, successive subtraction is an *empirical method*, so there is some question about whether it is really suitable for use in geometry. Before attempting to deal with this matter, let us see how the method is used in Euclid's arithmetic.

Euclid gives two arithmetical applications of *ἀνθυφαίρεσις*. They are to be found in the first two propositions of Book VII. It is best to begin by examining Proposition VII. 2, which poses the following problem: "Given two numbers not prime to one another, to find their greatest common measure."

Following his usual practice, Euclid uses two line segments, *AB* and *CD*, to illustrate the two numbers. *CD* is the shorter of the two because it is supposed to represent the smaller number. According to him, there are two cases to consider. The first is that the smaller number *CD* measures the greater one; in this case *CD* itself is the required greatest common measure. The second case is that *CD* does not measure the greater number. It then has to be subtracted successively from *AB* until a number is obtained which divides the original ones. This number will be their greatest common measure. Its existence is guaranteed by the definition of what it means for two numbers not to be prime to one another; so the only thing which needs to be shown is that the method of successive subtraction really works. Euclid proves that it does, in a perfectly correct and adequate manner. His illustrations, however, add little if anything to the argument.

Our present concern is not so much with the proof itself as with the extent to which the subtractions can actually be carried out. In particular, it should be noted that successive subtraction will not necessarily yield the greatest common measure of the two line segments. It can only be applied to these *in theory*. As far as Euclid is concerned, there is a crucial difference between numbers and the line segments which represent them; the former are composed of units, whereas the latter are not. Thus successive subtraction is not a *practical and concrete method* except when applied to whole numbers.

We can see this even more clearly by considering Proposition VII. 1, which states: "Two unequal numbers being set out, and the less being continually subtracted in turn from the greater, if the number which is left never measures the one before it until a unit is left, the original numbers will be prime to one another."

There is no difficulty about this theorem if we keep in mind that Euclid is concerned only with *whole numbers* and that, according to the Greeks, these are built up out of *units which cannot be divided into any smaller parts*. So the successive subtraction of two numbers which are relatively prime will terminate once the unit has been reached. There is no possibility of continuing the process any further than this in arithmetic. When it comes to line segments, the above method is only applicable to those which, like numbers, are multiples of some given unit. The segments which play a role in the theory of music are of this kind. A musical interval can always be expressed as a ratio between whole numbers; hence the monochord itself can be regarded as some multiple of a unit, and its precise length (in terms of this unit) can be determined in an empirical way by successive subtraction. On the other hand, the line segments used by Euclid to illustrate numbers cannot be viewed as specific multiples of some unit because they are supposed to represent *arbitrary* numbers. This means that successive subtraction, insofar as it applies to such segments, is not a practical or empirical method, but an abstract and ideal one.

It should perhaps be mentioned that some scholars have taken exception to Euclid's curious custom of illustrating his arithmetical propositions with the help of line segments. Thaer, for example, remarked in his German translation of the *Elements*[19] that the illustrations in Books V and VII were similar in kind, but that those in the latter book were not of much help in understanding the theorems; he therefore preferred to replace them by ones which *made use of broken lines* and could thus represent specific numerical examples.

---

[19] Thaer, *Die Elemente von Euklid*, Part 3, p. 73.

We have yet to explain how it came about that line segments were used to symbolize numbers. There is no doubt in my mind that this practice originated in the field of music; for it was there that (whole) numbers were identified with lengths of string on a monochord or canon. The theory of music discussed in Part 2 must therefore have preceded the arithmetical propositions found in the *Elements*. In fact, it even influenced the terminology of arithmetic. Euclid does not talk about the 'greatest common divisor' of two numbers, but always about their κοινὸν μέτρον (common *measure*); furthermore, he uses the verb μετρεῖν ('to measure' or 'to be measured') to mean 'divide'. These expressions, like the terms μέρος *(part)* and μέρη *(parts)* of a number (see Definitions VII. 3 and 4), are of geometric (or quasi-geometric) origin. They hark back to the time when successive subtraction was an empirical method which was applied to lengths of string as part of a musical experiment.

Let us now turn to the question of how Euclid applied successive subtraction *(ἀνθυφαίρεσις)* in geometry. There is only one geometrical theorem (X. 2) which mentions this method. Before quoting it, I would like to recall the following two facts. (1) In the theory of music, the *common measure* of two lengths of string was found by using ἀνθυφαίρεσις. (2) In geometry, line segments which had no *common measure* were said to be *linearly incommensurable*.

It follows from (1) and (2) that successive subtraction is a process which, when applied to two incommensurable segments, will not terminate. This conclusion is stated in Proposition X. 2, which runs as follows:

"If, when the less of two unequal magnitudes is continually subtracted in turn from the greater, that which is left never measures the one before it, the magnitudes will be incommensurable."

There is an obvious sense in which this theorem is a counterpart to the arithmetical proposition (VII. 1) quoted above. The major difference between the two stems from the fact that there are no *'least magnitudes'* in geometry.[20] Hence ἀνθυφαίρεσις, when applied to two arbitrary line segments, need not always terminate; whereas in arithmetic

---

[20] *Procli Diadochi in Primum Euclidis Elementorum Librum Commentarii* ed. G. Friedlein (Lipsiae 1873) 60.11: ἐν γεωμετρίᾳ γὰρ τὸ ἐλάχιστον ὅλως οὐκ ἔστιν . . .

it must do so, even in the case of numbers which are relatively prime. Proposition X. 2 gives a criterion of incommensurability in terms of successive subtraction, namely, if this process goes on to infinity, the two magnitudes to which it is being applied are incommensurable.

*This criterion*, however, because it depends upon establishing that a certain process is infinite, *is a purely theoretical one; it cannot be applied in practice.* In my opinion this accounts for the fact that the ancients, as has often been noted, seem never to have made use of Proposition X. 2 in their mathematical proofs.[21]

One possible objection to the above is the following. Certain irrational magnitudes, namely those which are solutions to equations of the form $ax^2 + bx + c = 0$ (where $a$, $b$, and $c$ are integers), can be written as *periodic* 'infinite continued fractions'. Furthermore, for a given magnitude of this sort it is not difficult to verify that the 'fraction' in question is periodic. Hence it appears that an infinite process can sometimes serve as a practical criterion of incommensurability and, what is more, that the Pythagoreans may well have known about this. For on the one hand, as Becker correctly observed (see Part 2, n. 7), the method used to obtain such 'fractions' is nothing more than a modern version of the Euclidean algorithm (or ἀνϑυφαίρεσις), and on the other they can easily be shown to be periodic with the help of so-called '*side-*' and '*diagonal-numbers*'. (I myself have argued above (see p. 92) that the enumeration of these numbers, which was devised by the Pythagoreans, can be interpreted as an ancient method of approximating the value of $\sqrt{2}$.)

The following digression about side- and diagonal-numbers is intended to answer this objection.

We have already discussed (see pp. 89 ff.) the passage from Plato's *Republic* (546c) where the number 7 is described as the 'rational diagonal' belonging to the square with sides of length 5. In our explanation of this curious phrase it was mentioned that if a square has sides 5 units long, the length of its diagonal will be approximately 7 units; hence 7 is the 'expressible diagonal' corresponding to the side 5. (The

---

[21] A. Frajese, *Attraverso La storia della matematica* (Rome 1962), pp. 201ff. See also *Gli Elementi di Euclide* by A. Frajese and L. Maccioni (Turin 1970), pp. 599–600, where the following observation is made: "una fatto strano: né presso Euclide, né presso altri matematici greci (almeno per quanto ce ne è giunto) si trova la facile dimostrazione della incommensurabilità del lato e della diagonale del quadrato secondo il metodo dell' algoritmo euclideo della X.2."

diagonal, $d$, of such a square is actually greater than 7; for by Pythagoras' theorem $d^2 = 5^2 + 5^2$, and so $d = \sqrt{50}$.) My reason for recalling this passage is that it furnishes an example of a pair of side- and diagonal-numbers, namely 5 and 7.

The interest of the Pythagoreans in such numbers centered around two problems. These were: (1) how to obtain squares with integral sides whose diagonals were approximately equal to integral lengths; and (2) how the so-called 'expressible' and 'inexpressible' diagonals of a given square were related. There are a great many texts which attest to the fact that the Pythagoreans were especially concerned with (2). Proclus, for example, writes (*In Platonis Rem Publicam Commentarii*, II. 23):

The Pythagoreans show with the help of numbers *(διὰ τῶν ἀριθμῶν οἱ Πυθαγόρειοι δεικνύουσιν)* that the [squares of the] expressible diagonals, the counterparts of the inexpressible ones *(ὅτι αἱ ταῖς ἀρρήτοις διαμέτροις παρακείμεναι ῥηταί)*, are greater or less by a unit than twice [the squares on the corresponding sides] *(μονάδι μείζους εἰσὶν ἢ ἐλάττους διπλάσιον)*.

(A literal translation of Proclus' words would falsify their meaning; therefore I have given a paraphrase of them above.)

Proclus illustrates his assertion with two examples; one deals with the pair 5 and 7, and the other with the numbers 2 and 3. In the first case the square of the expressible diagonal is 49 ( $= 7^2$ ), which is *one less* than twice the square on the side $(2 \cdot (5^2) = 50)$; in the second, the square on the diagonal $(9 = 3^2)$ is *one more* than twice the square on the side $(8 = 2 \cdot 2^2)$.

We now want to investigate how the Pythagoreans discovered a method of generating pairs of side- and diagonal-numbers. It seems to me that modern research has devoted insufficient attention to this interesting historical question. For the most part scholars have contented themselves with recording what Theon of Smyrna and Proclus had to say about the matter. The former wrote (*Expositio rerum mathematicarum ad legendum Platonem utilium*, ed. E. Hiller, pp. 43–4): "The unit is the origin of every number, whether it be a side or a diagonal one. So we take two units, a side-unit and a diagonal-unit; we then form a new side by adding the diagonal-unit to the side-unit, and a new diagonal by adding twice the side-unit to the diagonal-unit."

According to the above, the sequence $(a_i, d_i)$ of side- and diagonal-numbers begins with $a_1 = 1$ and $d_1 = 1$; $a_2 = a_1 + d_1 = 2$ and $d_2 = 2a_1 + d_1$; $a_3 = a_2 + d_2 = 5$ and $d_3 = 2a_2 + d_2 = 7$, and in general

$$a_{n+1} = a_n + d_n \text{ and } d_{n+1} = 2a_n + d_n .$$

Proclus (*In Platonem Rem Publicam Commentarii*, II. 27) mentions these same formulae for the calculation of side- and diagonal-numbers. He writes: ἡ μὲν διάμετρος προσλαβοῦσα τὴν πλευράν, ἧς ἐστιν διάμετρος, γίνεται πλευρά, ἡ δὲ πλευρὰ ἑαυτῇ συντεθεῖσα καὶ προσλαβοῦσα τὴν διάμετρον τὴν ἑαυτῆς γίνεται διάμετρος, and goes on to say that Euclid proves these formulae in Book II (Proposition II. 10) of the *Elements*. Let us now take a look at the first eight terms in the sequence of side- and diagonal-numbers; these are listed in the table below:

|       | Side-number | Diagonal-number |        |
|-------|-------------|-----------------|--------|
| I     | 1           | 1               | (+1)   |
| II    | 2           | 3               | (−1)   |
| III   | 5           | 7               | (+1)   |
| IV    | 12          | 17              | (−1)   |
| V     | 29          | 41              | (+1)   |
| VI    | 70          | 99              | (−1)   |
| VII   | 169         | 239             | (+1)   |
| VIII  | 408         | 577             | (−1)   |

(In the right-hand column I have indicated whether the square of the expressible diagonal has to be increased or diminished by one in order to obtain the square of the inexpressible diagonal, (i.e. twice the square on the corresponding side.)

Our present interest in these numbers stems from the following (see E. Stamates, *Eukleidou Geometria*, Vol. 2, Athens 1953, pp. 9ff.). As we read down the table, it is obvious that the actual diagonals of the squares on the sides in column 2, which of course are incommensurable in length, approximate more and more closely to the corresponding diagonal-numbers in column 3. Furthermore, the ancient Pythagoreans must have known this, since our table does not contain anything which goes beyond their knowledge of the subject.

Let us now modify the table in a way which ancient mathematicians would never have thought of doing, by forming ratios from each pair of side- and diagonal-numbers and expressing these as decimal fractions. This will give us the following:

| I | 1 : 1 | 1 |
| II | 3 : 2 | 1.500000 . . . . |
| III | 7 : 5 | 1.400000 . . . . |
| IV | 17 : 12 | 1.4166666 . . . . |
| V | 41 : 29 | 1.4137931 . . . . |
| VI | 99 : 70 | 1.4142857 . . . . |
| VII | 239 : 169 | 1.4142011 . . . . |
| VIII | 577 : 408 | 1.4142156 . . . . |

We need only glance at the infinite decimals in the right-hand column to see that they are approximations to the value of $\sqrt{2}$. Even more interesting is the fact that, as we read down the column, the odd rows tend towards this value from below, whereas the even ones tend towards it from above. This means, in particular, that the entire sequence is bounded below by 1, the number in the first row, and bounded above by 1.5, the number in the second. This sequence (and its continuation to infinity) can also be illustrated on the real line; 1 and 1.5 mark out the end points of the interval within which the point $\sqrt{2}$ lies, and the numbers in the odd [even] rows approach this point from the left [right] without ever reaching it.

It is difficult to say whether or not the Pythagoreans knew that their side- and diagonal-numbers could be converted into a sequence which tends towards the limit $\sqrt{2}$. In the first place decimal fractions were unknown in antiquity, and in the second place the Greeks of the classical period did not have a concept which corresponded exactly to our $\sqrt{2}$. The ancient counterpart to the notion of $\sqrt{2}$ was something like *'the linearly incommensurable diagonal of the unit square'* or *'the mean proportional between the unit length and its double'*. Yet there is a sense in which our modern interpretation accurately represents ancient ideas about side- and diagonal-numbers. This can be seen from such facts as that the decimal fractions in our second table are greater than $\sqrt{2}$ in exactly those cases where the square of the corresponding expressible diagonal is greater by one than twice the square on the side (similarly,

they are less than $\sqrt{2}$ whenever the square of the corresponding diago-
nal-number is one less than twice the square on the side).

If we now transform our first table by writing each pair of side- and
diagonal-numbers as a *fraction* and then expanding it as a *continued
fraction*, the result is even more interesting. Rows III and IV, for
example, will look like this:

$$\text{III} \qquad \frac{7}{5} = 1 + \cfrac{1}{2 + \cfrac{1}{2}}$$

$$\text{IV} \qquad \frac{17}{12} = 1 + \cfrac{1}{2 + \cfrac{1}{2 + \cfrac{1}{2}}}$$

As we move down the columns, we obtain successively longer continued
fractions, all of which have the same form. $\Big[$ If the $n$th such fraction is
$f_n$, then $f_{n+1} = 1 + \dfrac{1}{1 + f_n} \cdot \Big]$ We are thus led to ask whether the
Pythagoreans knew about this method of representing the ratios be-
tween side- and diagonal-numbers.

It should be remembered that, as far as the Greeks were concerned,
fractions played no role in pure mathematics until the time of Archimedes. (Of course, this is not to deny that they were used in the practical applications of mathematics. Furthermore, the three fractions
*hemiolion*, *epitriton* and *epogdoon* played a very important part in the
theory of music.) This suggests that the answer to our question is
negative. Since there is no evidence that the Pythagoreans made use of
ordinary fractions, it hardly seems likely that they would have known
about continued ones. However, there is something more which needs
to be said about this matter. As was mentioned above, the Euclidean
algorithm *(ἀνθυφαίρεσις)* is nothing but an ancient version of the
method by which we calculate the ratio between two numbers as a
continued fraction. Moreover, the basic idea behind the method for
generating the continued fractions in our third table is easy to grasp,
especially if one is familiar with Zeno's argument about a body in
motion (see Aristotle's *Physics* 29.239b2, also p. 198 above). Just as
Zeno's argument procedes by dividing the distance between the body

and its destination into two halves, halving one of these halves, and so on, each fraction in the sequence of side- and diagonal-numbers is obtained by adding a half to the bottommost denominator of the preceding one (beginning with the unit). All things considered, therefore, it seems to me that the continued fractions in our third table do correspond, to some extent, to the way in which the ancients thought about side- and diagonal-numbers.

I want now to go back to the claim which I made above, namely, that successive subtraction, because it does not terminate when applied to two incommensurable magnitudes, can only serve as a *theoretical criterion of incommensurability*. I do not believe that the validity of this claim is affected in any way by our discussion of side- and diagonal-numbers. It should not be forgotten that the claim was made about *geometrical* magnitudes, and for all practical purposes these cannot be subtracted successively from one another more than a finite number of times. Indeed, the very notion of successive subtraction going on to infinity is a theoretical one which cannot be realized in practice. (Similarly the infinite sequence of side- and diagonal-numbers is part of a *theory* which has only the most tenuous connection with the practical applications of mathematics.) My conclusion, therefore, is that neither successive subtraction nor its modern variant, 'continued fractions', can serve as a practical criterion of incommensurability in geometry. Incommensurability is a *theoretical* concept; it is not an empirical property of geometrical quantities.

Of course Euclid is quite right in asserting that "*If, when the less of two unequal magnitudes is continually subtracted in turn from the greater, that which is left never measures the one before it, the magnitudes will be incommensurable*", and his proof of this statement is equally correct. Neither the statement nor the proof, however, helps us to decide in a given case whether the process of successive subtraction, when applied to two magnitudes, terminates or not. I want to argue that there is no *practical* method which enables us to reach such a decision in every case. We might hope to find intuitive evidence and arguments which, although they would not be accepted as conclusive proof, would convince a mathematician that the process was never going to terminate. The non-mathematician, on the other hand, might well regard these same arguments as a refutation of the very fact which they were intended to demonstrate. This can be seen from the following:

**14** Szabó

Rademacher and Töplitz, in their interesting little book *Von Zahlen und Figuren*,[22] use Proposition X. 2 to show that the side and diagonal of a square are linearly incommensurable. Their argument runs as follows:

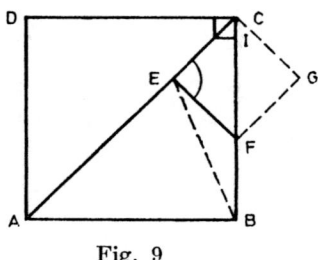

Fig. 9

Given a square $ABCD$ (see *Fig. 9*), we want to apply the method of successive subtraction to its side *(AB)* and diagonal *(AC)*. So let us mark off the length $AE$ ( $= AB$) along $AC$ and draw the perpendicular, $EF$, from $E$ to the side $CB$; then $EC = EF = BF$. (The proof of this equality is omitted for the sake of brevity.) Hence $CF = AB - (AC - - AB)$; i.e. $CF$ is the length obtained by subtracting the difference between the side and diagonal from the side. Now $EF$ and $CF$ respectively form the side and diagonal of a new square, $EFGC$. Therefore, we can repeat the whole procedure for this smaller square (i.e. we subtract $IF = EF$ from $CF$ and draw the perpendicular from $I$ to the side $FC$) and obtain yet another square (with side $IC$). The side of this third square can then be subtracted from its diagonal and so on. Continuing in this way, we get smaller and smaller squares. Their sides, however, will never measure their diagonals exactly. Thus it is clear that, although the remainders become smaller and smaller, the process of successive subtraction never terminates. Hence we may infer by Proposition X. 2 that the two magnitudes (the side and diagonal of the original square) are incommensurable in length.

This 'proof', which Rademacher and Töplitz have reconstructed, is not to be found in any of the ancient texts which have come down to us.

[22] H. Rademacher and O. Töplitz, *Von Zahlen und Figuren*, Berlin 1930, p. 258; cf. also O. Töplitz, *Die Entwicklung der Infinitesimalrechnung*, Berlin 1949, pp. 2–6.

It is deficient in one respect (namely, the fact that the process can be continued to infinity is not proved rigorously), but is nevertheless quite convincing. Each step is so clear and straightforward that Rademacher and Töplitz were led to wonder why no ancient mathematician ever found a similar argument for the existence of incommensurability. If we are to answer this question from a historical point of view, we will have to tackle two separate but interrelated problems; the first is to explain how a *mathematician* would interpret the preceding construction, and the second is to give an account of how a *non-mathematician* would react to it.

It should be stressed at the outset, of course, that mathematical and non-mathematical patterns of reasoning are *radically different*. Aristotle makes this point in a passage from the *Metaphysics* which touches on the problem of incommensurability.[23] He remarks that non-mathematicians simply do not believe in the existence of lengths which have no common measure and are therefore perplexed by the incommensurability of the side and diagonal of a square; on the other hand, those who are 'well versed in geometry' would be even more perplexed if it could be shown that the side and diagonal of a square do have a common measure. The difference between these two ways of thinking is well illustrated by the difference between the mathematical and non-mathematical interpretations of 'infinite successive subtraction'.

A mathematician would probably explain the construction sketched above in the following way:

(1) The *first step* shows that the difference between the side and diagonal of a square can be viewed as the side of a new square.

(2) The *second step* shows that the difference between the side and diagonal of this new square is equal to the result of performing the second operation in the process of successive subtraction.

(3) The *third step* shows that this second difference can be viewed as the side of yet another square and that performing the next operation in the process of successive subtraction corresponds to subtracting the side of this square from its diagonal.

There is no need to carry the construction any further since it obviously can be continued by repeating the first three steps in order. Furthermore, there is no limit to the number of times this can be done.

---

[23] *Metaphysics* A2.983a13ff.

14*

As the number of steps increases, of course, the resulting squares de-
crease in size, so that it soon becomes practically impossible to construct any more of them. But this is a matter of no great concern to us.
Such a sequence is clearly *not* constructible *in practice;* for the process
of construction never terminates and, in addition, all the sets encountered in the external world are *finite.*[24] As mathematicians, however, we
are not interested in the question of whether this construction actually
can be carried out. It is sufficient for our purposes to know as a result of
*mathematical reflection* that we are faced with a case of infinite successive subtraction. The construction is useful for proving incommensurability precisely because it is an *ideal* (or *theoretical*) one which does not
terminate after a finite number of steps.

A non-mathematician, on the other hand, would be inclined to take
a different view and to reason as follows. Successive subtraction is a
*practical device* for finding the common measure of two quantities.
Furthermore, the application of a *practical* method leads to *practical*
rather than theoretical results. Hence successive subtraction, even
when it is applied to the side and diagonal of a square, does not go on to
infinity. By following the procedure described above, we can quite
easily construct a square which is so small that its side is *practically* as
long as its diagonal; in other words, if we carry out sufficiently many
steps, we will eventually find a length which, for all *practical* purposes,
is a 'common measure' of two incommensurable magnitudes.

In my opinion, the non-mathematician is right to maintain that the
process of successive subtraction will always terminate in *practice.*
This is the reason why Proposition X. 2 can only provide a theoretical
criterion of incommensurability. It also explains why no ancient mathematician ever used this proposition to prove the existence of incommensurability. (What prompted Euclid to include such a criterion in Book
X of the *Elements* is an interesting question in its own right. To answer

[24] One should keep in mind Fraenkel's thoughtful remark (*Abstract Set
Theory*, Amsterdam, 1953, p. 9): "As a matter of fact, the recent research in
physics has in increasing measure convinced us that the exploration of nature
cannot lead to either infinitely large or infinitely small magnitudes. The assumption of a finite extent of the physical space, as well as the assumption
of an only finite divisibility of matter and energy (so that the smallest particles
of matter and energy are finite), completely harmonize with experience. It thus
seems that the external world can afford us nothing but finite sets."

it, however, we would need to undertake a detailed investigation of the book in question, a task which lies outside the scope of the present work.)

Nevertheless, what Rademacher and Töplitz have succeeded in reconstructing is a correct 'proof'. Furthermore, it is one which accords well with the spirit of ancient mathematics. It should be noticed, however, that diagrams cannot illustrate the most important part of the argument. Figure 9, for example, *shows* only the first three steps of the construction; it does *not* establish that these steps can be repeated infinitely many times. Hence the proof is not justified by visual evidence but only by theoretical considerations.

When the Greeks discovered the existence of linear incommensurability, they found themselves confronted with a mathematical fact which could not be proved conclusively by practical methods. Such methods actually serve to establish the exact opposite, namely that any two quantities have a common measure. From a practical point of view, Proposition X. 2 is useless as a criterion of incommensurability. It is not surprising, therefore, that this criterion was never used in antiquity. It was necessary to turn away from visual evidence and to reject empiricism in mathematics before incommensurable magnitudes could be proved to exist.

*

Let us now see how pre-Euclidean mathematicians proved that the side and diagonal of a square have no common measure. Their proof formerly appeared at the end of Book X, but it is relegated to an appendix (27) in Heiberg's edition of the *Elements*. As Becker has remarked, modern editors (beginning with E. F. August in 1829) omitted the proposition in question from the main body of the text because they could not see any connection between it and the rest of Book X. Their action, although justified, should not be allowed to obscure the fact that it is a proposition of great antiquity which was discovered long before the time of Euclid and was thought worth preserving, if not by Euclid himself, at least by one of his earliest editors.[25] We know that both the

[25] O. Becker, 'Die Lehre vom Geraden und Ungeraden . . .' (see n. 15 above).

proposition and its proof antedate Euclid because they are mentioned on more than one occasion in the works of Aristotle.[26] Indeed, Becker has argued quite convincingly that the so-called 'theory of even and odd' (which was developed no later than the middle of the fifth century) was directed, in part at least, towards proving this theorem.[27] It seems likely, therefore, that we are dealing with one of the oldest proofs in Greek mathematics. According to Aristotle, mathematicians "prove that the diagonal of a square is incommensurable with its side by showing that, if it is assumed to be commensurable, *odd numbers will be equal to even*".[28] Becker observed that every possible variant of the proof must fit this description,[29] and he attempted to reconstruct a simplified version of it, using only the theory of even and odd. The proof itself runs as follows:

We begin by assuming that the diagonal of a square is commensurable with its side, and we will show that this assumption cannot possibly be true because it leads to a contradiction. Hence we will be able to conclude that the two segments in question are incommensurable.

Now the diagonal *(d)* and two adjacent sides *(a* and *a)* of the square form a right-angled isosceles triangle. So, by Pythagoras' theorem, we have

$$d^2 = 2a^2.$$

Since $a$ and $d$ are commensurable (by assumption), we can take them to be numbers; in fact, we can take them to be numbers which are relatively prime. (For, if they are not relatively prime, we need merely divide them by their common factors.) This means that at most one of them can be even. The above equation shows that $d$ is even. We can conclude, therefore, that *a must be an odd number.*

But if $d$ is even, there is some number $m$ such that $d = 2m$. So our previous equation can be written

$$(2m)^2 = 4m^2 = 2a^2.$$

Dividing both sides by 2, we obtain

$$2m^2 = a^2.$$

[26] See Part 1, n. 2 above.
[27] O. Becker, op. cit., n. 25 above.
[28] Ibid., p. 544, n. 11.
[29] *Prior Analytics*, I.23.41a26 and I.44.50a37.

This last equation shows that *a must be an even number*. Thus our initial assumption has led us to the contradictory conclusion that *a is both even and odd*. Therefore the assumption must be false; in other words, its negation must be true.

As we can see, the preceding argument is a *reductio ad absurdum* or *indirect proof*. I would like to mention three facts about it, which, in my opinion, are important from a historical point of view.

(1) The proof has nothing to do with *visual* or *empirical* evidence. Although we can easily draw a picture of the two quantities whose incommensurability is to be proved, the argument is a purely theoretical one which does not depend upon the visible properties of such a representation. It appeals *to reason alone*.

(2) In view of what was said above, it seems likely that the mathematicians of antiquity adopted this theoretical method of proof because their earlier attempts at proving the existence of incommensurability were *not sufficiently clear or convincing*. The most that one can hope to establish by empirical methods is that no common measure has been found; but this does not refute the naive common-sense opinion according to which any two magnitudes are commensurable. To prove conclusively that the side and diagonal of a square are incommensurable, it must have been necessary to devise new techniques and to reject 'practical' arguments in favour of abstract ones.

(3) The authors of the proof outlined above must clearly have had some prior notion of incommensurability. The creation of this concept was in intself a bold step, for the existence of incommensurability is not a fact which could have been discovered by accident, nor is it one which could have been accepted without some astonishment. There is a considerable difference between acknowledging that no common measure has been found for two given geometrical magnitudes and establishing that *no such measure can exist*. The concept of incommensurability, however, applies only to the latter situation; in a sense, therefore, it presupposes a method of showing that incommensurable magnitudes really exist.

In any event, I believe that the proof we have been discussing illustrates very well the close connection between the introduction of new techniques of proof, on the one hand, and the rejection of empiricism and visual evidence on the other. Just as Plato emphasized that numbers do not have visible or tangible bodies and that they can only be comprehended by *pure reason*,[30] so in the present case it must be stressed that the side and diagonal of a square are not lengths which actually can be seen but are merely ideal entities. Geometrical magnitudes can only be incommensurable *in theory*.

It seems, therefore, that the discovery of linear incommensurability was related to the development of proof by contradiction and to the rejection of empiricism in mathematics.

### 3.3 THE ORIGIN OF ANTI-EMPIRICISM AND INDIRECT PROOF

Let us now turn to the question of how Greek mathematicians arrived at the idea that numbers and geometrical figures were ideal entities which had to be treated in a completely abstract way. Those who are familiar with mathematical reasoning will find this a very simple and obvious idea. It should not be forgotten, however, that mathematics was at one time a *practical* science whose subject matter was empirical and whose principles were established by *empirical methods*. Accountants and surveyors used to take pride in the fact that their calculations could actually be verified by means of empirical experiments. In view of all this, the sudden turn away from empiricism seems somewhat surprising.

The appearance of indirect proofs in mathematics is even more surprising. In the preceding chapter we saw how one such proof, which did not depend upon empirical considerations or upon visual evidence, was used to establish the existence of incommensurability, a fact which could never have been proved conclusively by empirical means or by mere inspection of the diagrams used to illustrate constructions. It should be remarked that Euclid often uses indirect proofs in the *Elements*. Indeed, as we shall see later, by Plato's time arguments of this sort had come to be regarded as typical of mathematical proof.[31]

---

[30] See nn. 13 and 14 above.

[31] In the *Theaetetus* (see 162e) Socrates deplores the use of vague arguments based on plausibility and probability to settle matters of importance and states that a mathematician "who argued from probabilities and likelihoods in geometry, would not be worth an ace". He then goes on to give a vigorous proof by contradiction. From this I infer that the use of indirect proof was thought to be characteristic of mathematics in Plato's time.

Now I want to argue that neither anti-empiricism nor the method of indirect proof could have arisen spontaneously in mathematics. I do not believe, for example, that mathematicians were prompted solely by their dealing with numbers and geometrical figures to change radically their way of thinking and to adopt an interesting new method of proof;[32] they must have been subjected to some influence from outside mathematics. There is no other way to explain why a predominantly empirical mathematical tradition suddenly and for no apparent reason became anti-empirical and anti-visual. Similarly there is no way to explain how the device of indirect proof could have been discovered, if we restrict our attention to that body of purely practical and empirical knowledge which was at one time the whole of mathematics. My view is that these two features of Greek mathematics, its rejection of empiricism and its characteristic use of indirect proof, are attributable to the decisive influence of the *Eleatic* school of philosophy.

This book is not the place for a detailed historical investigation of *Eleatic philosophy*. The best I can do here is to refer the reader to those works which have contributed most to my forming the following opinion about the Eleatics.[33]

Their philosophy is distinguished by its rejection of practical empirical knowledge and of sense perception in general. Parmenides empha-

---

[32] I have not seen this conjecture stated clearly anywhere, yet I think that mathematicians in particular have been disposed to view these achievements as ones which belong exclusively to mathematics. This can be seen from Reidemeister's remarks (in his book, *Das exakte Denken der Griechen*, Hamburg 1949, pp. 10ff) about Proposition IX. 30. Having outlined the proposition, he asserts that its boldest and most original step is the inference that a certain number has to be even *because it cannot be odd*. From this he concludes that *the idea of reasoning consistently with concepts came about as a result of dealing with numbers*.

[33] In addition to the important book by G. Calogero *(Studi sull' Eleatismo*, Rome 1932) I have to mention *Parmenides und die Geschichte der griechischen Mathematik* by *K. Reinhardt* (2nd edition, Frankfurt am Main 1959; 1st edition, Bonn 1916). I would also like to draw the readers attention to four of my own papers which, contrary to my initial expectations, led me into an investigation of the early history of Greek mathematics. These appear in *Acta Antiqua Academiae Scientiarum Hungaricae* (Budapest) and are as follows:

(1) 'Zur Geschichte der griechischen Dialektik' **1** (1953) pp. 377–410;
(2) 'Zur Geschichte der Dialektik des Denkens' **2** (1954) 17–62;
(3) 'Zum Verständnis der Eleaten' **2** (1954) 243–89;
(4) 'Eleatica' **3** (1955) 67–103.

sizes that truth cannot be grasped by means of sense perception, which is misleading, but only be reason *(λόγῳ)*. To get a clear idea of what he means by 'reason', let us take a look at one of his arguments; it asserts that what is *(τὸ ὄν)* cannot have *come into being* and runs as follows: Suppose that what is did come into being, then it could only have come from what *is* or from what *is not;* there is no third possibility. Now if it had come from what is, it would already have been existent before it came into being; hence to say that it came into being in this way would make no sense. If, on the other hand, the claim is made that what is came from what is not, this leads immediately to a contradiction. What is can never have been the opposite of itself, what is not, and hence could not have come into being in this way either.

It is apparent that indirect arguments played a very important part in Eleatic philosophy. Without them it would not have been possible to establish such central doctrines as that there is *no motion, no change, no becoming, no perishing, no space* and *no time.* Of course, these doctrines contradict the evidence of our senses and are incompatible with empiricism, nonetheless the Eleatics, bolstered by their belief that reason was the only guide to truth, accepted them. Furthermore, the whole of Eleatic dialectic is nothing but an ingenious application of the method of indirect proof, which is why Aristotle considered Zeno to be the inventor of dialectic.[34] However, there is no real difference between Zeno's dialectic and the arguments of Parmenides. The most noteworthy feature of both is their use of indirect proof.

As I remarked above, I believe that the influence of Eleatic philosophy was responsible for the rejection of empiricism and visual evidence in Greek mathematics, as well as for the introduction of indirect proof. Indeed, I hope to show in the following chapters that the construction of Greek mathematics as a deductive system was a result of this same influence and that, had it not been for the philosophy of Parmenides and Zeno, it would not have been possible to build up so ingenious a system as Euclid's *Elements.* I will attempt to give a detailed defense of this view below. First, however, I would like to summarize the evidence in favour of it which has emerged from our discussion up to now.

---

[34] *The Fragments of Aristotle* (ed. V. Rose, Lipsiae 1886), fragment 65 (Diogenes Laertius 9.25): φησὶ δ᾽ Ἀριστοτέλης ἐν τῷ σοφιστῇ εὑρετὴν αὐτὸν (scil. *Ζηνώνα)* γενέσθαι διαλεκτικῆς κτλ.

(1) The earliest application of indirect proof which I have encountered in my studies of Greek language and culture occurs in the didactic poem of Parmenides. On the other hand, it is agreed nowadays that the earliest known mathematical applications of this technique are all without exception of a later date. This simple observation about chronology suggests the following conjecture. Unless the method of indirect proof was developed independently in mathematics and in Eleatic philosophy, it must have originated exclusively among the Eleatics. They could hardly have adopted it *from mathematics* since it does not appear to have been used there until a comparatively late date.

(2) I have already emphasized that there is no way to explain how the method of indirect proof could have been developed from the practical and empirical knowledge which at one time made up mathematics. Any attempt to show how this technique might have arisen within mathematics is bound to end in failure. However, there is no difficulty in explaining how the teachings of the Eleatics led to the introduction of indirect proof, for it was a method designed originally to refute the cosmogony of Anaximenes.[35]

(3) In Greek mathematics the first applications of indirect argument *coincided* with the appearance of an anti-empirical and anti-visual trend. This is a striking fact which is somewhat puzzling from a mathematical point of view. Neither of these two innovations would be conceivable in empirical mathematics, and there is no reason why indirect proof should have been accompanied by anti-empiricism, even if mathematics had been a theoretical science. Although the two cannot be separated from each other in the proof that the side and diagonal of a square are incommensurable, many of the other indirect arguments given by Euclid do not result in conclusions which contradict the evidence of our senses. If we want to find a meaningful and necessary relationship between indirect proof and anti-empiricism, we will have to look outside mathematics and, in fact, at Eleatic philosophy. The Eleatics *had* to be anti-empiricist, i.e. they were *forced* to reject sense perception and practical experience; had they not done so, they would

---

[35] On this see my paper 'Zur Verständnis der Eleaten' mentioned in n. 33 above, pp. 247ff. Popper also observes that Eleatic philosophy originated in cosmogony (see pp. 18–20 of *Problems in the Philosophy of Mathematics* edited by I. Lakatos, Amsterdam 1967).

not have been able to accept the validity of the conclusions which they reached by way of indirect argument. The connection which they established between the two was then carried over into mathematics.

### 3.4 EUCLID'S FOUNDATIONS

In the chapters which follow I will attempt to show that the deductive system of mathematics presented in the *Elements* could not have been constructed without Eleatic philosophy. My claim is not only that Greek mathematicians in the time before Euclid adopted an *anti-empirical attitude* and the *method of indirect proof* from the Eleatics, but also that their efforts to transform their mathematical knowledge into a system built up according to certain basic principles must have been prompted by the teachings of the Eleatics. I want to corroborate this conjecture by examining in detail a series of historical problems which have to do with Euclid's work. Let us begin, therefore, by taking a closer look at the problem of Euclid's foundations.

\*

The books of Euclid's *Elements* were compiled around 300 B.C. Right at the beginning of the first one we find an interesting list of unproved assertions which are classified into three groups as follows:

1. *definitiones,*
2. *postulata,*
3. *communes animi conceptiones.*

After listing these assertions, Euclid turns his attention to the propositions which go to make up the remainder of the book. He never discusses the significance of the three groups. We are not told why he found it necessary to place them at the beginning of his work, nor how the members of one group differ from those of another. All we know from the *Elements* is that *postulata* and *communes animi conceptiones* appear only at the beginning of Book I, whereas further *definitiones* precede most of the other books. VIII, IX, XII and XIII are the only books which contain no new definitions. This is clearly because all the definitions required for the arithmetical books (VII, VIII and IX) are listed at the beginning of Book VII, and all the concepts used in Books XII and XIII are ones which have been introduced previously.

If we want to find out why Euclid prefaces his treatment of the mathematical propositions by listing three groups of unproved assertions, we must turn to the work of Proclus Diadochus, who produced an admirable commentary on Book I of the *Elements* in the 5th century A.D. Proclus writes:[36]

"Since this science of geometry is based, we say, on hypothesis *(ἐξ ὑποθέσεως εἶναι)* and proves its later propositions *(τὰ ἐφεξῆς ἀποδει-κνύναι)* from determinate first principles *(ἀπὸ ἀρχῶν ὡρισμένων)* . . . he who prepares an introduction to geometry should present separately the *principles* of the science *(χωρὶς μὲν παραδοῦναι τὰς ἀρχὰς τῆς ἐπιστήμης)* and the *conclusions* which follow from the principles *(τὰ συμπεράσματα)*, giving no argument for the principles but only for the theorems that are derived from them. For no science demonstrates its own first principles or presents a reason for them; rather each holds them as self-evident, that is, as more evident than their consequences. The science knows them through themselves, and the later propositions through them. . . . Whoever throws into the same pot his principles and their consequences disarranges his understanding completely by mixing up things that do not belong together. For a principle and what follows from it are by nature different from each other."

This quotation appears at least to tell us why Euclid had to start out by listing three groups of unproved assertions. These groups are clearly composed of the *principles* which Proclus refers to as ἀρχαί or ὡρισμέναι ἀρχαί. (In the continuation of the above quotation[37] he calls them κοιναὶ ἀρχαί as well. It is worth noting here what Proclus means by the word κοιναί in this context. The principles in Book I are said to be 'common' because they are valid, at least in part, for the remaining books as well. As we shall see later, this same word, when used in the phrase κοιναὶ ἔννοιαι, has an entirely different meaning.)

[36] Page 75.6ff. in Friedlein's edition of the commentary. The translation is by Glenn R. Morrow, op. cit., n. 5 above.
[37] Ibid., p 76. 3–4.

It is generally accepted that the set of principles assembled by Euclid is neither *necessary* nor *sufficient* for his development of mathematics. Tannery has pointed out that Euclid's definitions (I. 4 and 7) of *'straight line'* and *'plane area'* are completely superfluous as far as the books of the *Elements* are concerned. They are included among the principles but are never used at all.[38] He also remarked that the principles are not the only unproved assumptions which Euclid makes use of in his exposition; in the very first construction (Proposition I. 1), it is assumed without proof that the two circles which are drawn there will always intersect. Of course, a great many similar examples are to be found in the *Elements*.[39]

When Proclus speaks of Euclid dividing the whole of mathematical knowledge into unproved *hypotheses* on the one hand and *consequences* derived from these on the other, this has to be understood as a description of an 'ideal requirement'. The three groups of unproved assertions were intended to serve as a foundation for mathematics; they were to be stated at the outset, and everything else had to be derived from them. The extent to which Euclid succeeded in realizing this intention, however, is a separate question.

Whether Euclid's principles are in fact necessary and sufficient to obtain the whole of his geometry is not the only problem concerning them which interests us. We also want to know why they are listed under exactly *three* headings, what distinguishes the principles in one list from those in another, and, above all, how the idea of compiling such lists of principles arose in the first place.

One of the things which makes it difficult to answer these questions is the fact that those who wrote about mathematics and philosophy in antiquity did not always agree upon the way in which the Greek words corresponding to *postulata*, *definitiones*, etc. were to be used. Proclus, for example, quotes a sentence from Archimedes to show that the latter counts as *postulata* principles which he, Proclus, believes would be better described as *communes animi conceptiones*.[40] He goes on to

---

[38] *Revue des Études grecques* **10** (1897), 14–8 (pp. 540–4 of *Mémoires Scientifiques* **2**, 540–4).

[39] 'Sur l'authenticité des axiomes d'Euclide', *Bulletin des Sciences mathématiques*, 2nd series **8** (1884), 162–75 (reprinted in *Mémoires Scientifiques* **2**, 48–63).

[40] Proclus, op. cit. (n. 20 above), p. 181. 16–24.

observe that the term which he uses for *communes animi conceptiones* is employed by other authors in a wider and more comprehensive sense. Therefore it seems that, in post-Euclidean times at least, there was no unanimity about how various mathematical principles were to be classified. Indeed, one modern scholar[41] has remarked that total confusion and vagueness prevailed amongst mathematicians after Aristotle as far as the terminology relating to ἀρχαί was concerned, and that even the later commentators on Euclid and Archimedes who sought to introduce a precise Aristotelian terminology were unable to clear it up.

This confusion also shows in the fact that Proclus, when discussing individual ἀρχαί, sometimes uses terms which differ from the ones found in the *Elements*. We shall consider a particular case of this in some detail later on. For the time being, however, I just want to point out the following facts.

In the *Elements definitions* are always called ὅροι. Yet Proclus, even though he must have been familiar with Euclid's term for definitions and in fact did not hesitate to use it on occasion,[42] seems to have preferred the expression ὑποθέσεις.[43] This tells us that Proclus employed the mathematical term 'hypotheseis' with two distinct meanings, although he never made a point of bringing this fact to his reader's attention. It is clear that he sometimes intended it to be understood as a *general* term for 'unproved mathematical assumptions'. The passage quoted above,[44] for example, begins with a sentence in which the word ὑπόθεσις has to be taken in this latter sense: "Since this science of geometry is *based*, we say, *on hypothesis (ἐξ ὑποθέσεως εἶναι)* . . . ". Furthermore, Proclus frequently refers to the unproved foundations of mathematics, the principles *(archai)*, simply as ὑποθέσεις.[45] All this, however, should not be allowed to obscure the fact that in other places he uses the same word to denote *only* the definitions and not the foundations of mathematics in general.

[41] K. von Fritz, 'Die ἀρχαί in der griechischen Mathematik', *Archiv für Begriffsgeschichte*, Bonn, 1 (1955), 101.

[42] See, for example, p. 81. 26 of Proclus, op. cit. (n. 20 above).

[43] Ibid., p. 76. 4–6 and p. 178. 1–2.

[44] See n. 36 above.

[45] Proclus, op. cit. (n. 20 above), p. 77. 2: πολλάκις δὲ καὶ πάντα ταῦτα (scil. τὰς ἀρχὰς τῆς ἐπιστήμης) καλοῦσιν ὑποθέσεις.

This ambiguity in the meaning of the term ὑπόθεσις seems to go back quite a long way in mathematics. My reason for thinking so is the following.

Plato sometimes uses the verb ὑποτίθημι in the sense of 'to determine', 'to give a definition'.[46] His use of the word in the following quotation is perhaps even more interesting:[47]

> "I fancy that you know that those who study geometry and calculation and similar subjects, *take as hypotheses (ὑποθέμενοι) the odd* and *the even*, and *figures*, and *three kinds of angles*, and other similar things in each different inquiry. They make them into hypotheses *(ποιησάμενοι ὑποθέσεις αὐτά)* as though they knew them, and will give no further account of them either to themselves or to others on the ground that they are plain to everyone. Starting from these, they go on till they arrive by agreement at the original object of their inquiry."

It is apparent that Plato, when he talks in this passage about *hypotheses (ὑποθέσεις)*, is referring to the fundamental principles which are usually called *archai* by Proclus. However, his choice of examples is not without interest. He mentions first the concepts of 'odd' and 'even', which are defined in Book VII (Definitions 6 and 7) of the *Elements*, and then the 'three kinds of angles' and geometrical 'figures', which are dealt with in Book I (the former by Definitions 10, 11 and 12, and the latter by Definitions 15, 18, 19, 20, 21, 22, etc.). We can see that all the *hypotheses* mentioned in the above quotation are in fact *definitions;* in other words, for Plato, just as for Proclus, the term ὑπόθεσις had two distinct meanings.

The use of ὑποθέσεις to mean 'definitions' was not confined to Plato's time, for the verb ὑποτίθεσθαι (in the form ὑποτιθέμεθα) serves more than once in Archimedes' *On Conoids and Spheroids* to introduce a *definition*.[48] It must be admitted, however, that Archimedes' terminology, as

---

[46] See, for example, *Charmides* 160d; see also 159c.

[47] *The Republic* VI.510 c–d; translation by A. D. Lindsay (Everyman's Library, 1950).

[48] Archimedes, *Opera omnia*, ed. J. L. Heiberg, Lipsiae 1910, vol. 1, pp. 248 and 252.

far as the *foundations* of mathematics are concerned, is notoriously inconsistent.[49]

The fact that in ancient times the same word could refer both to *definitions* and to the *foundations* of mathematics in general is in itself very revealing. We may conclude from it that the distinctions which Euclid drew between various kinds of mathematical principles were not required originally. He doubtless had his own reasons for classifying his assumptions under three headings, but these need not concern us for the present. Now, as Tannery has remarked, no distinction is made between *definitions* and *postulates* in the oldest complete work on Greek mathematics which has been handed down to us (he was referring to *On a Moving Sphere* by Autolycus of Pitane).[50] We can, however, say something more about this matter than Tannery did. He pointed out that the second of the two 'definitions' to be found in Autolycus' work is really a *postulate*,[51] but failed to add that there is no justification for referring to either of the two assertions in question as definitions. They are not prefaced by any such title in the original manuscript and are called *horoi* (i.e. 'definitions') only because the text was embellished in an arbitrary and superfluous manner by Conrad Rauchfuss (also known as Dasypodius), one of its early editors. Unfortunately Hultsch, when he came to prepare his own edition of the works of Autolycus, did not remove the occurrences of ὅροι which his predecessor had inserted.[52]

Perhaps to begin with it was not considered absolutely necessary to identify the assumptions which stood at the beginning of a mathematical work by giving them some kind of title. Certainly Autolycus does not have a special name for his assumptions; he simply states them

---

[49] K. von Fritz, op. cit. (n. 41 above), p. 57.

[50] *Autolyci, De sphaera quae movetur liber. De ortibus et occasibus libri duo*, ed. F. Hultsch, Lipsiae 1885, see the index under ὅροι: "Inter definitiones et axiomata ab Autolyco nullum discrimen factum esse docet P. Tannery, etc."

[51] Tannery (*Mémoires Scientifiques* 2, 58), after quoting both of them in a French translation, goes on to say:

Il est clair que cette seconde définition est, en réalité, un *postulat*, et nous apprenons ainsi qu'immédiatement avant Euclide, il n'était pas de règle de distinguer avec précision les définitions et les axiomes, que tout était ou au moins pouvait être rangé sous la même rubrique. Il y a là un fait considérable.

[52] In Hultsch's edition (see n. 50 above) the assumptions are entitled ὅροι in both of Autolycus' works. On both occasions, however, we find the words 'add. Da' (i.e. *addidit Dasypodius*) in his critical apparatus.

without proof at the outset. Furthermore, fundamental mathematical principles can sometimes be recognized by the way in which they are formulated. As Tannery has observed, if Euclid had omitted the heading αἰτήματα from his list of *postulates*, we would still have been able to recognize them for what they were because the list begins with the expression ᾐτήσθω.[53] Similarly in books on mathematics the definitions and other fundamental principles could have been prefaced by the word ὑποκείσθω. This expression is to be found in the works of later mathematicians; Archimedes, for example, uses it to introduce the principles of fluid mechanics in his book *On Floating Bodies*.[54] Furthermore, we encounter the same verb used in a very similar way in the most ancient part of Pythagorean arithmetic, the so-called theory of odd and even.[55] I have in mind the phrase ὅπερ οὐχ ὑπόκειται which occurs in the proof of Proposition IX. 34 when a conclusion contrary to hypothesis has been reached.

### 3.5 ARISTOTLE AND FOUNDATIONS OF MATHEMATICS

In the preceding chapter we saw that ὑποθέσεις was used to denote *definitions* as well as the unproved *foundations* of mathematics. For the time being, we will not look into the threefold classification of mathematical principles which is found in Euclid. We shall turn our attention instead to the question of how the idea arose that a mathematical discussion had to be based on unproved and unprovable foundations. In other words, we want to discover the origins of the idea that mathematics as a deductive science needed a foundation of definitions and axioms.

Before attempting to answer this question by applying the same methods which have been used so often in the first two parts of this book, I would like to summarize briefly the approach which recent scholars have taken to it.

---

[53] P. Tannery, *Mémoires Scientifiques* 2, 58: "Il est parfaitement possible qu'Euclide n'ait nullement inscrit: αἰτήματα en tête de ses postulats géométriques; mais en tout cas il les a nettement distingués des définitions précédentes en les commençant 'ᾐτήσθω' (qu'il soit demandé)."

[54] K. von Fritz, 'Die ἀρχαί in der griechischen Mathematik', *Archiv für Begriffsgeschichte* (Bonn) 1 (1955), 57.

[55] See n. 12 above.

Von Fritz has observed that before the time of Aristotle a lively
controversy prevailed about whether there could be such a thing as
*proven knowledge* in the strictest sense of the words; some maintained
that the search for it would lead to an *infinite regress*, while others
claimed that this difficulty could be overcome by proving the various
propositions which constituted a system of knowledge from one
another.[56] Aristotle opposed both these views. His doctrine was that
every science had to be derived from true but unprovable first principles, and he attempted to establish the various properties which these
principles would need to possess. Nonetheless, a remark which Proclus
made about Apollonius of Perga[57] (namely that Apollonius attempted
to prove Euclid's first axiom) seems to indicate that as late as the end
of the third century futile attempts were still being made to prove first
principles. Therefore, it would appear that Aristotle's view was not
universally accepted even a hundred years after his death.[58]

Von Fritz's view is that Aristotle deserves a great deal of the credit
for laying down the foundations of mathematics in a systematic way,
or at least for making known the possibilities and limits of such a foundation. Of course he does not want to deny that a considerable amount
of mathematical material was available to Aristotle, nor that Aristotle
actually made use of it. (Thus he thinks it possible, and even likely,
that a start was made on laying the foundations of mathematics in
pre-Aristotelian times.) Yet he insists[59] that Aristotle's discoveries were
the ones which made it possible, *in principle* at least, to place mathematics on a secure foundation, and offers the following justification for
this claim – It was Aristotle who attempted to show that every science
has to start out from principles which, although unprovable, are nonetheless both true and certain. Furthermore, definitions, postulates and
axioms cannot be regarded as a foundation for mathematics until it is
acknowledged that these need not be proved; indeed, unless one realizes
that fundamental principles are unprovable, one cannot even conceive
of laying a foundation for mathematics. Anyone who believes that all
the various propositions which go to make up a scientific system can be

---

[56] *Posterior Analytics*, I.3.72b5ff., and I.19–23.83b, 32–84ab.
[57] Proclus, op. cit. (n. 20 above), p. 183. 13–4.
[58] K. von Fritz, op. cit. (n. 54 above), pp. 64–5.
[59] Ibid., p. 98.

15*

proved from one another will not be able to see any difference between a theorem and a fundamental principle; he will regard both as provable and will not understand why mathematicians have to preface the statements which they intend to prove by a list of 'fundamental' propositions. On the other hand, mathematical assumptions will appear as arbitrarily chosen starting points, and not as real foundations at all, to those who think that proofs rest on an infinite regress.[60]

As we can see, this account assigns a very important role to Aristotle; it seems almost to suggest that he was the first person to set about providing a true foundation for mathematics. Yet, if no one before Aristotle had a clear understanding of what was involved in laying these foundations, the significance of definitions and axioms in pre-Aristotelian mathematics requires some explanation. This matter is not discussed by von Fritz, nor does he give any indication of how sophisticated *techniques of proof* could have been developed at a time when the basic foundations of mathematics *had not yet been clarified*. I believe that these techniques, which had been all but perfected before Aristotle's time, presuppose a more or less clear understanding of foundational questions. Von Fritz, on the other hand, seems to hold the view that rigorous methods of inference were developed *prior* to the first systematic attempts at securing the foundations of mathematics.[61]

I want to take a different approach to the question of how the idea arose that mathematics had to be organized into a deductive system. First, however, I would like to give a brief indication of my opinion about the efforts which have been made to trace the origins of systematic mathematics back to Aristotle.

Most of the historical research which has been done on the development of the axiomatic method in Greek mathematics does in fact take Aristotle as its starting point.[62] It is easy to understand why this should be so. Aristotle frequently refers to contemporary mathematics in his

[60] Aristotle, of course, thought that there was a simple and natural starting point for mathematics; he did not believe that the fundamental principles could be selected arbitrarily.

[61] Op. cit. (n. 54 above), p. 90.

[62] See, for example, the introduction to Heath's translation of the *Elements*, 'Die ἀρχαί in der griechischen Mathematik' by K. von Fritz (*Archiv für Begriffsgeschichte*, Vol. I, 1955) and Becker's paper in *Archiv für Begriffsgeschichte*, 4 (1959), 210ff.

writings; he quotes examples from mathematics to illustrate his own ideas, and sometimes he even tries to explain the way in which mathematicians reason. Most significant of all, however, is the fact that the first coherent discussion of the problems connected with axiomatizing mathematics is to be found in his *Prior Analytics*. Aristotle, of course, is a source of some very valuable information to the modern student of the history of mathematics. Furthermore, some questions about the history of axiomatics simply cannot be answered without studying his texts.

Yet the statements which he makes about axiomatics are often arbitrary and historically inaccurate; hence they need to be treated with some caution. From the point view of mathematics, it must be admitted that Aristotle's theories about the nature of axiomatics and especially the terminology which he sought to establish for the fundamental principles of science seem almost to be products of his own imagination. Consider, for example, how the various kinds of first principles *(ἀρχαί)* are distinguished in the second chapter of the *Posterior Analytics*.[63]

A distinction is first drawn between θέσις and ἀξίωμα. A *thesis*, Aristotle says, is like all first principles unprovable; furthermore, it need not be known by one who wants to learn something (about the science in question). *Axioms*, on the other hand, do have to be known by those who desire scientific understanding. He goes on to say that a thesis is either a ὑπόθεσις or a ὁρισμός; a *hypothesis* states that something is or is not, whereas a *horismos* (definition) does not. Definitions (like *hypotheseis)* are assumptions or stipulations; unlike *hypotheseis*, however, they do not stipulate that something *is*. (A definition stipulates, for example, that the *unit* is an indivisible quantity, but not that the *unit* exists.) Thus Aristotle's general classification of principles is as follows:

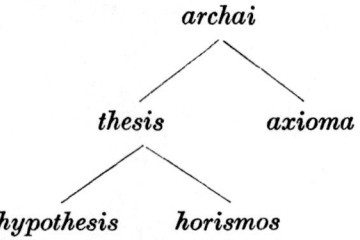

*archai*

*thesis*          *axioma*

*hypothesis*     *horismos*

[63] von Fritz, 'Die ἀρχαί in der griechischen Mathematik', *Archiv für Begriffsgeschichte* **1** (1955), 25.

Let us now see whether we can reconcile Aristotle's classification, as explained above, with Euclid's. It can be *shown conclusively* that, *axiomata* correspond to *communes animi conceptiones*. Furthermore it would seem that Aristotle's *hypotheseis* are the same as Euclid's *aitemata*. If we accept this latter identification, however, we come up against some considerable difficulties. In the first place it implies that Aristotle had a special collective name for definitions and postulates (yet no other ancient author, as far as we know, used the word *theseis* in this sense), and in the second place we can easily establish that Aristotle ascribes *two* distinct meanings to the term *aitema*, neither of which has more than the most tenuous connection with Euclid's *aitemata* (or postulates).[64] Finally, it should be said that even the identification itself does not stand up to critical scrutiny.

Euclid's postulates *are not* existential statements of the same kind as Aristotle's *hypotheseis*. Aristotle asserts[65] that mathematicians assumed the existence of the unit in arithmetic and of the line and point in geometry, and that they proved or showed the existence of the even and odd in arithmetic and of the incommensurable and the triangle in geometry. However, there are no such existential postulates to be found in Euclid, and the *proofs* of existential statements in the *Elements* are *not at all* of the sort mentioned by Aristotle. Aristotle expects the existence of the even and odd 'to be proved', whereas Euclid *defines* even and odd numbers, but he does not go on to prove their existence. Clearly 'existence' meant something different to Aristotle than it did to Plato, the Eleatics and mathematicians in general. The arithmetical books of the *Elements* (VII–IX) are not preceded by any existential postulates; indeed they are not preceded by any postulates at all, but only by definitions. Similarly the existence of quantity *is not postulated* at the beginning of the planimetric books, contrary to what Aristotle seems to require. It is true that some sort of comparison can be made between Euclid's postulates on the one hand and Aristotle's ὑποθέσεις on the other, but the two are a long way from being identical. Aristotle's views about axiomatics and

---

[64] Ibid., p. 42.

[65] *Posterior Analytics*, I.10.76b; cf. K. von Fritz, op. cit. (n. 63 above), pp. 54–5.

the terminology which goes along with them simply *cannot be applied* to Euclid, unless they are first subjected to an arbitrary reinterpretation.

The terminology employed by other ancient mathematicians is even further removed from Aristotle's. Indeed, if we were to adopt Aristotle's ideas and to regard them as some kind of *norm*, we would be forced to agree with von Fritz, that when it came to naming the various groups of ἀρχαί, total confusion and vagueness prevailed amongst post-Aristotelian mathematicians.

For this reason I do not want to place too much reliance upon what Aristotle has to say about the terminology of axiomatics. As far as this subject is concerned, his views (or rather his proposals) are of more interest to the student of Aristotle than to the historian of Greek mathematics.

My own attempt to explain the origins of axiomatic mathematics will be based on a historical analysis of the key terms involved. Before beginning it, however, I would like to mention some important facts which have come to light as a result of previous research.

Tannery seems to have been the first person to notice that Euclid's foundations, especially the definitions in Book I of the *Elements*, contain some interesting absurdities. The definitions of 'straight line' and 'plane area' (I.4 and I.7 respectively), for example, are not used in any of Euclid's theorems. Furthermore, the terminology introduced by the definitions does not always agree with that found in the propositions and proofs. The terms ἑτερόμηκες, ῥόμβος and ῥομβοειδές (rectangle, rhombus and rhomboid), for example, are defined in I.22 but occur nowhere else in the *Elements;* Euclid speaks instead of παραλληλόγραμμα. Another example is furnished by Definition I.19, which is incompatible with Euclid's convention for naming polygons; for the definition leads one to expect that these figures will be given names which refer to the number of their sides *( τρίπλευρα, πολύπλευρα )*, whereas they turn out to be described by the number of their *angles* (τρίγωνον, πεντάγωνον, etc.).[66]

Now the conclusion which Tannery drew from this is undoubtedly the correct one, namely that the foundations in Book I did not originate

---

[66] P. Tannery, *Mémoires Scientifiques* **2**, pp. 48–9.

with Euclid.[67] The author of the *Elements* was not the first mathematician to preface his work with a list of fundamental assumptions. A great many of his predecessors, especially those who themselves compiled books of 'elements', did so as well. Much of what Euclid took over from these earlier mathematicians he left unaltered, regardless of whether it was necessary for his work or whether the terminology used was the same as his own. In fact the systematic development of deductive mathematics began at least two centuries before Euclid, and by the middle of the fifth century Hippocrates of Chios had compiled a book of 'elements'.[68] Of course this development simply could not have taken place without some mathematical principles being accepted as *fundamental*. Furthermore, there is no doubt that Hippocrates' *Elements* began with a list of *fundamental assumptions*, even though today we do not have any idea of what these were.

In my opinion, these observations justify the belief that the systematic development of mathematics based on first principles began long before the time of Aristotle.

### 3.6 'HYPOTHESEIS'

Let us now turn to the study of some mathematical terms. We shall begin with ὑποθέσεις. As was mentioned above, this word has two (mathematical) meanings in the writings of Plato and Proclus; it is used to denote the 'foundations of mathematics' in general and also to refer only to 'definitions'. The corresponding verb, ὑποτίθεσθαι, means 'to define' in both a mathematical and a non-mathematical sense; one can find examples of the former usage in Archimedes[69] and of the latter in Plato.[70] These meanings, of course, are easy to explain from an etymological point of view. Ὑπόθεσις from ὑπό and τίθεσθαι means that which is *placed under*, i.e. that which can serve

---

[67] Ibid., p. 55; cf. A. M. Frenkian, *Le postulat chez Euclid et chez les Modernes*, Paris 1940, p. 14.

[68] Proclus, 66. 7–8 (in Friedlein's edition).

[69] See n. 48 above.

[70] See n. 46 above.

as a *foundation* for something. But we are interested in more than etymology here. We want to find out exactly which uses of the word led to its being taken over into mathematics and to its becoming later a precise mathematical term. Let us therefore consider some examples which illustrate related senses of ὑπόθεσις.

It is advisable to look first at a passage from Plato's *Meno* which deals with the question of whether or not virtue can be taught.[71] Socrates wants to begin by investigating the nature of virtue; his companion, however, is too impatient. As a compromise, Socrates proposes that in considering the question he be allowed to make use of a *hypothesis* (συγχώρησον ἐξ ὑποθέσεως αὐτὸ σκοπεῖσθαι, Men. 86e 3). He then goes on to illustrate *the hypothetical method* by means of an example from geometry: "to make use of a hypothesis – the sort of thing, I mean, that geometers often use in their inquiries" *(λέγω δὲ τὸ ἐξ ὑποθέσεως ὧδε, ὥσπερ οἱ γεωμέτραι πολλάκις σκοποῦνται).*

These words are by themselves sufficient to show that ὑπόθεσις and *the hypothetical method* must have played a major role in the mathematics of Plato's day. Socrates is able to call on an example from geometry because, as he himself remarks, this same method is frequently used there.

The example which Socrates discusses briefly is also very informative. He considers the question of whether a certain area can be inscribed in a given circle, and claims that a geometer would most likely set about answering it in the following way: "I don't know yet whether it fulfills the conditions, but I think I have a ὑπόθεσις which will help us in the matter. It is this. If the area is such that . . .,"[72] then, I should say, one result follows *(ἄλλο τι συμβαίνειν μοι δοκεῖ);* if not, then the result is different." This response is *hypothetical* in the sense that it is *qualified.* The geometer bases his argument on an assumption; this assumption is his *hypothesis*, and he goes on to say what follows from it. If the assumption (i.e. the ὑπόθεσις) were changed, then the conclusion would also be different. We can see that Socrates

---

[71] Cf. O. Becker, 'Die Archai in der griechischen Mathematik', *Archiv für Begriffsgeschichte* **4** (1959), 210ff. The translation of the passage quoted from the *Meno* is by W. K. C. Guthrie.

[72] Cf. P. Tannery, *'Mémoires Scientifiques'* **1**, 39–45 and **2**, 400–6; cf. also T. L. Heath, *A History of Greek Mathematics*, Oxford 1921, 1, p. 300.

is sketching a geometrical argument which is based on a ὑπόθεσις *(assumption* or *supposition)*. The argument depends upon this assumption and is qualified by it.

The actual wording of Plato's text reveals an interesting fact about mathematical ὑποθέσεις. Socrates asserts that geometers conduct their investigations on the basis of assumptions *(ἐξ ὑποθέσεως σκοποῦνται)*. Similarly, Proclus[73] maintains that geometry is a science which rests on assumptions *(ἐξ ὑποθέσεως εἶναι)*. They both use the same expression, but the situations which they describe are different. Hence the word ὑπόθεσις does not have precisely the same sense in the two cases. Proclus means by it the *fundamental assumptions* of mathematics which have been laid down once and for all. (We saw above that in Book IV of the *Republic* (510c–d) Plato also refers to these *foundations*, and in particular to the definitions among them, as ὑποθέσεις.) Socrates, on the other hand, is not referring to *the definitive foundations of mathematics*, but merely to an ad hoc *assumption* which is to serve as the basis of a particular mathematical argument. Furthermore, it seems clear that in mathematics ὑπόθεσις meant an ad hoc assumption before it acquired its other meaning.

In my opinion, Socrates uses two other expressions which are worthy of our attention. One is the phrase with which he introduces the 'hypothetical method'; he asks to be *allowed* to make use of a hypothesis *(συγχώρησον ἐξ ὑποθέσεως αὐτὸ σκοπεῖσθαι)*. The other is his description of the consequences of a ὑπόθεσις "then, I should say, one result *follows;* if not, then the result is different *(ἄλλο τι συμβαίνειν μοι δοκεῖ)*". It is typical of contexts such as the above (Meno 86 3ff) that the term ὑπόθεσις is found together with the words συγχώρησον and συμβαίνειν. The one indicates the way in which a 'hypothesis' comes into being and the other the way in which it is used.

The *hypothetical method* can only be used in a dialogue if both parties *agree* to it. This is why Socrates has to *ask* his companion for *permission* to make use of it. Without such an *agreement* a 'hypothesis' cannot really serve as a foundation for any kind of joint inquiry. Hence in Plato's *dialectic*[74] *hypotheseis* are sometimes called ὁμολογήματα those things which are conceded or on which all parties agree.[75]

---

[73] Proclus (Friedlein's edition) 75. 6–14.

[74] In this book 'dialectic' means Plato's διαλεκτικὴ τέχνη.

[75] See, for example, Plato's *Theaetetus* 155a–b.

The word 'hypothesis' seems therefore to have been a *dialectical* term as well as a *mathematical* one. The fact that it came from dialectic also tells us something about the mathematical uses of its cognate verb. Archimedes, for example, introduces the principles of fluid mechanics in his work *On Floating Bodies* with the phrase ὑποκείσθω,[76] 'Let it be supposed'; it is as if he wanted to say 'let us allow ourselves to take as a foundation . . .'.[77] Since the word ὑπόθεσις and its synonyms[78] clearly derive from *dialectic*, we will have to investigate how they were used there if we want to gain a full understanding of their role in mathematics.

The verb συμβαίνειν, which Socrates uses in connection with the *consequences* of a 'hypothesis', is equally interesting. Plato's vocabulary contains a great many synonyms for it. One such is the verb συμφωνεῖν.[79] The consequences of an assumption are also called τὰ ἑπόμενα,[80] τὰ ἑξῆς,[81] etc. An examination of Plato's writings reveals that these phrases are sometimes used as precise mathematical terms[82] and sometimes as ordinary *dialectical* ones.

We see that the technical term ὑπόθεσις forms part of a whole complex of expressions which appear together in both *dialectic* and *the language of mathematics*.[83] Therefore, we will have to broaden our investigation to take account of these as well.

[76] *Opera omnia*, ed. Heiberg, Lipsiae 1910–15, II, p. 318. Archimedes uses the expressions ὑποκείσθω and ὑποτιθέμεθα as stylistic variants of one another (see n. 48). In Greek it can be said of a ὑπόθεσις that it ὑπόκειται.

[77] Euclid's list of postulates begins with the word ἠτήσθω ("let it be *required* that . . ."); it is *required* of the other participant in the argument that he accept the statements which follow (the *postulates*) as a basis for the discussion.

[78] Not only does Plato treat the expressions ὁμολόγημα and ὑπόθεσις as synonyms, he sometimes uses the word ἀρχή in a similar sense. See, for example, *Cratylus* 436d.

[79] See, for example, *Phaedo* 100a.

[80] *Cratylus* 436d.

[81] *Phaedo* 100c.

[82] For example, the expression τὰ συμβαίνοντα in *Meno* 87a.

[83] The expressions τὰ ἑπόμενα, τὰ ἑξῆς, etc. could, of course, also be used as mathematical terms. See Ž. Markovič, 'Les mathématiques chez Platon et Aristote', *Bulletin International de l'Académie Yougoslave des Sciences et des Beaux Arts, Classes des Sciences mathématiques et naturelles*, **32** (1939) 1–21.

### 3.7 THE 'ASSUMPTIONS' IN DIALECTIC

The *hypothesis* in a dialectical proceeding seems to represent a starting point chosen almost *at random*. The participants in a discussion agree upon some assertion which is to serve as the basis of their subsequent inquiry, and it then becomes their ὑπόθεσις.[84]

Some scholars[85] have attempted to explain the development of Aristotelian logic from dialectic along these lines. According to them, one of the participants in a dialectical debate would select some statement and set out to persuade the other of its correctness. In order to do this, he had to find premises which his opponent would accept as true and from which his original statement followed as a matter of logical necessity. These premises, chosen almost at random, were the *hypotheseis*.

In fact, Plato had something like this in mind when he talked about provisional assumptions in the *Republic*.[86] The passage in question runs as follows: "we may proceed on the assumption that what we have said is right *(ὑποθέμενοι ὡς τούτου οὕτως ἔχοντος)*, agreeing that *(ὁμολογήσαντες)* if at any time we have to change our opinion *(ἐάν ποτε ἄλλη φανῇ· ταῦτα ἢ ταύτῃ)*, all the consequences of our assumption shall be considered invalid *(πάντα ἡμῖν τὰ ἀπὸ τούτου συμβαίνοντα λελύμενα ἔσεσθαι)*."

A *hypothesis*, however, is very closely connected with what is inferred from it. Hence, far from being completely arbitrary, it must be chosen with the greatest care. This point is made in the *Cratylus* where we read:[87]

---

[84] The initial propositions of a dialectical debate were, in fact, sometimes chosen in a *completely arbitrary* manner; see Aristotle's *Topics* (VIII.1.155b29 and VIII.3.159a3ff). In the passages in question, Aristotle refers to these propositions as ἀξιώματα, but he is clearly using this word as a synonym for ὑπόθεσις (cf. K. von Fritz, op. cit. (n. 63 above), p. 31).

[85] K. von Fritz, op. cit. (n. 63 above), p. 20. See also the article 'Syllogistik' by E. Kapp in *RE* (Pauly–Wissowa, *Realencyclopädie der Klassischen Altertumswissenschaft*, IVA1048, 1058–9) and E. Kapp, *Greek Foundation of Traditional Logic* (Columbia University Press, New York 1942).

[86] *Republic* IV.437a. The translation is by A. D. Lindsay.

[87] *Cratylus* 436d. The translation is by Jowett.

"If he did begin in error *(εἰ γὰρ τὸ πρῶτον σφαλεὶς ὁ τιθέμενος)*, he may have forced the remainder into agreement with the original error and with himself *(τἆλλα ἤδη πρὸς τοῦτ' ἐβιάζετο καὶ αὐτῷ συμφωνεῖν ἠνάγκαζεν);* there would be nothing strange in this, any more than in geometric diagrams, which have often a slight and invisible flaw in the first part of the process *(ὥσπερ τῶν διαγραμμάτων ἐνίοτε τοῦ πρώτου σμικροῦ καὶ ἀδήλου ψεύδους γενομένου)*, and are consistently mistaken in the long deductions which follow. And this is the reason why every man should expend his chief thought and attention to the consideration of his first principles *(δεῖ δὴ περὶ τῆς ἀρχῆς παντὸς πράγματος παντὶ ἀνδρὶ τὸν πολὺν λόγον εἶναι καὶ τὴν πολλὴν σκέψιν)* — are they or are they not rightly laid down *(εἴτε ὀρθῶς, εἴτε μὴ ὑπόκειται)?*[88] And when he has duly sifted them *(ἐκείνης δ' ἐξετασθείσης ἱκανῶς)*, all the rest will follow."

There are a number of reasons why this passage is of interest. First of all it shows clearly that Plato attached great importance to the role of foundations in a system of ideas. For, as Socrates is made to explain, it is vital that the assumptions be chosen correctly since the smallest mistake in any one of them can lead to completely false consequences. This is an idea which recurs often in Plato's dialogues. In the *Phaedo*, for example, Simmias declares that, even though he cannot raise any objections against the argument which Socrates has just given, he still harbours some doubts, and Socrates responds in the following manner:[89] "Not only that, Simmias, . . . but our first assumptions *(τὰς ὑποθέσεις τὰς πρώτας)* ought to be more carefully examined, even though they seem to you to be certain *(εἰ πισταὶ ὑμῖν εἰσιν)*. And if you analyse them completely, you will, I think, follow and agree with the argument, . . .".

These words, which imply that 'first assumptions' have to be *certain*, inevitably bring to mind the πρῶται ἀρχαί of mathematics. As far as we know, Aristotle was the first author to discuss explicitly the nature and function of first principles in mathematics.[90] Yet Socrates' remarks about the 'first assumptions' of a dialectical argument give the impres-

[88] The verb ὑπόκειται is discussed in n. 76 above.
[89] *Phaedo* 107b. The translation is by H. N. Fowler (London 1953).
[90] Cf. Chapter 3.5 above.

sion that Plato was also fully aware of the problem of 'first assumptions' in mathematics.

This brings us to the second interesting feature of our quotation from the *Cratylus*. I am referring to the fact that it contains yet another example from geometry. Socrates points out that an incorrectly chosen *hypothesis* is like a geometrical diagram which contains some flaw; they both can lead to a whole succession of false consequences, all of which are consistent with the initial error.

It is surely no accident that Plato's discussions of *hypotheseis* frequently contain a reference to geometry or to mathematics in general. He was of the opinion that dialectical arguments should conform to the high standards set by the methods of mathematics. In ordinary conversation, an inconclusive argument which is at best plausible will often suffice to settle the point at issue. Mathematicians, however, did not accept such arguments;[91] they were concerned only with convincing proof, and their concern was reflected in their manner of choosing and handling *hypotheseis*. Consider, for example, the three interesting ὁμολογήματα which are found in the *Theaetetus*:[92]

(1) "Nothing can become greater or less, either in size or in number, so long as it remains equal to itself" *(μηδέποτε μηδὲν ἂν μεῖζον μηδὲ ἔλαττον γενέσθαι μήτε ὄγκῳ μήτε ἀριθμῷ, ἕως ἴσον εἴη αὐτὸ ἑαυτῷ).*

(2) "A thing to which nothing is added and from which nothing is taken away is neither increased nor diminished, but always remains the same in amount *(ᾧ μήτε προστιθοῖτο, μήτε ἀφαιροῖτο, τοῦτο μήτε αὐξάνεσθαί ποτε, μήτε φθίνειν, ἀεὶ δὲ ἴσον εἶναι).*

(3) "A thing which was not at an earlier moment cannot be at a later moment without becoming and being in process of becoming *(ὃ μὴ πρότερον ἦν, ὕστερον ἀλλὰ τοῦτο εἶναι ἄνευ τοῦ γενέσθαι καὶ γίγνεσθαι ἀδύνατον)."*

These fundamental assertions are very reminiscent of mathematical axioms.[93] Yet, even though they are drawn up with such precision and care, Plato merely calls them ὁμολογήματα, "concessions which the participants in a discussion have agreed to make". The name clearly indicates the *dialectical* origins of the kind of statement to

[91] See *Theaetetus* 162e.
[92] Ibid. 155a–b. The translation is by F. M. Cornford (London 1935).
[93] Ž. Markovič, op. cit. (n. 83 above), p. 2.

which it refers. All this suggests that mathematics and dialectic not only shared a common terminology but were also interconnected disciplines. In fact, it looks as if Plato was writing at a time when mathematics was *just a branch of dialectic*.

It becomes even more apparent that dialectical and mathematical methods were closely related, indeed that they were identical, when one examines the way in which *hypotheseis* are used in Plato's arguments. On one occasion Socrates describes his manner of thinking[94] in the following way:[95] "I assume in each case some principle which I consider strongest *(ὑποθέμενος ἑκάστοτε λόγον ὃν ἂν κρίνω ἐρρωμενέστατον εἶναι)*" [notice that he bases his inquiry on some 'strongest principle' in very much the same way as mathematicians used to base theirs on *foundations* which had been laid down once and for all], "and whatever seems to me to agree with this, I regard as *true (ἃ μὲν ἄν μοι δοκῇ τούτῳ συμφωνεῖν, τίθημι ὡς ἀληθῆ ὄντα)* and whatever disagrees with it, as *untrue (ἃ δ' ἂν μή, ὡς οὐκ ἀληθῆ)*". As we can see, in the above quotation consequences are said *to agree (συμφωνεῖν)* with the *hypothesis*, and not *to follow (συμβαίνειν)*[96] from it. Both these expressions belong to the ordinary vocabulary of dialectic mentioned in the preceding chapter. Of course Plato also uses the verb *διαφωνεῖν*, the opposite of *συμφωνεῖν*, in contexts where something is said *to disagree* with a *hypothesis*.[97] In fact, we shall soon see that the *disagreement* between statements expressed with the help of this last word is the earliest form of what we would call a logical *contradiction*.

### 3.8 HOW 'HYPOTHESEIS' WERE USED

Our next task is to investigate the role which *hypotheseis* played in dialectic. We shall examine in some detail an easily accessible example of dialectical argument taken from the works of Plato. It should perhaps be mentioned that Plato's dialogues contain a great many

---

[94] In the *Theaetetus* (189e–190) Plato describes *thinking* as a kind of conversation which the soul holds with itself.

[95] *Phaedo* 100a. The translation is by Fowler.

[96] The Latin word for this is *consequi*.

[97] *Phaedo* 101d.

examples of his method of argumentation, which incidentally was often used in mathematics as well. Furthermore, it really makes very little difference which one of them we choose. However, there is a certain argument in the *Theaetetus* which suits our purpose particularly well for the following reasons:

(1) Plato's own words suggest that it was patterned after a mathematical argument. (Immediately before giving it, Socrates emphasizes that one should not be satisfied with reasoning which is merely plausible and contrasts such reasoning with the vigorous proofs expected of mathematicians. [*Theaetetus* 162e])

(2) It can very easily be seen to have a counterpart in early mathematics.

(3) It illustrates very well not only how the *consequences* of a *hypothesis* were checked, but also how the *hypothesis* itself was subjected to scrutiny.

The main points of the argument in question can be summarized ¬s follows. The issue to be debated is whether or not knowledge and sense perception are identical (163a: εἰ ἄρα ἐστὶν ἐπιστήμη καὶ αἴσθησις ταὐτὸν ἢ ἕτερον). It is first assumed that they are indeed the same, and this assumption is taken to be the *hypothesis* of the inquiry.[98] If, however, *knowledge* (ἐπιστήμη) is the same as *sense perception* (αἴσθησις), then to *see* something must be to *know* it as well, since 'sight' is one kind of perception. Furthermore, what is not seen cannot be known. [Here and in what follows Plato treats sight as representative of all perception.] Yet a man who *sees* something, and hence (by assumption) who *knows* it, will still have knowledge of it after he has shut his eyes. In other words, he will know something which he can no longer see. Thus one is forced to conclude that he will both 'know' and 'not know' this thing at the same time.

Hence the initial assumption leads to consequences which are obviously *contradictory*. This means that the *hypothesis* or ὁμολόγημα from which these consequences can be derived must itself be false. As Socrates puts it: "Apparently, then, if you say that knowledge and perception are the same thing, it leads to an impossibility" (164b: τῶν ἀδυνάτων δή τι συμβαίνειν φαίνεται, ἐάν τις ἐπιστήμην καὶ αἴσθησιν ταὐτὸν φῇ εἶναι).

---

[98] See n. 78 above.

The sentence quoted above in both Greek and English concludes the argument. It contains some interesting terms, in particular the Greek word ἀδύνατον which is used to denote the 'impossibility' resulting from the *hypothesis*. Anyone who has read Euclid in the original knows that he almost always concludes his indirect proofs with this same word. The phrase he uses is: ὅπερ ἐστὶν ἀδύνατον.[99] Plato's argument is in fact no less of a correct *indirect proof* than the ones found in the *Elements*. Socrates wants to prove that knowledge and sense perception are not the same. Furthermore, he feels that his proof must be patterned after a mathematical one if it is to be convincing.[100] He therefore adopts the indirect method and starts out by assuming the opposite of what he wants to prove. Then he shows that this assumption leads to a clear contradiction (διαφωνεῖ). He is thus able to conclude that his initial assumption cannot possibly be correct because an *adynaton* appears amongst its consequences, and hence that its negation must be true. This is exactly the same kind of argument as the Pythagoreans used to prove that the diagonal of a square is incommensurable with its side, for they deduced the existence of a number which was both even and odd from the assumption that the two lengths in question were commensurable. Plato's dialogues, of course, contain a great many arguments which are similar to this one from the *Theaetetus*.[101]

The reader may by now have noticed that Plato attaches *several different meanings* to the dialectical/mathematical term ὑπόθεσις. Furthermore, he may be wondering why I have not bothered to distinguish between them. So, before embarking on a more detailed investigation of the argument outlined above, I would like to discuss this matter briefly. *Definitions* and the *foundations* of science in general, which mathematicians take to be self-evident and not in need of any further examination,[102] are called *hypotheseis* by Plato, as are

[99] Some propositions (for example, *Elements* I.6 and IX.20, 30, 33, and 34) conclude with the words ὅπερ ἄτοπον or ὅπερ οὐχ ὑπόκειται. Both these phrases are nothing more than stylistic variants of the one mentioned in the text, and they are also used by Plato in his discussions of dialectic.

[100] In mathematical contexts, Plato refers to 'conclusive proof' as ἀπόδειξις καὶ ἀνάγκη; see *Theaetetus* 162e.

[101] Becker (*Archiv für Begriffsgeschichte* 4 (1958), 212) has pointed out that a similar kind of argument is to be found in *Phaedo* 101d 4–5.

[102] *Republic* VI.510c–d.

those *first principles* in a dialectical debate which Socrates advises us to select on the basis of their strength.[103] On the other hand, Plato also uses this term to denote any assertion whose validity *is to be examined* in a dialectical debate. (The argument which we have just been considering and its mathematical counterpart, the ancient Pythagorean proof of incommensurability, are both built around *hypotheseis* of the second kind.) I do not wish to deny that one can draw a legitimate distinction along these lines, but I must emphasize that it is not essential for us to do so. Clearly the only difference between a 'fundamental' *hypothesis* and one which has merely been granted provisional acceptance is that the latter could lead to an *adynaton* (in which case it would have to be rejected as false) whereas the former does not (and hence can serve as a genuine 'foundation'). The statement that the 'one' is indivisible, for example, became accepted as a true (fundamental) *hypothesis* of arithmetic because its negation, just like the false *hypothesis* that the diagonal of a square is commensurable with its side, was found to lead to a contradiction. Discussion of this example is reserved for a later chapter.

Plato's argument is especially informative because it illustrates very well exactly what was involved in making use of a *hypothesis*. From what has been said up to now it should be apparent that *hypotheseis* were laid down with a view to examining their *consequences*. Examinations of this kind had one of two purposes. If the *hypothesis* was a genuine one whose truth was not in doubt, its consequences were examined in order to gain an understanding of the conclusions derivable from it. If, on the other hand, the *hypothesis* was a questionable assertion, then its consequences were examined to see whether they contained an *adynaton*, i.e. to see whether the *hypothesis* itself could be shown to be untenable. In any event, a crucial role was played by the consequences of the *hypothesis*.

The question which now presents itself is how such consequences were obtained in Plato's dialectic. Let us attempt to find out how the chain of assertions connecting a *hypothesis* with its consequences was constructed and what rules governed the choice of the individual links in this chain.

[103] *Phaedo* 100.

## 3.9 'HYPOTHESEIS' AND THE METHOD OF INDIRECT PROOF

If one looks at those passages in Plato where the dialectical method is used to investigate the consequences of some *hypothesis*, then one is forced to conclude that such investigations were concerned exclusively with the question of whether or not the individual assertions which were taken to be consequences *(τὰ συμβαίνοντα)*[104] agreed with one another *(συμφωνεῖν)*. In Plato's dialectic, however, the only way to decide this question was by applying a *negative criterion*. This meant that his arguments, instead of explaining directly why statements were in accord with one another (ὁμολογεῖν, συμφωνεῖν, etc.), invariably concentrated upon those assertions which *did not agree (διαφωνεῖν)*. Plato regarded this kind of mutual disagreement as a sure guide, and he *always* used it when he set out to examine the consequences of a *hypothesis*. More precisely: the consequences of a *hypothesis* were taken to be those statements which agreed with it, were compatible with it or did not contradict it;[105] however, if a contradiction was discovered amongst these consequences, this indicated that the *hypothesis* was false or that it led to an *adynaton*.[106]

The consequences of a *hypothesis* were checked only for consistency, and the criterion used was basically a negative one. It follows therefore that Plato's dialectic was wholly dependent upon the method of indirect proof. The consistency of a statement or of a whole system of ideas can be demonstrated in a negative way *only* by establishing

---

[104] *Meno* 87a.

[105] The verb συμβαίνειν was used in connection with statements which followed necessarily from some premises (whether these premises were true or false). This is the sense in which Aristotle uses this word (*Physics* Z9.239b30) when he discusses Zeno's argument about the flying arrow; he concedes that the conclusion ('the flying arrow is at rest') follows from the premise of the argument, but asserts that the premise is false.

[106] It seems to me that the principle of non-contradiction plays a major role in Plato's dialectic. Socrates is always portrayed as trying to find out whether the statements under discussion are consistent with some hypothesis which has already been accepted as true. If the hypothesis is a 'strong' one, all statements which contradict it are rejected as false (see *Phaedo* 100ff, especially 101a). In some cases, however, the hypothesis itself is a questionable assertion which has only been granted tentative acceptance, and it may actually be rejected, if it is found to contradict some other statement.

that the contrary statement or the opposing system contains a contradiction, i.e. only by giving an indirect proof.

It would have been inconceivable to examine the consequences of a *hypothesis* without using indirect arguments. *Indeed, by its very nature the kind of hypothesis which we have been discussing could not have played its part in Plato's dialectic without the method of indirect proof.* I consider this fact to be of prime importance. As we shall soon see, Plato himself explicitly acknowledged it and deliberately chose to express it through the character of Parmenides.

The observation that the use of *hypotheseis* in dialectical reasoning was inseparable from the method of indirect proof fits in very well with the fact that in the *Theaetetus* Plato describes his argument dealing with the relationship between knowledge and perception as one which is patterned after a mathematical proof; we know that indirect arguments figured very prominently in early Greek mathematics. Consider, for example, Book VII of the *Elements* and in particular the first thirty-six propositions of this book. They are by all accounts of great antiquity,[107] and no less than fifteen of them have indirect proofs. Similarly six or eight of the seventeen propositions which make up the ancient theory of even and odd[108] are established by means of indirect arguments. In a subsequent chapter I shall attempt to show that even the central and fundamental theorem of Pythagorean arithmetic, which states that the *one* is indivisible, was proved in an indirect manner. It seems therefore that the use of indirect proof can be regarded as a characteristic feature of mathematics. No doubt Plato had this in mind when he compared his argument (*Theaetetus* 162–4) to a mathematical one.

### 3.10 A QUESTION OF PRIORITY

A number of interesting facts have emerged from the evidence presented up to now. We saw that the mathematical term *hypothesis* and its synonyms, as well as a whole complex of related expressions, all came originally from dialectic. Viewed as mathematical terms, they

---

[107] B. L. van der Waerden, 'Die Arithmetik der Pythagoreer', *Math. Ann.* **120** (1947–49), 127–53.

[108] See n. 12.

are in a sense 'loan words' which were taken over from dialectic, their earliest field of application. If we wanted to interpret this as a reflection of some relationship between the two fields, we would have to say that mathematics was at least in one respect a branch of dialectic. It seems reasonable to suppose that mathematicians made such extensive use of ʿdialectical terms because *mathematics itself grew out of the more ancient subject of dialectic.*

It should be mentioned, however, that this historical conjecture appears to conflict with something else which we have learnt. Plato held up mathematical methods as an *example;* he wanted to introduce mathematical rigour into dialectic. This suggests that it was mathematics which preceded and influenced dialectic rather than the other way around. The close connection between the use of *hypotheseis* in dialectic and the method of indirect proof lends additional support to this latter view. Indirect arguments seem to have been a distinctive feature of mathematics and, according to Plato at least, were peculiar to it. Yet, if we accept that mathematics was the older discipline, we are faced with the problem of explaining how the mathematical terms mentioned above could have been derived from *dialectic.*

The question of priority is therefore of crucial importance, and we will have to take a serious look at it. We need to examine both the sort of *hypothesis* found in Plato's dialectic and the method of indirect proof with a view to finding out when and under what circumstances they were first used.

Let me digress at this point and give an illustration of how easy it is to be misled when attempting to answer these questions. Eudemus' account of the quadrature of lunes by Hippocrates of Chios might be thought to provide some support for the view that *hypotheseis* were already being used by Greek mathematicians of the fifth century. According to this account,[109] Hippocrates "set down an ἀρχή and accepted, as the first thing useful for his proof, that . . ." *(ἀρχὴν μὲν οὖν ἐποιήσατο καὶ πρῶτον ἔθετο τῶν πρὸς αὐτοὺς χρησιμῶν).* Although the word *hypothesis* does not occur in this sentence, the presence of such expressions as ἀρχή, πρῶτον and ἔθετο, which are either synonymous with it or related to it, seems to suggest that Eudemus is talking

[109] F. Rudio, *Der Bericht des Simplicius über die Quadraturen des Antiphon und des Hippokrates* (Leipzig 1907); P. Tannery, *Mémoires scientifiques* **1**, pp. 347-9; O. Becker, *Quellen und Studien* . . .ʿB, **3** (1936), pp. 427-9.

about a *hypothesis* of the kind we have been investigating. The words ἀρχή and πρῶτον are especially suggestive; they bring to mind not only the passages from Plato discussed previously[110] but also the πρῶται ἀρχαί of mathematics.

This interpretation, however, is completely mistaken. The continuation of the above quotation makes it plain that Eudemus is concerned with something altogether different. The *arche* or initial proposition (πρῶτον) of Hippocrates' argument turns out to be an ordinary geometrical theorem, which runs as follows: "Similar segments of circles have the same ratio to one another as the squares on their bases have." It does not fit Socrates' description of a hypothesis; i.e. it is not an assertion upon which mathematicians base their investigations and of which they give no further account.[111] As Eudemus himself goes on to say, Hippocrates succeeded in *proving* his initial proposition (τοῦτο δὲ ἐδείκνυεν ἐκ τοῦ κτλ) by showing that it followed from another one, namely "circles have the same ratio as the squares on their diameters".[112]

So those words which the casual reader of Eudemus' account might interpret as references to the πρῶται ἀρχαί of mathematics have in fact a rather different significance. They are used to denote the *lemmas* which Hippocrates, probably right at the *beginning* of his work, found necessary for his arguments. Nothing which Eudemus has to say is relevant to our present interest in the use of *hypotheseis*. On the other hand, we cannot determine whether or not Hippocrates used the method of indirect proof without examining in detail the text which contains Eudemus' words. Hence this question must remain unanswered here.

The fragments of early Greek mathematics which modern scholarship has recently been able to recover also seem to indicate that *hypotheseis* and the method of indirect proof were used in the fifth century. This evidence must be treated with caution, however, if one is to avoid getting involved in a circular argument. Although it is generally agreed nowadays that most of the propositions in Book VII of the *Elements* date back to the fifth century B.C. and that the

[110] *Cratylus* 436d and *Phaedo* 107b.
[111] *Republic* VI, 510c–d.
[112] Cf. O. Becker, *Archiv für Begriffsgeschichte* 4 (1958), 218ff.

so-called theory of even and odd is likewise of great antiquity (probably originating in the middle or even in the first half of the fifth century),[113] we are not entitled to infer from this that the Euclidean *proofs* of these ancient propositions, especially those which use the *indirect method*, are the original ones.

It must be admitted that we do not have sufficient evidence to give a definitive historical analysis of each one of the proofs concerned. Several of them are almost certainly not in their original forms;[114] they have obviously been reworked at some later date, partly no doubt for the purpose of fitting the propositions which they serve to establish into a new framework. However, this fact should not be taken to mean that those proofs which do not bear clear traces of subsequent reworking are necessarily as old as the propositions which correspond to them.

When it comes to those ancient propositions which are demonstrated by indirect means, there is only one which can definitely be said to have come down to us with its proof more or less intact. I am referring to the early Pythagorean theorem which states that the diagonal of a square is incommensurable with its side. (*Elements* X, Appendix 27 in Heiberg's edition; it is not included in Heath's translation, but is discussed by him in his introduction to Book X). In the case of this proposition we can be sure that the proof used in ancient times was essentially the same as the one found in Euclid because Aristotle refers to it in his *Prior Analytics*.[115] No such evidence exists, as far as I know, in any other case which involves an indirect argument.

Nonetheless, in my opinion, it is no accident that so many propositions of early Greek mathematics are proved indirectly. Their proofs must have come down to us substantially unchanged. Even in fifth century mathematics the indirect method must have been the most widely used technique of proof. I would not be so rash as to make these assertions solely on the basis of the passages from Plato quoted in previous chapters. In fact there is other evidence, which tells us

[113] O. Becker, *Grundlagen der Mathematik in geschichtlicher Entwicklung*, Freiburg–Munich 1954, p. 38.

[114] See O. Becker, *Quellen und Studien* ... B, **3** (1936) 533.

[115] Ibid., p. 544, n. 11; Aristotle, *Prior Analytics* I.23. 41a26 and I.4450a37.

exactly when and under what circumstances *hypothesis* and indirect proofs were first used. To establish the claims which I have made, all that needs to be done is to let this evidence speak for itself.

### 3.11 ZENO, THE INVENTOR OF DIALECTIC

In the *Parmenides* Plato makes Zeno the Eleatic explain in his own words the main points of his work.[116] From this it emerges that Zeno's treatise was intended to be a defence of the teachings of Parmenides (βοήθειά τις τῷ Παρμενίδου λόγῳ). Parmenides' doctrine of the *One* had been ridiculed by some on the grounds that it led to consequences which contradicted it (ὡς εἰ ἕν ἐστι, πολλὰ καὶ γελοῖα συμβαίνει πάσχειν τῷ λόγῳ). Zeno therefore set out to show that the *hypothesis* of those who opposed Parmenides ('πολλά ἐστιν') led to even more ludicrous consequences, when it was investigated in the right way (ὡς ἔτι γελοιότερον πάσχοι ἂν αὐτῶν ἡ ὑπόθεσις, εἰ πολλά ἐστιν, ἢ ἡ τοῦ ἓν εἶναι, εἴ τις ἱκανῶς ἐπεξίοι).

It follows from the above as well as from what Simplicius has to say[117] that Zeno was engaged in contrasting one *hypothesis* with another. According to Simplicius, these were ἡ ὑπόθεσις ἡ λέγουσα πολλά ἐστιν ('the *hypothesis* which states that what exists is *many*') and ἡ τοῦ ἓν εἶναι (the *hypothesis* 'which states that what exists is *one*'). Zeno then examined the statements which agreed with each of these *hypotheseis* or, in the words of our text, he checked to see which of these propositions led to a contradiction (ἐναντία αὐτῷ λέγει) so that he would be able to reject as false the one which did. This means that Zeno's method was identical with the one used in Plato's dialectic. In view of the fact that Aristotle described Zeno as the founder or inventor of dialectic,[118] we should not be too surprised at the presence of a common method.

There is absolutely no doubt that Plato's account of Zeno is historically accurate. Instances in which Zeno made use of *hypotheseis* and the method of indirect proof are so well known from the surviving fragments of his work that there is little point in enumerating them

[116] Plato, *Parmenides* 128.

[117] Simplicius, *Commentary on the Physics of Aristotle* 134.2 (the passage is concerned with *Physics* A3.187a1).

[118] *Aristotelis Fragm.* (ed. V. Rose), Lipsiae 1886, fragment 65.

here. Eleatic dialectic was built on these two things, and Plato's dialectic, as I have remarked on many previous occasions,[119] is nothing but a later and perhaps more highly developed form of Eleatic dialectic.

So we have been led to conclude that the earliest known applications of *hypotheseis* and the method of indirect proof were made by members of the Eleatic school. This in itself would enable us to answer our earlier question about the relative ages of dialectic and mathematics, *were it not for a noteworthy observation which has often been made about Zeno's dialectic.* I have in mind the conjecture that indirect proof, or *reductio ad absurdum*, was developed by the earliest Greek mathematicians and that Zeno learned about it from their work. If this were so, it would mean that indirect argument was originally mathematical or geometrical in character.[120]

It is apparent that the question of whether or not dialectic antedates mathematics, which cropped up in connection with our study of Plato's arguments, cannot be answered in a satisfactory manner simply by referring back to the Eleatics. The conjecture that Zeno took over the method of indirect proof from mathematics is not without some plausibility. After all, Plato emphasizes that when he uses *hypotheseis* and indirect proofs he is following the example of mathematicians; so perhaps Zeno did the same thing. I myself have stressed that Zeno's dialectic was a forerunner of Plato's; so perhaps the Eleatics were also forerunners of Plato in that they learned from mathematics. All this indicates that we are faced with a problem which requires serious and critical examination.

The first thing to be kept in mind is that we have no direct knowledge of Greek mathematics in the age of Zeno. What little we do know about the mathematics of this period is based entirely upon modern attempts to reconstruct it. This fact should in itself make us a little wary of the claim that indirect proof was used in mathematics *before the time of Zeno.* There is absolutely no evidence to support it. To insist that Zeno learned this technique from mathematics is to explain what is known by something which is unknown.

On the other hand, it is very easy to locate the source from which Zeno obtained the method of indirect proof if one does not discount

[119] See my paper 'Eleatica' referred to in n. 33 above.
[120] A. Rey, *La Jeunesse de la science grecque*, Paris 1933, p. 202.

completely the ancient tradition. Zeno was a pupil of Parmenides, and his greatest debt undoubtedly was to his teacher. There was no need for him to learn about indirect proof from some 'ancient mathematician'; this method is to be found fully developed in the didactic poem of Parmenides. It was Parmenides who proved his theses by refuting their negations.[121] The discovery of indirect proof was perhaps his greatest and most lasting contribution to philosophy. We are even in a position to be more precise about the circumstances of the discovery. It probably came about as a result of Parmenides' critique of Milesian cosmogony or, to be more exact, of Anaximenes' cosmogony.[122] Anaximenes held that the world came into being from a primary substance by means of 'rarefaction' and 'condensation'.[123] For example, he maintained that *water* could be reduced to *air;* *water*, before it came into being (i.e. before it was condensed), was 'air' or '*non-water*'. On the other hand, this line of reasoning could not be applied to the ὄν. As Parmenides realised, it would be self-contradictory to say of the ὄν that, before it came into being, it was a μὴ ὄν; this would be inconceivable and hence impossible. *What is* must always have existed; it could not have come into being from *what is not*. (Using the terminology of Zeno and Plato, we could describe the situation as follows: the *hypothesis* that what is 'came into being' leads to a contradiction or *adynaton*, for it implies that *what is* was once *non-existent*.) Thus the notion of a *logical contradiction* was formulated by Parmenides in the course of his criticism of Milesian cosmogony, and this in turn led him to discover the method of indirect proof.

### 3.12 PLATO AND THE ELEATICS

We have now shown that both the use of *hypotheseis* and the method of indirect proof can be traced back to the Eleatic school. Our next task is to see what light this knowledge throws on the origins of

---

[121] See the papers of mine listed in n. 33.

[122] Cf. K. Reinhardt, *Parmenides und die Geschichte der griechischen Philosophie* and my paper 'Zum Verständnis der Eleaten' (see n. 33 above).

[123] Hippolytus, Ref. I.7. 3 (Dies and Kranz, Fragmente I⁸.13 Anaximenes A7).

axiomatic mathematics. First, however, let us take another look at Plato's method of reasoning and pick out some features of it which relate to his use of *hypotheseis*. In so doing, we will also learn something about early Greek mathematics.

There is no doubt that Plato usually chooses to express his own views through the character of Socrates. It is well known, however, that Socrates is portrayed in the dialogues as being extremely distrustful of sense perception; i.e., he is suspicious of all knowledge obtained by means of the senses. In the *Phaedo*, for example, we read:[124]

"When the soul makes use of the body for any inquiry, either through seeing or hearing or any other of the senses — for inquiry through the body means inquiry through the senses — then it is dragged by the body to things *which never remain the same (οὐδέποτε κατὰ ταὐτὰ ἔχοντα)*, and it wanders about and is confused and dizzy like a drunken man because it lays hold on such things.

But when the soul inquires alone by itself, it departs into the realm of the pure, the everlasting *(ἀεὶ ὄν)*, the immortal and the changeless *(ἀθάνατον καὶ ὡσαύτως ἔχον)*."

In the above quotation the two kinds of knowledge are contrasted with one another. Plato's distinction between knowledge obtained with the help of the senses and pure, non-sensory knowledge, as well as his preference for the latter sort, is of course part of the legacy of Eleatic philosophy.[125] It was Parmenides who gave the following warning to his disciple:[126]

"You must debar your thought from this way of search, nor let ordinary experience in its variety force you along this way, the eye, sightless as it is, and the ear, full of sound, and the tongue,

---

[124] *Phaedo* 79c–d. The translation is by Fowler.

[125] This is not the place to discuss whether it is possible to acquire knowledge without the aid of the senses, but see my paper 'Zur Geschichte der Dialektik des Denkens' (referred to in n. 33 above).

[126] Diels and Kranz, Fragmente, I.28 Parmenides, b7; the translation is by Kathleen Freeman (*Ancilla to the Pre-Socratic Philosophers*, Oxford 1948). See also the fragment of Melissus, Diels and Kranz, Fragmente, I.30b8.

to rule; but judge by means of the Reason *(λόγῳ)* the much-contested proof which is expounded by me."

Thus the kind of argument from *hypotheseis* which is the subject of our present inquiry presupposes the total rejection of sense perception. This remark is true not only for the Eleatics (Zeno, for example, could not have maintained that his 'paradoxes' were valid without at the same time denying the truthfulness of sense perception) but also for Plato. The criterion of consistency, by which arguments from *hypotheseis* were judged, can only be applied *in the domain of pure thought*. Objects perceived through the senses are always changing; they become transformed into their opposites and, hence, are *contradictory*. Only such ideal entities as the concepts of 'Identity' *(τὸ ἴσον)* and 'the Beautiful' *(τὸ καλόν)*[127] are unchanging and never turn into their opposites. For the Eleatics, therefore, and for all who came after them, argument from *hypotheseis* and the rejection of sense-perception formed a single organic whole. (It is worth recalling here that the method of indirect proof made its first appearance in Greek mathematics at the same time as mathematicians began to turn away from visual evidence and to adopt an anti-empirical viewpoint. A full discussion of this matter is to be found in Chapter 3.3 above.)

Plato himself provides us with incontrovertible evidence for the view that he was following the example of the Eleatics in rejecting sense-perception and arguing indirectly from *hypotheseis*. Parmenides, in the dialogue of the same name, addresses the following words of praise to Socrates:[128] "There was one thing you said to him which impressed me very much: you would not allow the survey to be confined to visible things or to range only over that field; it was to extend to those objects which are specially apprehended by reason . . .".

These words, which Plato put into Parmenides' mouth, are consonant with Parmenides' own words quoted above and with Plato's dialectical method. Their presence in the dialogue indicates that Plato

---

[127] Cf. *Phaedo* 78d: αὐτὸ τὸ ἴσον, αὐτὸ τὸ καλόν, αὐτὸ ἕκαστον ὃ ἔστιν τὸ ὂν μὴ πότε μεταβολήν καὶ ἡντινοῦν ἐνδέχεται . . . ἀεὶ αὐτῶν ἕκαστον ὃ ἔστι . . . ὡσαύτως κατὰ ταὐτὰ ἔχει καὶ οὐδέποτε οὐδαμῇ οὐδαμῶς ἀλλοίωσιν οὐδεμίαν ἐνδέχεται.

[128] Plato, *Parmenides* 135e. The translation is adapted from one by F. M. Cornford (see n. 134 below).

had no wish to conceal his debt to the Eleatics. The continuation of this passage is also of considerable interest. Parmenides goes on to discuss how one should set about arguing from *hypotheseis*. He asserts that in a given situation it is not enough to consider some fixed *hypothesis* and its consequences; one must also put forward the opposing *hypothesis* and subject its consequences to careful examination.[129] He cites Zeno as an example of someone whose investigations always conform to this pattern.

The reason why Parmenides insists that both *hypotheseis* have to be considered should be obvious. Since all sense perception, and with it all physical experience, must on principle be rejected as *false* or *unreliable*, no empirical test can be devised to check whether or not an assertion is correct. The validity of a statement can be demonstrated in only one way. The consequences of both the statement in question and its negation must be examined to see which of them contains a contradiction. Once this has been determined, the statement must be rejected or accepted, depending upon whether it or its negation leads to the inconsistency. Every indirect proof, when carried out in full, utilizes this method of investigating a pair of *hypotheseis*; furthermore, the whole of Plato's dialectic is based upon it.

There is no doubt that by the fifth century B.C. the procedure described above was already being widely used in arithmetic. The proposition that the *One* is indivisible, for example, could not have been established without showing that its negation led to a contradiction, i.e. it could not have been established without using a pair of *hypotheseis* in the manner prescribed by Parmenides.

### 3.13 'HYPOTHESEIS' AND THE FOUNDATIONS OF MATHEMATICS

Since we have now ascertained that *hypotheseis* and the method of indirect proof came from Eleatic philosophy, we are finally in a position to decide whether dialectic or mathematics is the older science. Quite obviously it was dialectic which came first. The mere fact that

---

[129] Ibid. 135e–136: χρὴ δὲ καὶ τόδε ἔτι πρὸς τούτῳ ποιεῖν, μὴ μόνον εἰ ἔστιν ἕκαστον ὑποτιθέμενον σκοπεῖν τὰ συμβαίνοντα ἐκ τῆς ὑποθέσεως, ἀλλὰ καὶ εἰ μὴ ἔστι τὸ αὐτὸ τοῦτο ὑποτίθεσθαι, εἰ βούλει μᾶλλον γυμνασθῆναι.

*all those terms which relate to the foundations of mathematics are of dialectical origin* should have led us to this conclusion. Dialectic did not borrow any of its vocabulary from mathematics; instead, perfectly ordinary expressions from dialectic were transformed into technical mathematical terms. Hence early Greek mathematics, at least when it is viewed as an elaborately constructed system of knowledge, can properly be called a branch of dialectic.

In view of this, the fact that Plato held up the mathematical method as an example to be imitated is of considerable historical interest. It indicates that by his time *hypotheseis* and the method of indirect proof had been completely incorporated into mathematics and were no longer associated only with dialectic. In fact it seems that the period separating Parmenides and Zeno from Plato saw the Eleatic teaching about contradiction and consistency develop in two different directions. On the one hand it degenerated into sophistry,[130] and on the other it blossomed into the mathematics of the fifth century. This latter subject, of which we have at best a fragmentary knowledge, gave birth to Euclid's *Elements*. There was, of course, no question in Plato's mind about which of these two developments he should applaud.

We have not yet exhausted the interest of the fact that Plato refers to mathematics in connection with his use of *hypotheseis*. Not only does it allow us to conclude that *hypotheseis* played an important role in mathematics at that time, but it also enables us to make quite a plausible conjecture about the kind of *hypothesis* which was used in pre-Platonic mathematics.

The question at issue is whether ὑποϑέσεις in early Greek mathematics were for the most part *just ad hoc assumptions* like the geometrical *hypothesis* mentioned in the *Meno* (86e3), or, to put it another way, whether any attempts were made in pre-Platonic times to lay down a *general foundation* for mathematics.

In answering this question we must take account of the following facts:

(1) According to Proclus, the first person to compile a book of 'Elements' was Hippocrates of Chios, a fifth century mathematician.[131]

---

[130] See my paper 'Zur Geschichte der griechischen Dialektik' (referred to in n. 33 above).

[131] Proclus (ed. Friedlein), 66.7f.

It is difficult to believe that a work of this kind would not have been prefaced by some list of principles.

(2) We have already observed (in Chapter 3.5 above) that a number of Euclid's geometrical definitions are redundant; either they play no part in his proofs, or the names which they serve to introduce are never used. Furthermore, it can be shown (by examining Plato's vocabulary, for example) that these are of pre-Euclidean origin. This means that in all probability they are the remnants of some list of basic mathematical principles which was compiled before the time of Euclid.

(3) Plato mentions a few mathematical *hypotheseis* (concerning 'the odd and the even', for example, and 'the three kinds of angle')[132] which turn up in Euclid's list of definitions. He also remarks that mathematicians "will give no further account of them (i.e. these *hypotheseis*) either to themselves or to others". This suggests that in pre-Platonic times mathematicians were wont to preface their discussions with *a special list of such hypotheseis*, just as Autolycus of Pitane, Euclid and all the other mathematicians of a later era did. Only by doing this could they have indicated that these propositions, unlike all the remaining ones, *would not be proved*.

So there is no doubt that even before Plato's time efforts were made to collect together a group of fundamental mathematical principles. Since Plato's dialectic was patterned after mathematical reasoning, we can infer from what Socrates says in the *Phaedo* (100a) that mathematicians began each of their investigations by assuming some principle which they judged to be the strongest. When it was realized that by and large the same principles were needed for every mathematical investigation, the idea of collecting them together at the beginning of a work could easily have suggested itself.

I believe that we can go even further than this and conjecture that the earliest *foundations* were composed only of *definitions*. After all, we know that the word ὑπόθεσις (as a mathematical term) had two meanings; it could denote either a *fundamental principle* or just a *definition*. This in itself seems to indicate that *fundamental principles* were at one time identified with *definitions*. Furthermore, it is obvious that the very first *hypotheseis* on which the partners in a dialectical

---

[132] *Republic* VI.510c–d.

debate have to agree take the form of definitions. One needs to establish that whatever is being discussed *exists* and that it can always be distinguished *from what it is not;* it cannot be both itself and its opposite at the same time. The following quotation from Plato's Euthyphro illustrates this last point:[133] "Is not the holy (τὸ ὅσιον) always one and the same thing in every action, and, again, is not the unholy (τὸ ἀνόσιον) always opposite to the holy, and like itself?"

The Greek word for 'to define' (ὁρίζεσθαι) actually means *to mark off.* A definition was intended to mark off the Form or *Eidos* of an object from that which it *was not* and in this way secure the consistency of the Form in question. The earliest definitions, therefore, were designed to distinguish an object from its opposite. The first person to make use of one was Parmenides, when he drew a sharp distinction between the ὄν (What is) and the μὴ ὄν (What is not). As far as the Eleatics were concerned, there could have been no dialectic without this procedure of marking off or defining. Similarly, Plato asserts that it is impossible to engage in a dialectical debate without first laying down definitions:[134]

"If . . . a man refuses to admit that *Eide* of things exist or to distinguish a definite *Eidos* in every case, he will have nothing on which to fix his thought, so long as he will not allow *that each thing has a character which is always the same;* and in so doing he will completely destroy the significance of all dialectical debate."

Plato's meaning is clear: if one does not lay down definitions at the outset to ensure that the subject of one's discourse must always remain the same in character (i.e. to ensure that it can never become its opposite), it will turn out to be *contradictory* (i.e. to be both 'itself' and 'not itself'); but one cannot sensibly apply the dialectical method to anything which is not consistent.

---

[133] *Euthyphro* 5d; cf. *Phaedo* 102e.
[134] *Parmenides* 135b–c: εἴ γέ τις δή . . . αὖ μὴ ἐάσει εἴδη τῶν ὄντων εἶναι . . . μηδέ τι ὁριεῖται εἶδος ἑνὸς ἑκάστου, οὐδὲ ὅποι τρέψαι τὴν διάνοιαν ἕξει, μὴ ἐῶν ἰδέαν τῶν ὄντων ἑκάστου τὴν αὐτὴν ἀεὶ εἶναι, καὶ οὕτως τὴν τοῦ διαλέγεσθαι δύναμιν παντάπασι διαφθερεῖ. – The translation is taken (with some minor alternations) from the one by F. M. Cornford (*Plato and Parmenides*, London 1939).

Since early Greek mathematics can be regarded as a branch of dialectic, the first step in a mathematical investigation must also have been to lay down *definitions*. Such definitions, in fact, provided a totally new *foundation* for pure mathematics (and especially for arithmetic). They made it possible to guarantee consistency in a new and unexpected way. This claim will be substantiated in the next chapter, which is concerned with *hypotheseis* in mathematics.

### 3.14 THE DEFINITION OF 'UNIT'

By a stroke of good luck it is relatively easy to investigate the definitions at the beginning of *Elements* Book VII. As we know, the so-called theory of *even* and *odd*, which is to be found at the end of Book IX (IX.21–36), represents the earliest known example of deductive mathematics; it was doubtless included in the *Elements* only on account of its historical interest. Becker rediscovered this theory in 1936[135] and estimated its date to be the middle or some time in the first half of the fifth century.[136] Now one needs to know what it means for a number to be 'even' or 'odd' before one can construct such a theory. In fact, one needs to know more than this, and it is inconceivable that the architects of the theory were ignorant of Definitions VII.6–9 and 12. As Becker himself realized,[137] however, this implies that the definitions mentioned must also have been formulated no later than the middle of the fifth century.

In my opinion this last remark applies to some other definitions as well. It is scarcely possible to distinguish between even and odd numbers without giving some thought to the concept of *number*. Hence the definition of 'number' (VII.2) is also presupposed by the theory. Numbers, however, are defined as multitudes composed of *units;*[138] so Definition VII.1, which introduces the concept of 'unit', goes together with VII.2. As a matter of fact the concept of 'unit' also appears in the definition of 'odd number' and in many of the propositions and proofs which go to make up the theory of even and

---

[135] See n. 12 above.
[136] See n. 113 above.
[137] Ibid.
[138] Ἀριθμός ἐστιν τὸ ἐκ μονάδων συγκείμενον πλῆθος.

odd. Therefore my conjecture is that Definitions VII.1 and 2 are an integral part of this theory and that they were known at the time when it was first developed.

It could be argued that the Euclidean definitions of 'number' and 'unit' might have been formulated as a result of some philosophical activity which did not take place until long after the middle of the fifth century. Most of the propositions included in the theory of even and odd are very trivial and simple; they must have been accepted without proof by anyone who had the slightest knowledge of arithmetic, and they do not seem to require that the concepts of 'number' and 'unit' be given exact definitions.

Contrary to what one might think, however, the preceding argument does more to *confirm* my conjecture than to discredit it. It has been observed that the proofs in the theory are carried out in painful detail.[139] This indicates that whoever compiled them was less interested in content than in *form*. His intention must have been to show how complete proofs could be given for these simple propositions. Without the definitions of 'unit' and 'number', however, the theory of even and odd is not complete. Moreover, there are other reasons for thinking that these two definitions are of great antiquity.

Book VII of the *Elements* begins with the following definition.[140] "An unit is that by virtue of which each of the things that exist is called one."

This is such a succinct statement that at first glance it is difficult to comprehend what lies behind it. Plato talks on one occasion about the so-called 'theory of the one'.[141] We would do well, therefore, to try and discover as much as we can about this ancient 'theory'. In the *Republic* Plato tells us the following about the unit:[142]

"You know how it is with skilled mathematicians? If any one in an argument attempts to dissect the *one*, they laugh at him and will not allow it. If you cut it up, they multiply it, taking good care that the *one* shall never appear not as *one* but as *many parts*. . . .

---

[139] See B. L. van der Waerden, *Math. Ann.* **120** (1947/49), 139.

[140] Μονάς ἐστιν, καθ᾽ἣν ἕκαστον τῶν ὄντων ἓν λέγεται.

[141] *Republic* VII.525: ἡ περὶ τὸ ἓν μάθησις.

[142] *Republic* VII.525d–526a. The translation is by A. D. Lindsay.

If you were to say to them, "My wonderful friend, what sort of numbers are you discussing in which the 'one' answers your claims,[143] in which each 'one' is equal to every other 'one' without the smallest difference and contains within itself no parts?" How do you think they would reply?

I fancy by saying that they are talking about those numbers which can be apprehended by the understanding alone, but in no other way."

So, according to the above, the 'one' is indivisible. Of course any unit can be divided in practice, and from time immemorial Greek tradesmen and engineers used fractions in their calculations. The ancient science of arithmetic, however, admitted none of these things. It took no account of fractions until the time of Archimedes. As Plato says, instead of dividing the unit, mathematicians multiplied it. Theon of Smyrna explains how these words are to be understood:[144] "When the unit is divided in the domain of visible things, it is certainly reduced as a body and divided into parts which are smaller than the body itself, but it is increased in numbers, because many things take the place of one."

This quotation not only illuminates the meaning of Plato's remark, but also helps us to understand the significance of Euclid's definition of *a unit* as "*that by virtue of which each of the things that exist is called one*". This fact in itself makes it a lot easier for us to ascertain the age of the definition. Plato and Euclid share a common conception of what it means to be a 'unit'; hence the Euclidean definition of this notion must have been formulated in pre-Platonic times. Indeed, I believe that it dates back to the very beginning of deductive mathematics. My reasons for thinking so are given below.

Plato's explanation of why the unit in arithmetic was regarded as indivisible is very instructive. He asserts that mathematicians wanted to exclude the possibility of the one being something other than itself, i.e. of it ever appearing to be many *(εὐλαβούμενοι μή ποτε φανῇ τὸ ἓν μὴ ἓν ἀλλὰ πολλὰ μόρια)*. If the *one* were divisible, it would no longer

[143] ... *οἷον ὑμεῖς ἀξιοῦτε* ... We shall have more to say about the verb *ἀξιόω* when we come to discuss the *communes animi conceptiones*.

[144] P. 18 in Hiller's edition (Lipsiae 1878); the passage is quoted by van der Waerden in *Science Awakening*, p. 115.

17*

be one, but *many*. Hence the concept of 'unit' would be both itself and its opposite at the same time. The idea that the unit is indivisible became accepted once it was realized that the opposing idea (because it involved the notion of a divisible unit) was self-contradictory; thus the statement that the unit is divisible had to be rejected as false, which meant that its negation was taken to be true.

The preceding interpretation of Plato's words suggests that the following conclusions can be drawn from them:

(1) It is now clear what lies behind the Euclidean definition of 'unit'. It is definitely not the vacuous descriptive statement which it appears to be at first sight. Hidden in this definition is the theoretical justification for excluding fractions from the ancient Greek science of arithmetic. It represents the conclusion of a carefully considered argument and implicitly determines *whether or not the one is divisible*.

(2) We are also in a position to say how the Greeks arrived at the idea that the *one* is indivisible and at the Euclidean definition of 'unit'. The statement that the one is divisible was found to *contain an inconsistency*, and this discovery was taken to mean that its negation had to be true. So the proposition that the one is indivisible was obtained as the conclusion of an indirect argument. We know that indirect proofs in early Greek mathematics always took the same form: the inconsistency of some assertion was demonstrated in the course of an argument; this assertion had then to be abandoned, and its negation was considered to have been established. As we can see, this also describes the way in which the Euclidean definition of 'unit' was worked out, for it, too, was built upon an indirect argument. In view of this fact, it seems highly unlikely that rigorous methods of proof were developed in Greek mathematics *before* a start was made on laying the axiomatic foundations.[145] Our example suggests rather that these two developments must have taken place at about the same time.

A close examination of Plato's description of the 'one' in arithmetic leads us to a further conclusion. He states that the one is indivisible and exists only in the mind since "each 'one' is equal to every other 'one' without the smallest difference and contains within itself no parts". All this, especially if the last phrase is read in the original

---

[145] See nn. 2 and 61 above.

Greek *(ἴσον τε ἕκαστον πᾶν παντὶ καὶ οὐδὲ σμικρὸν διαφέρον, μόριόν τε ἔχον ἐν ἑαυτῷ οὐδέν)*, inevitably brings Parmenides to mind. The Eleatics ascribed exactly these same characteristics to 'Being'. The following passage, selected almost at random from Parmenides' poem, serves to illustrate this point:[146] "Nor is Being divisible, since it is all alike. Nor is there anything there which could prevent it from holding together, nor any lesser thing, but all is full of Being. Therefore it is altogether continuous, for Being is close to Being."

It is no accident that the Eleatics often spoke as if *Being (τὸ ὄν)* and *the One (τὸ ἕν)* were interchangeable concepts.

It is fair to say, therefore, that the Euclidean definition of 'unit' is nothing but a concise summary of the Eleatic doctrine of 'Being'. The definition was obtained by the same kind of indirect reasoning as Parmenides used to develop his theory of 'Being'.[147] This is what Plato had in mind when he mentioned in passing the "theory of the One" *(ἡ περὶ τὸ ἓν μάθησις)*.[148]

### 3.15 ARITHMETIC AND THE TEACHING OF THE ELEATICS

Our investigation of the definition of 'unit' has led us straight to Eleatic philosophy. This definition, which traditionally has been attributed to Pythagoras,[149] turns out to be, in a sense, a recapitulation of Eleatic doctrine. However, we know that there are many other features of early Greek mathematics which undoubtedly reveal the influence of the Eleatics. Hence it is natural to speculate that the earliest system of Greek arithmetic may have been nothing more than an extension of the teaching of the Eleatics. I hope to show below that this conjecture is indeed correct. Before doing so, however,

[146] Diels and Kranz, *Fragmente* I.28 Parmenides B fragment 8.22ff. The translation is by Kathleen Freeman.

[147] Parmenides also proved his thesis (that the ὄν existed) by refuting its negation. See also O. Gigon, *Der Ursprung der griechischen Philosophie*, Basel 1945, p. 250.

[148] *Republic* VII.525.

[149] Cf. Sextus Empiricus. *Adversus math.* X.260–1: ὁ Πυθαγόρας ἀρχὴν ἔφησεν εἶναι τῶν ὄντων τὴν μονάδα, ἧς κατὰ μετοχὴν ἕκαστον τῶν ὄντων ἓν λέγεται. The concluding words of this quotation are almost the same as Euclid's definition of 'unit'.

I will have to discuss an erroneous view which was quite popular among earlier scholars.

In much of the scientific literature the relationship between the Eleatics and the Pythagoreans is described in terms which differ significantly from those used above. It is claimed that the Eleatics and Pythagoreans were opposed to one another. This view, which basically goes back to Tannery,[150] is usually associated with the following three theses:

(1) The teaching of Parmenides, despite the fact that it resulted from a close study of Pythagorean doctrines, was directed against the Pythagoreans' *pluralistic conception* of the universe.

(2) The Pythagoreans did not abandon their belief in the multiplicity of things and became the sharpest and most effective critics of the views of Parmenides.

(3) The arguments of Parmenides' pupil, Zeno, were directed against the Pythagoreans and, in particular, against their belief in the existence of plurality; they were designed to show that this belief was incompatible with the existence of motion.

Many distinguished scholars have subscribed to (3) in particular.[151] Nowadays, however, all three theses are treated with a certain amount of scepticism.[152] In fact, *they are not supported by any evidence worth*

---

[150] *Pour l'histoire de la science hellène*, Paris 1 887, pp. 249–50: "Parménide avait écrit son poème dans un milieu où, comme penseurs, les pythagoriciens seuls étaient en honneur; il avait reproduit plus ou moins exactement leur enseignement exotérique . . . mais en tout cas, il avait nié la vérité de leur thèse dualiste. — Les attaques contre son poème durent donc venir surtout de pythagoriciens, et c'est eux que Zénon prit à partie." and pp. 248–9: "Zénon n'a nullement nié le mouvement . . . il a seulement affirmé son incompatibilité avec la croyance à la pluralité."

[151] For references, see C. Thaer, 'Antike Mathematik 1906–1930', in *Bursians Jahresberichte über die Fortschritte der klass. Altertumswissenschaft* **283** (1943), 44ff. For an opposing view of Zeno's arguments, see B. L. van der Waerden, *Math. Ann.* **117** (1940), 143ff.

[152] G. Vlastos, in his review (*Gnomon* 1953) of a book by J. E. Raven (*Pythagoreans and Eleatics. An account of the interaction between the two opposed schools during the fifth and early fourth centuries B.C.*, Cambridge 1948), quotes the following remark made by W. A. Heidel ('The Pythagoreans and Greek Mathematics', *Am. J. Ph.* **61** (1940) 21): "there is not, so far as I know, a single hint in our sources that the Greeks were aware of the purpose of Zeno to criticize the fundamental doctrines of the Pythagoreans".

*taking seriously.* Nonetheless, the basic idea behind them deserves a closer examination to see *exactly what kind of 'opposition' there was between the Eleatics and Pythagoreans (i.e. arithmeticians).*

As we know, the Eleatics maintained that the only thing which existed was 'the One', 'What is'; they denied that there was any plurality and claimed to be able to prove that the very notion of plurality was self-contradictory.[153] To deny *plurality*, however, is to do away with arithmetic. Hence a student of arithmetic, although he would have had no trouble in accepting the notion of 'unit', could not have followed the Eleatics in their rejection of *multiplicity.* He would somehow have had to retain this latter concept. In fact, Euclid's second arithmetical definition (VII.2) is designed expressly to save the concept of plurality. For it states: "A *number (ἀριθμός)* is a *multitude composed of units (τὸ ἐκ μονάδων συγκείμενον πλῆθος).*"

This definition indicates that there was in fact an essential difference between the Eleatic and Pythagorean treatments of 'plurality'. Nonetheless, it would be wrong to say that they were opposed to one another. The Pythagoreans did not dispute the Eleatic doctrine of the 'One'; all they did was *to develop it further.* According to Plato,[154] they multiplied the 'One'. My reasons for claiming that this represents an extension of Eleatic teaching are given below.

(1) Arithmeticians treated 'numbers' in exactly the same way that the Eleatics treated 'Being'. We know from Plato[155] that numbers were regarded not as physical objects which could be apprehended by the senses, but as abstract ones which could be approached only by way of pure thought.[156] Similarly, Parmenides warned his disciple against trying to grasp 'Being' by means of the senses and insisted that it could only be comprehended by pure reason.[157] Since 'numbers' were construed as *mere abstractions*, they represented instances of a new kind of 'plurality' whose existence was less easy to challenge. The Eleatics were able to show that, because *plurality on a practical level*

---

[153] See W. Capelle, *Die Vorsokratiker*, Leipzig 1935, pp. 173ff: 'Zenons Beweise gegen die Annahme der Vielheit der Dinge'.

[154] Plato, *Republic* VII 525e.

[155] Ibid., 525d.

[156] Ibid., 526.

[157] See n. 126 above.

was a self-contradictory notion, there could not be a 'multi-plicity of things'. Their arguments, however, did not apply to the concept of 'number' which was obtained by multiplying the abstract 'One'.

(2) It becomes even more apparent that the Euclidean definition of 'number' did in fact develop out of the Eleatic doctrine of the 'One' (or 'Being'), if we keep in mind the reason why arithmeticians found it necessary to multiply the 'One'. Plato explicitly tells us that this was done to avoid dividing the unit. Fractions were replaced by *numerical ratios* in Pythagorean arithmetic.[158] Hence the new concept of 'number', which was introduced by Definition VII.2, made it possible for the Eleatic doctrine of indivisibility to be retained.

We are perfectly justified in speaking here of a *new concept of number*, for it is not the naive and traditional notion which is defined by VII.2. The Pythagorean's 'numbers' differed from the usual kind not only because they did away with fractions, but also because they did not include the *unit*. Counting begins with *one*, and, according to the usual way of thinking, *one* is also a number; but this is incompatible with Definitions VII.1 and 2. Numbers, if they are multitudes composed of units, can also be decomposed into units. The unit itself, on the other hand, cannot be decomposed at all. Hence it had to be excluded from the domain of numbers. This in turn led to the use of such cumbersome phrases as 'If $a$ is a number or one . . .'.[159]

It is interesting to note how difficult the earliest Greek mathematicians found it to treat 'one' consistently as a non-number. Let us consider just one example of this, which is provided by a pair of propositions in Book VII of the *Elements*. The first of these propositions runs as follows:

VII.9: "If a number be part of a number, and another be the same part of another, alternately also, whatever part or parts the first is of the third, the same part, or the same parts, will the second also be of the fourth."

So the theorem states that if $a : b = c : d$, where $a$, $b$, $c$ and $d$ are whole numbers, then $a : c = b : d$. The fact that the unit was not considered to be a number meant that the preceding theorem had

[158] B. L. van der Waerden, *Erwachende Wissenschaft*, p. 189.
[159] Ibid., p. 180, n. 1.

also to be formulated for the case in which $a = 1$. This is accomplished by Proposition VII.15, which reads:

VII.15: "If an unit measures any number, and *another number* measures any other number the same number of times, alternately also, the unit will measure the *third number* the same number of times that the second measures the *fourth*."

There is no doubt that this latter theorem differs from the former in just one respect; it talks about the 'unit' instead of a *'first number'*. This is the only reason why it received a separate statement and proof. One should notice, however, that the terms *'another number'*, *'third number'* and *'fourth* (number)' are inappropriate. If the unit is really *not a number*, then the formula $1 : b - c : d$ contains *only three numbers*, and it is nonsense to speak of a *'fourth'*. By his unfortunate choice of words, the author of VII.15 reveals the fact that, although he only formulated this proposition because the unit *was not a number* (according to Definition VII.2), he could not rid himself of the everyday view that it was.

We can summarize the results of this last chapter by saying that the first two definitions in Euclid's arithmetical books clearly betray the influence of the Eleatics. The problem of divisibility (or the Eleatic doctrine of indivisibility) had a decisive effect upon the form which these 'fundamental principles of arithmetic' took. This conclusion is further confirmed by the contents of the next chapter.


### 3.16 THE DIVISIBILITY OF NUMBERS

We have seen that the Eleatic theory of the indivisible 'One' caused the Pythagoreans to introduce numbers as multiples of a unit. They were then able to get rid of fractions in favour of *numerical ratios*. More was required, however, if they were to solve completely and in a consistent manner the problem posed by Eleatic philosophy. Although it could now be maintained that the One itself was indivisible, the old problem of divisibility acquired a new meaning when applied to multiples of the One. The difficulty arose from the fact that a *number* 'composed of units' would also have to be *decomposable*.

Arithmeticians, therefore, found themselves confronted with a new problem. If they had merely been faced with *a plurality of things in the real world*, they might perhaps have been able to deny that it was divisible. The Eleatics, after all, had discredited this notion of divisibility by showing that it was self-contradictory. (Hence in Eleatic philosophy 'Being' was one and indivisible; there was no plurality nor any divisibility.) The situation was completely different, however, now that the Eleatic One, which existed only in the mind, had been multiplied to form a new and abstract 'plurality'. This made it necessary to take a fresh look at the problem of *divisibility*. It appears that the investigation of this problem became the central and fundamental concern of Greek arithmetic in its earliest stage of development.

The so-called theory of *even* and *odd*, which is the earliest piece of deductive mathematics known to us,[160] seems to have arisen from the problem of division, or from the problem of *halving*. The simplest kind of division, of course, is division into two parts. The Greek word διαιρεῖν, which in arithmetic meant 'to halve', was used in the ancient philosophical terminology of the Eleatics to express *any* kind of decomposition or division. Parmenides, for example, writes in his poem,[161] οὐδὲ διαιρετόν ἐστιν (Nor is Being divisible), by which he means that Being cannot be split up in any way. There is little doubt that at first διαιρεῖν simply meant *'to divide'* or *'to decompose'*; the special meaning which it acquired in arithmetic *(to halve)* developed from this earlier and more general one. It is significant that Euclid, when he uses the verb διαιρεῖν in his definitions (VII.6 and 7), only finds it necessary to add the qualifying adverb δίχα (twofold). On the other hand, Plato (*Laws* 895e), who wrote somewhat earlier, defines an even number as ἄρτιος ἀριθμὸς ὁ διαιρούμενος εἰς ἴσα δύο μέρη (a number divisible into two equal parts). The latter's choice of words indicates that διαιρεῖν, when it first came to be used in arithmetic, was associated with its Eleatic meaning and had to be qualified in a precise way. Later on, however, the arithmetical use of this verb was confined to the theory of even and odd; as a result of this it no longer needed such precise qualification, and the adverb δίχα was thought to be sufficient.

The etymological considerations in the preceding paragraph clearly indicate that it took some time before the arithmetical problem of

---

[160] See n. 12 above.
[161] Diels and Kranz, *Fragmente*, I.28b, fragment 8.22.

which numbers can be halved was separated from the general problem of 'divisibility', which the Eleatics had formulated for Being (the One). We know from Definitions VII.6 and 7 that the Pythagorean theory of *even* and *odd* partitioned numbers into two groups: those which could be halved (even numbers) and those which could not (odd ones). Historically speaking, this marked the first step of an investigation into the divisibility properties of numbers (i.e. multiples of an indivisible unit). The distinction between even and odd numbers ultimately led to a theory of them.

It is clear that the *problem of divisibility* also suggested the distinction between *prime* and *composite* numbers. Mathematicians must quickly have realized that every number could be divided by the unit; this was guaranteed by the definition of number. They were then led to distinguish between those numbers which were divisible *only* by the unit and those which were not (i.e. those which were divisible by some other number as well). The former were called 'prime' (Definition VII.12), and the latter were called 'non-prime' or 'composite' (Definition VII.14).

The problem of divisibility also lies behind such definitions as: *Definition* VII.3: "A number is a part of a number *(μέρος)*, the less of the greater, when it measures the greater"; and *Definition* VII.4: "but parts *(μέρη)* when it does not measure it".

Let me mention in passing that μετρεῖν, the most important expression in these definitions (VII.12, 14, 3 and 4), is clearly of *geometric origin*. This leads me to believe that the definitions themselves belong to a part of arithmetic which, as I remarked previously (see p. 170), arose out of the *theory of music*. However, the theory of odd and even, does not employ terms of this sort; there is absolutely no reason to think that it was built upon the results of earlier investigations in the field of music.

*

The last three chapters were intended to show that the foundations of arithmetic which are laid down in Book VII of the *Elements* can be regarded as an extension of Eleatic teaching. We now want to turn our attention to another group of Euclid's mathematical principles.

## 3.17 THE PROBLEM OF THE 'AITEMATA'

One of our major concerns in the present investigation is to discover exactly what significance should be attached to the fact that the mathematical principles at the beginning of *Elements* Book I are grouped under three headings, and to give a historical explanation of how this threefold division came into being. Up to now we have only been able to throw some incidental light on these questions. We already know something about how the ὑποθέσεις or ἀρχαί, the foundations of mathematics, were discovered. Furthermore, we have succeeded in clarifying to some extent the problems connected with the earliest mathematical definitions *(ὅροι* or simply *ὑποθέσεις)*, especially the arithmetical ones. We have as yet, however, hardly touched upon the much more difficult problem of Euclid's αἰτήματα *(postulata)* and κοιναὶ ἔννοιαι *(communes animi conceptiones)*. Let us first take up the matter of the so-called postulates *(αἰτήματα)*.

At the moment all we know is that the word αἴτημα, as a mathematical term, appears to have been merely a synonym for ὑπόθεσις. (This remark does not apply to Euclid's usage, which I propose to ignore for the time being.) I have already mentioned that in one of his writings Archimedes more than once uses the word ὑποτίθεσθαι (in the form ὑποτιθέμεθα) to introduce definitions;[162] in another of his works, the principles of fluid mechanics are introduced by the expression ὑποκείσθω (let it be taken as a basis).[163] On the other hand, similar principles of mechanics are introduced in a third work by the word αἰτούμεθα.[164] Proclus comments on this last passage that Archimedes would have done better to speak of 'axioms',[165] although Archimedes himself, after listing the principles in question, refers to them as τούτων ὑποκειμένων.[166] We can infer from this that Archimedes did not distinguish between the mathematical terms ὑπόθεσις, ὑποκείμενον and αἴτημα.

The term αἴτημα (a 'request' or 'demand'; in Latin *postulatum*), just like its synonyms *hypothesis* and *hypokeimenon*, is obviously of dia-

---

[162] 'De conoidibus et sphaeroidibus', *Opera Omnia* (ed. Heiberg, Lipsiae 1910–15) I, pp. 248 and 252.

[163] 'De corporibus fluitantibus', *Opera* (see n. 162 above), II 318.

[164] 'De planorum aequilibriis', *Opera* (see n. 162 above), II 124.

[165] Proclus, (Friedlein, Lipsiae 1873), 181.17ff.

[166] *Opera* (see n. 162 above), II.126.

lectical origin. It comes from the verb αἰτέω ('to ask, to request or to demand). We know that one of the participants in a dialectical debate was obliged to start out by making a very important *request*. Socrates, in the passage from the *Meno* discussed previously (86e3), had to *request* that he be allowed to use the hypothetical method of reasoning. The words which he uses are *"allow me (συγχώρησον), in considering whether or not it can be taught, to make use of a hypothesis"*. (A joint investigation could not be based on an assumption or *hypothesis* unless *both* participants agreed to it. Hence one of them had to *ask* for the agreement of the other.) Dialectical debates are full of such expressions as *'ask', 'demand', 'request'* and even *'take' (λαμβάνειν)* on the one side, and *'give', 'allow'* and *'disallow'* on the other.[167]

An αἴτημα appears to have been some kind of initial proposition which one participant in a dialectical argument *demanded* that the other allow. For this reason, such propositions were called *demands (postulata)*. If *the demand* was conceded, i.e. if the second participant accepted that the statement being put forward was correct, it meant that both partners were in agreement about the starting point of their debate. In such cases the statement upon which they agreed could also be called a ὁμολόγημα or a ὑπόθεσις. This explains why Archimedes was able to use αἴτημα (or the verb αἰτούμεθα in his mathematical writings as a synonym for ὑπόθεσις or ὑποκείσθω.) It may even have been the case that in dialectic the word αἴτημα was used for the most part to denote an assumption.

Yet the mere fact that Euclid chose to call a special group of unproved principles αἰτήματα leads one to suspect that this name, at least originally, was intended to pick out principles which had some *distinguishing feature*. A comparison between the names ὁμολογή-ματα (those things upon which there is agreement) and αἰτήματα (demands) immediately reveals a difference between them; ὁμολόγημα seems to stress that the two partners have reached *a true agreement*,

---

[167] Consider, for example, the underlined words in the following passages: *Phaedo* 100b5: ὑποθέμενος εἶναί τι καλὸν αὐτὸ καθ᾽αὐτὸ . . . ἃ εἰ μοι δίδως τε καὶ συγχωρεῖς εἶναι ταῦτα . . . ᾽Αλλὰ μήν, ἔφη ὁ Κέβης, ὡς διδόντος σοι οὐκ ἂν φθά-νοις περαίνων, and Aristotle, *Physics* Z9.239b30 (the subject under discussion is one of Zeno's arguments): ὅτι ἡ οἰστὸς φερομένη ἕστηκεν. συμβαίνει δὲ παρὰ τὸ λαμβάνειν τὸν χρόνον συγκεῖσθαι ἐκ τῶν νῦν. μὴ διδομένου γὰρ τούτου, οὐκ ἔσται ὁ συλλογισμός.

whereas αἴτημα indicates merely that something has been *demanded*
by one of them. Therefore it seems possible that the word αἴτημα,
when it was first used as a dialectical term, was intended to signify a
*demand* made by one of the partners rather than an *agreement (ὁμολόγημα)* reached by both of them.

In fact, this is exactly the difference which appears to exist, under
certain circumstances at least, between the meaning of αἴτημα on
the one hand and that of ὑπόθεσις or ὁμολόγημα on the other, when
these words are considered as dialectical terms. Proclus, for example,
gives the following account of Aristotle's opinion[168] about this subject:[169] "Whenever, on the other hand, the statement is unknown and
nevertheless is taken as true *without the student's conceding it*, then
he (i.e. Aristotle) says, we call it a *postulate (αἴτημα):* for example,
that all right angles are equal." It has correctly been remarked that
Aristotle is not talking about *mathematical postulates* in the passage
to which Proclus (somewhat *inaccurately*) refers.[170] Proclus is wrong
to cite Euclid's fourth postulate as an example of what Aristotle had
in mind. The latter intended to define αἴτημα as a dialectical term and
not as a mathematical one. This need not concern us at the moment,
however, since our present goal is simply to obtain a general understanding of what αἴτημα meant in dialectic.

From Proclus' words it appears that Aristotle would uphold the
following view: In the terminology of dialectic an αἴτημα could, under
certain circumstances, refer to an assertion which one partner in a
debate wanted to assume (whose acceptance he *demanded*), but which
the other partner refused to concede. It is true that Proclus speaks of
a 'student' and a 'teacher', but this fact is of little significance. The
relationship between student and teacher is obviously the same as
the one which holds between two persons engaged in a dialectical
debate.

The question which now presents itself is whether the meaning of
αἴτημα in dialectic casts any light on the historical problem of Euclid's

---

[168] The passage which Proclus had in mind was obviously *Posterior Analytics*
I.10.76b 27–34.

[169] Proclus (ed. Friedlein), 76, 17ff. The translation is by Glenn R. Morrow
(Princeton 1970).

[170] K. von Fritz, op. cit. (n. 2 above), p. 42 where he refers to p. 540 of *Aristotle's
prior and posterior Analytics* by D. Ross (Oxford 1949).

*αἰτήματα*. Perhaps these postulates should be interpreted as mathematical 'demands' which some participants in a scientific discussion were unwilling to accept immediately and without reservation. We will only be able to answer this question by taking a closer look at the *αἰτήματα* themselves. First, however, we must consider two other matters; one is the treatment which Euclid's postulates have received from previous historians, and the other is the question of how they are to be dated. These will be dealt with in the next two chapters, after which we will return to our original question.


### 3.18 EUCLID'S POSTULATES

Euclid's five *αἰτήματα* read as follows:

"Let the following be postulated:
1. To draw a straight line from any point to any point.
2. To produce a finite straight line continuously in a straight line.
3. To describe a circle with any center and distance.
4. That all right angles are equal to one another.
5. That, if a straight line falling on two straight lines make the interior angles on the same side less than two right angles, the two straight lines, if produced indefinitely, meet on that side on which are the angles less than the two right angles."

Since the appearance of Zeuthen's pioneering work, there has been little disagreement amongst scholars about the meaning and significance of these postulates.[171] Becker, for example, describes them in the following way:[172]

[171] H. G. Zeuthen, 'Die geometrische Konstruktion als Existenzbeweis', *Math. Ann.* **47** (1896), pp. 222–8. In his *Geschichte der Mathematik im Altertum und Mittelalter* (Kopenhagen 1896), Zeuthen wrote (on p. 120) that, since the problems of the ancients were in essence propositions about existence and their solutions were proofs of existence, the postulates were *assertions about the existence of certain forms*, whose acceptance was demanded without proof. For a different view, see A. Frajese, 'Sur la signification des postulats euclidiens'. *Archives Internationales d'Histoire des Sciences* (1951) pp. 382–92, and *La matematica nel mondo antico*, Roma 1951, pp. 92–7.

[172] O. Becker, *Das Mathematische Denken der Antike* (Göttingen 1957), p. 19.

"The *Postulates* serve to guarantee the existence of certain basic forms, namely straight lines, circles and their points of intersection, out of which the other figures are built up in a constructive manner. The famous fifth postulate, for example, guarantees the existence of a point at which two convergent lines will meet."

However, there is more disagreement about the dating of these postulates. Tannery, for example, started out by observing that Aristotle seems not to have known about mathematical postulates and to have regarded αἴτημα exclusively as a dialectical term. He concluded from this that Euclid should be given credit for the first three postulates at least.[173] But when it came to the fourth and fifth ones Tannery conjectured that they did *not* originate with Euclid and argued somewhat surprisingly that they were *unworthy* of the author of the *Elements*.[174] Heath, on the other hand, was of a different opinion: "I think that the great postulate 5 is due to Euclid himself, and it seems probable that Postulate 4 is also his, if not Postulates 1–3 as well".[175]

Later scholars, although they may disagree with Tannery on some points of detail, seem by and large to accept his contention that Euclid should be given credit for at least some of the postulates. Hoffmann is no exception; in his history of mathematics he remarks that the postulates may be regarded as an important methodological contribution made by Euclid himself.[176]

It is only in very recent times that noticeable efforts have been made to trace these postulates back to the pre-Euclidean period. Von Fritz, for example, has claimed on the basis of a passage in Proclus (p. 179 in Friedlein's edition) that the constructive postulates (2–3) were introduced by Speusippus, a pupil of Plato.[177] I do not accept this conjecture, nor can I find anything in Proclus to support it;[178] nevertheless I think it noteworthy that such claims have been made.

[173] P. Tannery, *Mémoires scientifiques* **2**, 48–63.

[174] Cf. A. Frenkian, op. cit., p. 15 (n. 67 above), p. 15; see also C. Thaer, *Die Elemente von Euclid*, p. 22

[175] T. L. Heath, *A History of Greek Mathematics* (Oxford 1921), Vol. 1, p. 375.

[176] J. E. Hoffman, 'Geschichte der Mathematik, Part 1', *Sammlung Göschen*, (Berlin) **226** (1953), 32.

[177] K. von Fritz, op. cit. (n. 2 above), p. 97.

[178] Cf. O. Becker's paper in *Archiv für Begriffsgeschichte* **4** (1959), 213.

There have also been other attempts to clarify the prehistory of Euclid's postulates. Becker, for example, has succeeded in reconstructing what could very well be an early form of the fifth postulate (and hence of the fourth as well).[179] Not even the most plausible attempt of this sort, however, is sufficient to establish that postulates were used in pre-Euclidean times. This must be taken for granted, i.e. one must at the very least assume that the *kind* of mathematical principle which Euclid calls a postulate was known in pre-Euclidean times, if such reconstructions are to have any historical interest.

I would now like to show that such an assumption can be justified. In what follows I shall attempt to explain how postulates originated. Euclid's fourth and fifth $a\check{\imath}\tau\eta\mu a$ will not be discussed at all, since they form a separate topic by themselves. Instead I want to concentrate on his first three, the postulates which deal with constructions, and address myself to the question of their age.

### 3.19 THE CONSTRUCTIONS OF OENOPIDES

In my opinion, an answer to the question mentioned at the end of the last chapter has been available for quite some time. Since it has gone almost unnoticed, however, the facts pertaining to it need perhaps to be stated in a more perspicuous and precise manner.

It will be recalled that Becker wrote that Euclid's postulates were designed to guarantee the existence of certain basic forms from which the other figures could be constructed. The passage in which he made this statement continues as follows:[180]

"The constructive view of mathematical existence goes back to Oenopides in the fifth century. He was the first person to carry out elementary constructions, like that of the perpendicular to a line, using only a ruler and compasses. It appears that in the school of Plato the use of any other instruments was avoided whenever possible."

[179] Ibid., pp. 212–8.
[180] O. Becker, *Das mathematische Denken der Antike*, pp. 19ff.

The prehistory of Euclid's postulates cannot be discussed satisfactorily without making some reference to Oenopides. His name is mentioned twice in Proclus' commentary on Euclid. The two passages in which it occurs deal with certain constructions and are of such importance that they are both quoted below.

One of Euclid's problems reads: *Elements* I.12: "To a given infinite straight line, from a given point which is not on it, to draw a perpendicular straight line."

Proclus makes the following remark about this proposition:[181]

"This problem was first investigated by Oenopides, who thought it useful in astronomy. In archaic fashion, however, he calls the perpendicular a line drawn '*gnomonwise*' because the gnomon also is at right angles to the horizon." (The pointer on a sundial was called a *gnomon*.)

The other passage which concerns Oenopides is a comment on *Proposition* I.23: "On a given straight line and at a point on it to construct a rectilineal angle equal to a given rectilineal angle."

It runs as follows:[182] "This too is a problem, and the credit for its discovery belongs rather to Oenopides, as Eudemus tells us."

Both these constructions are so simple that it is difficult at first to understand why Proclus bothered to mention the name of their discoverer. Geometry was no longer in its infancy at the time of Oenopides (around 440 B.C.); it can scarcely be claimed that these simple constructions were unknown until then.[183] Oenopides, after all, was no more than a few years older than Hippocrates of Chios, and Hippocrates' geometry was a highly developed science, remarkable not only for the number of theorems which it contained but also for the rigour and precision with which these were proved.[184] Therefore it is clear that Proclus' words require a different interpretation.

There is one to be found in the works of Heath, whose opinion about this matter can be summed up in three sentences. Although each one

---

[181] Proclus (ed. Friedlein) 283.7ff. The translation is by Glenn R. Morrow (Princeton 1970).

[182] Ibid., 333, 5–6.

[183] A. D. Steele, 'Über die Rolle von Zirkel und Lineal in der griechischen Mathematik'. *Quellen und Studien* . . . B, **3** (1936), 287ff., 304.

[184] B. L. van der Waerden, *Erwachende Wissenschaft*, pp. 213–4.

of them says much the same thing as the others, all three are worth quoting:[185]

(a) "The point may be that he (Oenopides) was the first to solve the problem *by means of the ruler and compasses only.*"

(b) "Oenopides may have been the first to give the theoretical, rather than the practical, construction for the problem."

(c) "It may therefore be that Oenopides' significance lay in improvements of method from the point of view of theory."

I find these conjectures very plausible. The only comment which I would like to make concerns the phrase *"by means of ruler and compasses only"*. Proclus does not mention these instruments in the passages quoted above. It might therefore be thought that (a) is rather a fanciful conjecture. In my opinion, however, this is not the case; Heath's assertion is quite justified, although the manner in which it is expressed requires some explanation.

There is no doubt that the problems (*Elements* I.12 and 23) which Oenopides is supposed to have solved are interesting *only from a theoretical point of view. Practical* solutions to them were known from time immemorial. Furthermore, rulers and compasses were used by craftsmen long before the time of Oenopides. The purpose of Euclid's two propositions was *theoretical*, not practical, in other words, they were intended to demonstrate how the problems in question could be solved *on the basis of the first three postulates.*

Postulates 1–3 are equivalent in theory to allowing the use of ruler and compasses.[186] A ruler enables one 'to draw a straight line from any point to any point' and 'to produce a finite straight line continuously in a straight line'. Similarly it is easy 'to describe a circle with any centre and distance' by using a pair of compasses.

Thus to say that Oenopides solved the problem 'by means of the ruler and compasses' is just to say that he solved it *by making conscious use of Euclid's first three postulates.* Indeed, he may even have been the originator of these postulates, if he really was the first person to give theoretical constructions for Propositions I.12 and I.23. This is

---

[185] T. L. Heath, *A History of Greek Mathematics* (Oxford 1921) Vol 1, p. 175; cf. Frajese's paper in *Archimede* 6 (1967), pp. 285–94.

[186] Cf. J. E. Hofmann, 'Geschichte der Mathematik', Part 1, *Sammlung Göschen* (Berlin) 226 (1953), 32.

the only sensible interpretation of Heath's conjecture that "Oenopides' significance lay in improvements of method from the point of view of theory".

Anyone who disagrees with the conclusions reached above and believes that Postulates 1–3 are of more recent origin must reject everything which Proclus tells us about Oenopides as mistaken. It hardly seems possible to find a satisfactory explanation of Proclus' words which is not along the lines suggested by Heath.

### 3.20 THE FIRST THREE POSTULATES IN THE 'ELEMENTS'

In view of what we learned in the preceding chapter, it seems reasonable to conjecture that Euclid's first three postulates originated in the fifth century and that their author may have been Oenopides. The earliness of this date should not surprise us. We shall see in a subsequent chapter that Euclid's so-called κοιναὶ ἔννοιαι (communes animi conceptiones) are no less ancient.

Let us now turn to the question of why it was thought necessary to lay down postulates of any kind. It seems to me that the way in which this question has been tackled in the past is not altogether satisfactory. We are told quite correctly that in ancient geometry the only figures which could be said to *exist* were those obtained in a constructive manner, and that the postulates were intended to guarantee the existence of certain basic forms (straight lines, circles and points of intersection). However, these facts do not explain how this conception of mathematical *existence* arose, nor do they help us to understand why the unproved statements which guaranteed existence were called αἰτήματα. We know that the term αἴτημα came from dialectic where it was used to denote a 'demand' about which the second partner in a dialogue had *reservations*. Let us see whether there is any connection between this early meaning of the word and Euclid's postulates.

At first glance, Postulates 1–3 appear to be such simple, self-evident and easily fulfilled 'demands' that one is tempted to disregard the literal meaning of their name. There is no sense in insisting that the postulates are 'demands' or 'assumptions' of the kind described above just because their name was used to convey this meaning under certain circumstances in dialectic. Nonetheless there are, in my opinion, good reasons for thinking that the postulates were αἰτήματα in the original dialectical sense of the word. In order to be able to justify this claim, I must begin by quoting a passage from Proclus' commentary on Euclid:

> "The drawing of a line from any point to any point follows from the conception of the line as the *flowing* of a point and of the straight line as its uniform and undeviating *flowing*. For if we think of the point as *moving* uniformly over the shortest path, we shall come to the other point and so shall have got the first postulate without any complicated process of thought. And if we take a straight line as limited by a point and similarly imagine its extremity as *moving* uniformly over the shortest route, the second postulate will have been established by a simple and facile reflection. And if we think of a finite line as having one extremity stationary and the other extremity *moving* about this stationary point, we shall have produced the third postulate."[187]

(The translation of the above is by Glenn R. Morrow, but the italics are mine.)

I do not want to maintain that Proclus, in this short explanation of Postulates 1–3, meant to define a 'line' as *'the flowing of a point'* and a 'straight line' as *'its uniform and undeviating flowing'*; there are certainly no such definitions to be found in Euclid. At the moment, I am much more interested in the fact that he used certain simple kinds of *movement* to explain Euclid's first three postulates. As a matter of fact, the 'demands' made by these postulates cannot be satisfied *without movement*. It is questionable, however, whether 'motion' itself is really as simple and unproblematic as it appears to be from Proclus' words. Our commentator was clearly aware of this difficulty, for he went on to say:[188]

> "If someone should inquire how we can introduce *motions* into immovable geometrical objects and *move* things that are without

[187] Proclus (ed. Friedlein), 185.8ff.
[188] Ibid., 187.25ff.

parts – *operations that are altogether impossible* – we shall ask
that he be not annoyed . . . But let us think of this *motion* not as
bodily, but as imaginary [in the Greek text: κίνησις φανταστική],
and admit not that things without parts move with bodily motions,
but rather that they are subject to the ways of the imagination.
For 'nous', though partless, is moved, but not spatially; and imagination has its own motion corresponding to its own partlessness.
In attending to bodily motions, we lose sight of the motions that
exist among things without extendedness, etc.''

Luckily we do not have to worry overmuch about interpreting
Proclus' attempt to reassure his readers. It is sufficient for our purposes merely to note the fact that the above quotation, in a sense,
can be said to reproduce a *dialectical debate*. The debate is of interest
because one of the participants in it acts as a spokesman for those
who oppose the idea that the kind of *motion* required is 'conceivable'.
So we see that the unproved propositions which Euclid called 'postulates', and which he regarded as necessary for the foundations of geometry, are in the eyes of these opponents mere *demands (αἰτήματα)*,
which cannot be accepted without reservation.

Proclus seems to have thought that the opponents of *motion* in
geometry objected only to the idea of a moving point (i.e. to the idea
of that which has no parts, and hence no bodily existence, being in
motion). He readily concedes, therefore, that the *bodily* motion of
*that which has no body* is inconceivable and suggests as an alternative
that points move in an 'imaginary' rather than a 'bodily' way. Of course
this is little more than abstruse speculation on his part and is not to
be taken too seriously. However, we are left wondering, whether he
has dealt with all of the objections which were raised in dialectical
debate against the idea of *movement* in geometry.

As a matter of fact, we know that in the years which preceded the
time of Oenopides, the putative author of Postulates 1–3, there
flourished a very influential group of men who succeeded brilliantly
in showing that motion of any kind was contradictory and, hence,
*inconceivable*. Of course I am referring to the Eleatics and, in particular,
to Zeno, who gave the most incisive formulation of Parmenides' ideas.[189]

---

[189] See n. 33 above.

If we bear this in mind, it is easy to understand why Euclid's first three postulates had to be laid down. The only way to make geometrical constructions theoretically possible is to admit at least those kinds of *movement* which are indispensable to the production of the very simplest geometrical forms (namely straight lines, circles and their points of intersection). In fact, Postulates 1–3 guarantee the *existence* of certain basic geometrical forms; they ensure that it is possible to carry out the simplest constructions or, in other words, they *require* those kinds of *movement* which are necessary for this purpose. They really are *demands (αἰτήματα)* and not *agreements (ὁμολογήματα);* for they postulate *motion*, and anyone who adhered consistently to Eleatic teaching would not have been able to accept statements of this kind as a basis for further discussion. Therein lies the difference between αἴτημα on the one hand, and ὑπόθεσις and ὁμολόγημα on the other.

*

I hope that in the last four chapters I have succeeded in establishing the following important facts:

(1) Euclid's postulates or *aitemata* have the same origins as the general foundations of mathematics *(hypotheseis* or *archai)* and mathematical definitions *(hypotheseis* or *horoi)*, which is to say that they came from dialectic.

(2) The name *aitemata* refers to the fact that they are assumptions which were not accepted without reservation (by one of the participants in a dialogue).

Furthermore, we saw that when mathematicians in pre-Euclidean times laid down these αἰτήματα they extended the teaching of the Eleatics no less than when they introduced the basic arithmetical definitions (of 'unit', number', 'even number', 'odd number', 'prime number', 'composite number', etc.) discussed earlier. To lay the foundations of deductive mathematics, it was necessary to do more than just adopt the ideas of the Eleatics (the method of indirect proof, anti-empiricism, etc.); hence the earliest mathematicians, when they began this task, were obliged sometimes to *depart* from Eleatic teaching. The definitions in arithmetic are one manifestation of this, as are the postulates (amongst other things) in geometry.

## 3.21 THE 'KOINAI ENNOIAI'

We have remarked more than once that the fundamental principles which are stated without proof at the beginning of *Elements* Book I are listed in *three separate groups*. Two of these, the *definitions* and the *postulates*, have already been discussed. So let us now turn our attention to the third group, which Euclid calls κοιναὶ ἔννοιαι. This name is usually translated into Latin by the phrase '*communes animi conceptiones*'.

Let me begin by pointing out an interesting fact which it is important to keep in mind. Proclus' excellent commentary is full of quotations from Euclid's text; these quotations are very accurate and, as far as I know, have invariably been found to reproduce Euclid's words exactly. It is all the more surprising, therefore, that Proclus' names for the three kinds of mathematical principle are not always the same as Euclid's. The terms which they use are listed below for purposes of comparison:

|  | *Euclid* | *Proclus* |
|---|---|---|
| Definitions | ὅροι | ὑποθέσεις or ὅροι |
| Postulates | αἰτήματα | αἰτήματα |
| *Communes animi conceptiones* | κοιναὶ ἔννοιαι | ἀξιώματα |

As we can see, it is only in the case of the *postulates* that Proclus consistently follows Euclid's terminology. He refers to the *definitions* sometimes as *hypotheseis*[190] and sometimes as *horoi*.[191] In my opinion, however, this fact is of no great significance. We know that in early times *hypothesis* meant *definition* (as well as 'fundamental mathematical principle'); furthermore, the word *horos* occurs with this meaning in the writings of Plato. It is apparent that both these names were in use even before the time of Euclid.

As far as we are concerned, the fact that Proclus always refers to the κοιναὶ ἔννοιαι as ἀξιώματα is of much greater importance.[192] If this were a delibarate departure from Euclid's text, I feel sure that Proclus

---

[190] See, for example, Proclus (ed. Friedlein) 76.4ff. and 178.1ff.
[191] Ibid., 81.26.
[192] Ibid., 76.6; 77.1; 178.2 et passim.

would have given his reasons for making it. Yet he uses the term ἀξίωμα as naturally as if he had taken it straight from Euclid. He even dissociates himself, so to speak, from those who use the term κοιναὶ ἔννοιαι, saying that "according to them, *axiom* and '*common notion*' mean the same thing" *(ταὐτὸν γάρ ἐστιν κατὰ τούτους ἀξίωμα καὶ ἔννοια κοινή)*.[193] It seems to me that this fact can only be explained in the following way.

The kind of mathematical principle which is called a κοινὴ ἔννοια in our text of Euclid obviously bore the name ἀξίωμα in pre-Euclidean times. Aristotle refers more than once to mathematical *axiomata*;[194] furthermore, he frequently uses this term to denote the unproved proposition which we know as Euclid's third κοινὴ ἔννοια (when equals are taken from equals the remainders are equal).[195] No doubt the name *axiomata* also appeared in Proclus' copy of Euclid, and this explains why it is used so consistently in his commentary (instead of κοιναὶ ἔννοιαι); on the other hand, it was replaced by the newer expression κοιναὶ ἔννοιαι in the texts on which only modern editions of Euclid are based. This conjecture presents us with two problems: the first is to explain the original meaning of ἀξίωμα, and the second is to discover why this term is absent from the texts of Euclid known to us.

It is perhaps worthwhile to recall at this point how the problem of the κοιναὶ ἔννοιαι has been treated by historians in the past. Tannery observed that the term κοινὴ ἔννοια does *not* occur *with its technical meaning* in the works of Plato or Aristotle and that the Stoics appear to have been the first to use it in this sense. He concluded therefore that the group of principles which are called by this name in the *Elements* had to be of *post-Euclidean origin*.[196] Now, if he had merely claimed that their name was post-Euclidean, I would have no quarrel with him. Tannery, however, was not so modest; he maintained that the principles themselves could not have been gathered together

---

[193] See Proclus (ed. Friedlein), pp. 193–4 (especially 194. 8–9).

[194] See, for example, *Metaphysics* Γ3.1005a20.

[195] Ibid., K4.1061b20; compare this passage with Γ3. 1005a19ff.

[196] P. Tannery, *Mémoires scientifique* **2**, 60: "Ce terme d'ἔννοια n'est nullement de la langue philosophique de l'époque; on le chercherait vainement avec une signification technique dans l'œuvre de Platon ou dans celle d'Aristote; il appartient aux stoïciens dont l'école commençait seulement au temps d'Euclide et dont il ne pouvait subir l'influence à Alexandrie."

until after the time of Apollonius of Perga.[197] It is remarkable that he ignored completely a piece of evidence which contradicts this view. I am referring to the fact (which was certainly known to Tannery) that Aristotle was obviously acquainted with the κοιναὶ ἔννοιαι, only *under another name.*[198] (Tannery was never the most consistent of scholars.) It is even more surprising to find him explaining why the name ἀξίωμα had to be changed to κοινὴ ἔννοια; according to him, the Stoics were responsible for giving the word ἀξίωμα an entirely new meaning.[199] (This suggests that the influence of the Stoics only brought about *a change of name.* Yet the paper in which Tannery offered this explanation also contains his assertion that the whole list of principles known as κοιναὶ ἔννοιαι could not have been compiled before the time of Apollonius.) I must confess that I would be more sympathetic towards Tannery's conclusions and think them of greater worth, if they had been properly thought out and were consistent with one another. They are, however, mutually contradictory, despite being based on observations which are at least in part correct.

### 3.22 THE WORD ἀξίωμα

We have conjectured that the mathematical principles which go by the name of κοιναὶ ἔννοιαι in Euclid's text (as we know it) were

[197] Ibid. **2**, 56: "Quant aux *notions communes,* elles ne seraient pas de lui [Euclide]; il les aurait employées comme allant de soi ou comme supposées par les définitions; l'attention ne serait portée sur cette question qu'à l'époque d'Apollonius, qui essaya de démontrer ces propositions et reconnut leur liaison avec la définition de l'égalité et des opérations de l'addition et de la soustraction géométriques. Les éditeurs successifs d'Euclide auraient pris depuis lors l'habitude d'insérer un recueil plus ou moins complet de ces notions suivant le point de vue auquel ils se plaçaient mais la tradition serait restée longtemps assez flottante à cet égard."

[198] Ibid. **2**, 62: "Pour le mot *axiome,* Aristote l'emploie à peu près dans le même sens que nous, et il nous apprend qu'il était en usage chez les mathématiciens, mais particulièrement pour désigner les *notions communes.* Il cite même comme exemple que, si on retranche des choses égales de choses égales, les restes sont égaux."

[199] Ibid., **2**, 62–3: "Il est à remarquer, que les stoïciens changèrent complètement le sens du mot ἀξίωμα, et appelèrent de ce nom une proposition quelconque, vraie ou fausse; c'est là qu'il faut chercher la raison de l'adoption d'une autre désignation dans nos textes d'Euclide."

originally called ἀξιώματα. Let us now assume that this was indeed their original name and turn our attention to an investigation of its meaning.

The original meaning of the term ἀξίωμα, although it is basically no more complicated than that of αἴτημα, can no longer be established so easily. Even in antiquity this expression and its significance as a mathematical term were explained erroneously. Furthermore, this 'pseudoexplanation' is still current in the literature today and has made it almost impossible to gain a correct understanding of the matter. We shall have to begin by examining the source of this mistake.

Proclus, after quoting the five Euclidean axioms which he takes to be authentic, commences his explanation of them with the following words:[200] ταῦτ᾽ ἐστὶ τὰ κατὰ πάντας ἀναπόδεικτα καλούμενα ἀξιώματα, καθόσον ὑπὸ πάντων οὕτως ἔχειν ἀξιοῦνται, καὶ διαμφισβητεῖ καὶ πρὸς ταῦτα οὐδείς. Morrow's translation of this sentence runs as follows:[201] "These are what are generally called indemonstrable axioms, inasmuch as they are deemed by everybody to be true and no one ever disputes them."

Although this translation conveys the substance of the Greek sentence quite accurately, it obscures the fact that Proclus explains the noun ἀξίωμα in terms of the verb ἀξιοῦται. We cannot really reproach him for doing so, since ἀξίωμα is in fact derived from ἀξιόω. It seems reasonable, therefore, to investigate the exact meaning of this noun by looking at the verb which is its root. In my opinion, however, this is *not* what Proclus does. Instead of conducting an objective search for the meaning of ἀξίωμα he starts out with a *prejudice* which he then attempts to justify. This is shown by the fact that he mentions *only one of the possible meanings of* ἀξιόω.

The prejudice to which I am referring is his belief that the word ἀξίωμα was used (in Euclidean mathematics at least) to denote a statement *whose truth could not be doubted*. As we shall see later, Aristotle seems to have been largely responsible for the widespread acceptance which this view gained in antiquity. Proclus himself, on another occasion, attributes to Aristotle the opinion that "everyone would be

[200] Proclus (ed. Friedlein), 193, 15–7.
[201] P. 152 in G. R. Morrow's translation (see n. 5 above).

disposed to accept it [the truth of an axiom] *even though some might dispute it* for the sake of argument".[202]

Accordingly there could be no real dispute about the correctness of Euclid's axioms, and in fact no one in antiquity (or at least no one who lived after the time of Aristotle) seems to have seriously entertained the idea that their validity could be questioned. Proclus was amongst those who agreed with Aristotle about the nature of axioms, and he attempted to bolster up his views by appealing to etymological considerations. It was fortunate for him, therefore, that the verb ἀξιόω, when followed by an infinitive, had as one of its meanings 'to think fit'.[203]

Von Fritz has recently given an explanation of the word ἀξίωμα which follows the traditional one very closely.[204] The only difference between the two is that in the modern explanation an effort is made to derive the *mathematical meaning* of ἀξίωμα from an even earlier meaning (to value, to think worthy) of the verb ἀξιόω. Von Fritz however, just like Proclus before him, tacitly assumed that the mathematical meaning of this term *(ἀξίωμα)* was unambiguous and well established.

I now want to list what I consider to be the weaknesses of this position:

(1) ἀξίωμα entered mathematical terminology *from dialectic*. It is therefore wrong to investigate the etymology of this word with the aim of finding out only how it acquired its supposed *mathematical meaning*. One cannot gain a precise understanding of what it signified in mathematics, without first being clear about the role which it played in dialectic. It is obvious that ἀξίωμα, even as a mathematical term, was originally used in its dialectical sense.

(2) Furthermore, the language of mathematics contains clear indications that the earliest mathematical usage of ἀξίωμα was ignored by Proclus. The meaning of this term seems to have been *distorted* in post-Aristotelian times, when it came to denote a statement which was *as a matter of course* 'thought fit' (to be accepted).

---

[202] Proclus (ed. Friedlein), 182, 17ff; the translation is by Morrow (see n. 5 above).

[203] For this meaning of the word ἀξιόω, see the references (Xenophon, *Cyr.* 2.2.17 and *An.* 5.5.9; Pindar, *Nem.* 10.39, etc.) quoted in Pape's distionary.

[204] K. von Fritz, op. cit. (n. 2 above), pp. 29ff.

In fact, it is very easy to establish the *dialectical* meaning of the word ἀξίωμα. Von Fritz, for example, has written:[205] "Even when he is not dealing with dialectic or logic, Aristotle uses the word ἀξίωμα simply to mean a *supposition, opinion, doctrine*, etc.; this usage is a natural outgrowth of the pre-Aristotelian meaning of the verb ἀξιόω." He also made the following pertinent remark, which throws some more light on the subject:[206]

"When discussing in the *Topics* the dialectical technique of question and answer, Aristotle frequently used the word ἀξιοῦν to talk about the statement which the questioner hopes that the respondent will accept. This acceptance is then described as τιθέναι (see, for example, 155b30ff, 159a14ff and many other passages). If ἀξιοῦν is followed by τιθέναι, the dialectical argument can proceed."

If we now want to find out what ἀξιοῦν meant in the context described above, there is no need for us to go back to the original meaning of this verb (to think worthy). It is much simpler to look in a dictionary (Pape, for example, or Liddel and Scott), where the verb ἀξιόω is listed as having 'to demand' or 'to require' amongst its meanings. This was such a common usage in antiquity that it scarcely needs to be illustrated by examples; I will content myself with citing only two, both taken from the writings of Plato. The first one is παρὰ τοῦ ἰατροῦ φάρμακον ἀξιοῦν (*Republic* III 406 – to *ask* the physician *for* medicine), and the second is ἀξιόω ὑμᾶς ἀλλήλους διδάσκειν (*Apology* 19d – I *ask you* to inform one another).

Of course I do not want to deny that this meaning of the verb ('to demand' or *'to require'*) may very well have developed from the earlier one, *'to think worthy'* or *'to think fit'*. Nevertheless, I think that it is misleading to concentrate on this earlier sense in a discussion of how subsequent meanings evolved. It suggests that ἀξιόω was used in connection with demands which were certainly justified and 'deemed worthy' (to be accepted), whereas this verb was in fact used just as often to express a *'false* demand' or a *'false* supposition'.[207]

[205] Ibid., p. 35.
[206] Ibid., p. 32, n. 32.
[207] For example, Herodotus 6.87; Plato, *Menexenus* 239e, etc.

Correspondingly the noun ἀξίωμα, when used as a dialectical term, was originally synonymous with αἴτημα and just meant a 'demand' or 'request'. Sometimes it had this meaning outside dialectic as well; Sophocles, for example, used it to denote the *decree* (or *demand*) of the Gods.[208] Indeed, ἀξίωμα or ἀξίωσις was even used on occasions to refer to a *'petition'* or a *'written request'*.[209] Of course this word also acquired a variety of slightly different meanings in dialectic, but these are now easy to explain. Ἀξίωμα, like αἴτημα, signified an *assertion* which one of the participants in a dialectical debate *demanded* that the other grant; hence it also came to mean 'assertion', 'supposition', 'doctrine', etc.

I should emphasize here that the word ἀξίωμα, when it was used as a dialectical term in pre-Aristotelian times, does not seem to have connoted a 'demand' which was easily satisfied or a readily acceptable 'assertion'; an axioma did not have to be a *'plausible* assumption'. In fact one gets the impression that the reverse was true; the 'assertions' which were called *axiomata* (or *aitemata*) in dialectic appear to have been ones which were not accepted without *reservation*. This point is illustrated by the following passage from Plato's *Republic* (VII.526), which was quoted earlier in a different context:

"You know how it is with skilled mathematicians? If any one in an argument attempts to dissect the *one*, they laugh at him, and will not allow it. If you cut it up they multiply it, taking good care that the *one* shall never appear as not *one* but as many parts ... If you were to say to them, "My wonderful friend, what sort of numbers are you discussing in which the 'one' answers your claims *(περὶ ποίων ἀριθμῶν διαλέγεσθε, ἐν οἷς ἕν οἷον ὑμεῖς ἀξιοῦτε ἐστιν)*, in which each 'one' is equal to every other 'one' without the smallest difference and contains within itself no parts?" How do you think they would reply? I fancy by saying that they are talking about those numbers which can be apprehended by the understanding alone, but in no other way."

According to this quotation, mathematicians *claimed* or *required* the existence of a very special kind of 'one'. (Plato uses the verb

[208] *Oid. Col.* 1451.
[209] See the entry under this word in Pape's dictionary.

ἀξιοῦν in his description of their *claim*.) It should be noticed, however, that this was not a *claim* which the other side in a dialectical debate (i.e. non-mathematicians) could have accepted very readily. The mathematician's 'one' was not a part of the sensible world; hence the non-mathematician must have had reservations about allowing its existence.

We can see that ἀξίωμα and αἴτημα were interchangeable expressions in dialectic; these two terms were *exact synonyms*, as were the verbs ἀξιόω and αἰτέω.

This last remark applies to mathematics as well, if we ignore for the moment Euclid's works and the literature relating to them. We need only look through their writings to see that mathematicians regarded ἀξίωμα and αἴτημα as synonyms. (This is especially true of Archimedes, for whom ὑπόθεσις, ὑποκείμενον, αἴτημα, ἀξίωμα, λαμβανόμενον, etc. were all equivalent expressions.[210]) Furthermore, Proclus, after noting that Archimedes introduces with the word αἰτέω a principle which might rather be called an axiom, goes on to say about first principles in general:[211] "Others designate all of them *axioms*, as they call a *theorem* everything that requires a demonstration."

Proclus was fully aware of the fact that the terminology of Greek mathematics was neither uniform nor consistent as far as the naming of principles was concerned. No doubt he searched through the works of other mathematicians, hoping to find at least the traces of a distinction comparable to the one drawn by Euclid between *axiomata* and *aitemata*. He must have searched in vain, however, since these terms were used simply as synonyms by most mathematicians.

### 3.23 PLATO'S ὁμολογήματα AND EUCLID'S ἀξιώματα

Our task now is to find out why Euclid's *axioms* were given a special name. We have conjectured that the term ἀξιώματα was originally used instead of κοιναὶ ἔννοιαι to denote this group of principles. So the

---

[210] Cf. K. von Fritz, op. cit. (n. 2 above), p. 57 and P. Tannery, *Mémoires Scientifiques* **2**, 79.

[211] Proclus (ed. Friedlein), 181.16ff.: Ἤδη οἱ μὲν πάντα αἰτήματα . . . οἱ δὲ πάντα ἀξιώματα προσαγορεύουσιν.

question at issue is whether Euclid, like Archimedes, considered *axiomata* to be a synonym of *hypotheseis* or whether the fact that certain principles in the *Elements* were called by this name tells us something about the nature and historical development of axiomatics.

There are nine axioms to be found in the modern text of Euclid, and, with the exception of the last, each one of them asserts some basic property of *equality*. Because the first eight form such a homogeneous group, Axiom 9 (Two straight lines do not enclose a space) is usually thought to be a later addition.[212] These principles may therefore be described as axioms for equality. (Even Axiom 8, which states that the whole is greater than the part, is a kind of *equality axiom*, albeit a *negative* one, for it denies that a 'part' can ever be equal to the 'whole'. We shall have more to say about this later.)

We now want to explain why these statements were called ἀξιώματα. The significance of their name has already been discussed; *axioma* was intended to denote an assertion which, although *required* at the beginning of an argument, was not accepted without reservation. Let us investigate the extent to which this meaning is applicable to Euclid's axioms.

It is worthwhile to begin by comparing his axioms with two 'quasi-mathematical' assertions about equality which are made in the *Theaetetus*.[213] The assertions in question are:

(1) "Nothing can become greater or less, either in size or in number, so long as it remains *equal to itself*."

and

(2) "A thing to which nothing is added and from which nothing is taken away is neither increased nor diminished, but always remains *equal to itself*."

On closer inspection, the above statements turn out to be a kind of double definition of what it means for something to be 'equal to itself'. The first one says that a thing which remains 'equal to itself' cannot become bigger or smaller in any way. The second is the converse of

[212] Euclides: *Elementa* (ed. J. L. Heiberg, Lipsiae 1883–1916) I.10, and O. Becker, *Grundlagen der Mathematik in geschichtlicher Entwicklung* (Freiburg–Munich), p. 90.
[213] Plato, *Theaetetus* 155a. The translation is taken (with one minor change) from F. M. Cornford, *Plato's Theory of Knowledge* (London, 1935).

the first and asserts that a thing which does not become bigger or smaller remains 'equal to itself'.

It is instructive to note that these two assertions about equality are not called ἀξιώματα; Plato refers to them as ὁμολογήματα. We already know that *homologema* and *hypothesis* were interchangeable in dialectical terminology; a ὁμολόγημα was a statement upon which both participants in a debate *were in agreement*. It seems reasonable to conjecture that Plato's assertions about equality were called *homologemata* because they were uncontroversial, whereas Euclid's were called *axiomata* because they could not be accepted without reservations. In what follows we shall endeavour to verify this conjecture.

It is, as a matter of fact, hardly conceivable that anyone could have objected to Plato's *homologemata* about 'equality' (or, more precisely, about the property of 'being equal to itself'), for both these statements are purely *formal*. They merely explain in other words the meaning of the phrase 'equal to itself'; hence they constitute a correct formal *definition*. Such *formal statements* could certainly have been regarded as the 'strongest' (in Socrates' sense of the word – cf. *Phaedo* 100a) by the ancient dialecticians. They were not obtained by abstracting from everyday experience, but by a process of pure reflection. Hence they must have been considered no less consistent than the prototype of all assertions of this kind, Parmenides' famous thesis τὸ ὄν ἐστί (What is, is).

Euclid's equality axioms, on the other hand, are statements of a rather different kind. They read as follows:

1. Things which are equal to the same thing are also equal to one another.
2. If equals be added to equals, the wholes are equal.
3. If equals be subtracted from equals, the remainders are equal.
[4. If equals be added to unequals, the wholes are unequal.
5. Things which are double of the same thing are equal to one another.
6. Things which are halves of the same thing are equal to one another.]
7. Things which coincide with one another are equal to one another.
8. The whole is greater than the part.

Not one of these statements can be described as merely formal. On the contrary, they all assert properties of a relation (equality) which must have been regarded (by the Eleatics at least) as 'self-contradictory'. It is by no means evident that *two* distinct things (i.e. two things which are not the same) can ever be *'equal to one another'*; furthermore, it makes no sense to speak of *two* things unless they can be distinguished from one another in some way. The Eleatics were willing to concede only that a thing could *equal itself*, not that it could *equal another thing*.

Euclid's axioms are assertions which are justified by practical experience and, in some cases, directly by sense-perception. Axiom 7, for example, states that "things which coincide with one another are equal to one another". It can literally be *seen* that plane figures which coincide are actually equal; hence this axiom is verified by sensory experience. Similarly, the simplest way to convince oneself of the truth of Axiom 8 ("The whole is greater than the part") is by means of the senses. One need only look at a geometric figure[214] to see that the whole is greater than any part.

According to Eleatic doctrine, knowledge of the truth could not be acquired through sense-perception.[215] It is apparent, therefore, that the statements which were originally called *axiomata* in the *Elements* could never have played the role of *homologemata* or *hypotheseis* in a dialectical argument. Their acceptance may well have been *demanded* at the beginning of a debate, but the other party would not agree to them just because their truth was guaranteed by sense-perception. Hence these assertions, instead of being called ὁμολογήματα, were given the name ἀξιώματα.

As a matter of fact, it is well known that the Eleatics also discussed the problem of *equality;* but their treatment of it was not the same as Euclid's. Amongst those who have written about this subject is Becker, from whose work the following quotation is taken:[216]

---

[214] It was originally thought that Euclid's axioms were valid only for geometry. See O. Becker, op. cit. (n. 212 above), p. 90.

[215] See n. 126 above.

[216] O. Becker, *Mathematische Existenz, Untersuchunger zur Logik und Ontologie mathematischer Phänomene* (Halle 1927), p. 144. Becker refers there to Tannery (*Revue Philosophique* **20** (1885), 385) and Russel *(The Principles of Mathematics* (Cambridge 1903), Vol. I §331, p. 350 and §§340–1, pp. 358–60).

"Zeno's paradox of Achilles and the tortoise can be given a set-theoretical interpretation. When reformulated in this way, the paradox runs as follows: If Achilles is to overtake the tortoise, who has been given a lead, there must be a one to one correspondence between the distance which he has to cover (viewed as a set of points) and the distance covered by the tortoise; furthermore the correspondence must be such that the time which it takes Achilles to reach any point is the same as the time which the tortoise takes to reach the corresponding point. This is just one way of expressing the well known set-theoretical fact that *two infinite sets may have the same cardinality even though one of them is a proper subset of the other.* In the present case the distance travelled by the tortoise is a 'proper subset' of the distance travelled by Achilles."

This observation of Becker's, in my opinion, is quite correct, and it can be used to establish that pre-Euclidean mathematicians probably laid down *axiomata* for equality in order to avoid the paradoxes of Zeno. This is what I hope to show in the next chapter.

### 3.24 "THE WHOLE IS GREATER THAN THE PART"

Euclid's eighth axiom states that "the whole is greater than the part".[217] The first thing to notice about this principle is that it is rarely used in the *Elements*. (An example of one of the few places where it occurs is the proof of Proposition I.16.) Although it is quoted very seldom, it is rather similar to a formula which appears at the end of quite a number of Euclid's indirect arguments. The proof of Proposition I.6, for example, concludes in the following way: the false supposition which is to be refuted is shown to imply that "__ *will be equal to* __, *the less to the greater: which is absurd*".

The phrase quoted above (in italics) was most probably used in geometrical arguments long before the time of Euclid. It is quite common not only in his works but also in those of Autolycus of Pitane,

---

[217] *Elements*, Book I, Communes animi conceptiones VIII: τὸ ὅλον τοῦ μέρους μεῖζόν ἐστιν.

19*

who lived somewhat earlier.[218] Heiberg, in his Latin translation of the
*Elements*, always referred back to Axiom 8 when he came to the formula
in question.[219] This is an appropriate reference to make, of course,
since both statements say very much the same thing. Nevertheless,
it is worthwhile to investigate the relationship between them.

Tannery was the first to point out that the axiom and the formula
are *not* identical, but he offered no real explanation of this fact. He
seems to have thought that the former was, somewhat unfortunately,
abstracted from the latter at a relatively late date.[220] Apart from
this, however, he had nothing much to say about the matter.

The first task which faces us is to try to decide whether Axiom 8
was derived from the formula, as Tannery believed, or whether it was
the axiom which came first. In my opinion, this question can be
answered with a fair degree of certainty. If both possibilities are
examined seriously, it soon becomes apparent that one of them is
much more likely than the other.

Two notions, 'the *whole*' and 'the *part*', are contrasted in the axiom
("the whole is greater than the part"). The formula, on the other
hand, involves two entirely different concepts; 'the whole' is replaced
by the *greater* and 'the part' by the *less*. It is easy to account for these
substitutions on the assumption that the formula was obtained from
the axiom, for it is always the case in practice that the whole of a
thing is the greater and its part the less. Hence it is quite possible
that the formula which we find in Euclid developed from an earlier
one which dealt with the narrower concepts of 'whole' and 'part'.
*So there are good reasons for thinking that the axiom could have given
rise to the formula.*

If, on the other hand, one starts out by assuming that the formula
in some unspecified way was discovered first, it is extremely difficult
to explain how the axiom might have arisen subsequently. The
'greater' is not always a 'whole', nor is the 'less' always a 'part'.
There seems to be absolutely no reason why these more general concepts should have been replaced by narrower ones. In my opinion,

---

[218] For example, in *De sphaera quae movetur liber*, (ed. Hultsch), Lipsiae 1885,
Proposition 3 reads: ἴση ἄρα ἐστὶν ἡ ΔΗ (a line segment) τῇ ΔΖ (another line
segment), ἡ ἐλάσσων τῇ μείζονι, ὅπερ ἐστὶν ἀδύνατον.

[219] See, for example, Vol. I, p. 25 et passim.

[220] *Mémoires scientifiques* **2**, 54.

this indicates quite clearly that the axiom preceded the formula. The Greeks could easily have transformed "the whole is greater than the part" into "the greater cannot equal the less", whereas the reverse procedure is almost inconceivable.

We now want to find out what led to the formulation of Axiom 8 and when this took place.

The statement that "the whole is greater than the part" is at first sight so obvious that one cannot help wondering why such a banal *truth* was counted as an *axioma* (i.e. as an assertion which could not be accepted without reservation). Furthermore, one's surprise is only increased by the discovery that Euclid makes very little use of this axiom; instead, he often uses a principle which may very well have been suggested by the axiom but is not identical with it. Euclid, it seems, would have done better to give definitions of 'greater' and 'less'; for they would have been more useful to him than Axiom 8. The fact that he neglected to do so shows how closely he followed his original or, more likely, an earlier tradition.

In my opinion, the inclusion of "the whole is greater than the part" amongst Euclid's ἀξιώματα can only be explained in the following way. At one time the truth of this assertion must have been called into question and, because the consequences of its negation were found to be unacceptable, it was laid down as an unprovable principle. As a matter of fact, the preceding explanation is the correct one, and we are even in a position to say whose attack on the validity of this seemingly self-evident statement led to its adoption as an axiom.

Aristotle tells us[221] of a noteworthy argument which Zeno of Elea devised to show that "the half-time is equal to its double".[222] Unfortunately Zeno's account of his reasoning has been lost, and we know of it only from Aristotle. The latter, however, was interested in refuting the argument; his interpretation of it must therefore be treated with some caution. The conclusion reached by Zeno is a very paradoxical one; it asserts that a 'half' is equal to its 'double'. This is tantamount to saying that a 'part' is equal to the 'whole'. Given a period of time, its half is of course a *part*, and double that half is the *whole*. So we see that this paradox directly contradicts Euclid's eighth axiom.

[221] Aristotle, *Physics* Z9.239b33.
[222] Ibid., ἴσον εἶναι χρόνον τῷ διπλασίῳ τὸν ἥμισυν.

Our previous observations already make it seem reasonable to conjecture that Axiom 8 was laid down to avoid paradoxical conclusions like the one mentioned above. The paradox, however, is concerned with the nature of *time*, whereas Axiom 8 is a purely *mathematical* (or geometrical) principle. Although Zeno's argument undoubtedly had some influence on mathematicians, we are not yet in a position to say whether it was the direct cause of their decision to formulate this axiom. To answer this last question, we will need to examine the details of the argument.

That part of Aristotle's text which deals with Zeno's argument presents many philological difficulties, some of which have still not been solved satisfactorily.[223] Nevertheless, we can quite easily reconstruct Zeno's train of thought. Our task is facilitated by a certain diagram (Fig. 10 below) which is to be found in Simplicius' commentary on Aristotle. This diagram (which Simplicius borrowed from an earlier commentator, Alexander of Aphrodisias) serves to illustrate the argument.[224]

$$AAAA$$
$$BBBB \ \rightarrow$$
$$\leftarrow CCCC$$

Fig. 10

According to Aristotle, Zeno reasoned as follows: Suppose that there are three groups of bodies in a stadium (arranged as in Fig. 10) and that one group *(AAAA)* is at rest, while the other two *(BBBB* and *CCCC)* are moving at the same speed but in opposite directions. Suppose, furthermore, that the *B*'s and *C*'s begin to move at the same time from the positions shown in the figure and that they continue in motion until the first *B* reaches the last *A* (on the far right) and the first *C* reaches the first *A* (on the far left). Since the *B*'s and *C*'s are travelling at the same speed, they will clearly reach their respective goals at the same time. Now, if we want to measure the distance covered by these two groups of bodies or, what amounts to the same thing, the *time* which it took them to cover this distance, the magnitude

[223] Cf. J. Lachelier, *Revue de Métaphysique et de Morale* **18** (1910), pp. 345–55.
[224] Simplicius, 1016ff., see also 1019.27. The passage and the diagram both appear in Diels and Kranz, *Fragmente*, I.29, Zeno a28.

in question can be represented by $AA$, for in both cases the first body has passed exactly two $A$'s. Instead of measuring this period of time relative to the bodies at rest, however, we could just as well measure it relative to those in motion. It would then be represented by $BBBB$ (or $CCCC$) because the first $C$ has passed four $B$'s (and vice versa). Thus we are faced with a paradox; a fixed period of time can be measured either by two or by four letters. Hence we may conclude that "the half-time is equal to its double".

Before discussing the argument in more detail, I would like to point out one notable feature of it which has been ignored by previous commentators. I am referring to the fact that, although the letters $A$, $B$ and $C$ in Fig. 10 are supposed to denote *bodies* which are either at rest or in motion, half way through the argument they are used to denote certain *lengths* as well. It is clear that, if the time which the first $B$ (or $C$) takes to pass two $A$'s is represented by $AA$, these two letters $(AA)$ are being used to denote not only two bodies at rest but also a certain *length*, namely the distance travelled by $B$ (or $C$) in passing two such bodies.

One of the reasons why this fact is important is that it enables us to date the diagram found in Simplicius. We have already mentioned that Simplicius took this figure from an earlier commentary by Alexander of Aphrodisias (who lived at the end of the second century B.C.). However, the figure was devised long before the time of Aristotle. In fact, it must date back to the first half of the fifth century, the age of Zeno, for its author chose to symbolize lengths and *sums of shorter lengths* by groups of letters. This was a common practice in the fifth century B.C., but it became obsolete shortly thereafter.[225]

Another inference which can be drawn from our observation about the letters in Fig. 10 is the following. Zeno refers to the 'half-time' and its 'double', and he represents the former by two letters and the latter by four. These letters are obviously intended to denote *lengths*. ($AA$, the half-time, corresponds to one half of a *length*, and $BBBB$, its double, corresponds to the whole.) It is apparent, therefore, that Zeno's paradox must have had some connection with *geometry*, for geometry is a science which deals with the properties of *lengths* (or

---

[225] Archytas (in Boëtius, *De institutione musica*, ed. G. Friedlein, Lipsiae 1867, p. 287) denotes the sum of two lengths, $D$ and $E$, by $DE$.

lines). In view of this, it becomes even more likely that Euclid's eighth axiom was formulated in response to the paradox.

Let us now examine the 'fallacy' in Zeno's argument. Aristotle was firmly convinced that the entire argument was incorrect, and his pupils tried even more zealously to refute it. Eudemus, the first historian of mathematics, was perhaps the harshest critic of Zeno's reasoning. According to Simplicius,[226] "Eudemus asserts that Zeno's argument is totally absurd because it contains a blatantly obvious fallacy." It seems that Zeno had little luck with his paradox in antiquity. Even today, many scholars echo the judgment of Aristotle and his school. It is worthwhile, therefore, to take a closer look at this 'fallacy' as well as at the ingenious idea which lies behind it.

Zeno was right in thinking that, when bodies move at a constant speed $(c)$, the distance which they cover $(s)$ and the time which it takes them to cover this distance $(t)$ can be represented by a single *length*, whose parts will correspond to matching distances and times. However, his mistake was to ignore the fact that these bodies will cover twice the distance $(2s)$ in the same time $(t)$, if their speed is doubled. It is clear that if the $B$'s and $C$'s are moving in opposite directions at a constant speed $c$, their speed relative to one another will be $2c$. (They are only travelling at speed $c$ relative to bodies at rest.) Consequently, the distance which they cover in a fixed period of time will also double $(2s = 2c \cdot t)$. This means that, even though the 'half-length' $AA$ represents both distance and time, the 'whole length' $BBBB$ (or $CCCC$) can only represent distance.

It seems that Zeno commits a twofold error in his reasoning. On the one hand, he fails to mention that a speed which is supposed to remain 'constant' suddenly doubles, and on the other he claims that a certain length represents both distance and time, when in reality it is only a faithful representation of distance. This is the 'fallacy' which Aristotle discovered in Zeno's argument.

In my opinion, however, the preceding account does not do full justice to the argument. It is misleading to talk about a 'speed' $c$ which is 'doubled' and becomes $2c$. For this concept is nowadays defined as the ratio of distance travelled to required travelling time or, in other words, as the distance covered in a fixed amount of time.

---

[226] Simplicius on Aristotle's *Physics*, 1019.32ff.

Such a definition presupposes that both distance and time can be broken up into units and that some measure of *length* or *duration* can be assigned to these units. Zeno, however, held an entirely different view. He is criticized by Aristotle for maintaining that time is composed of *instants* (νῦν, literally 'nows').[227] Similarly, he regarded space as a collection of *dimensionless points*.[228] Both the 'half' and the 'whole length' were considered by Zeno to be *infinite sets of points*. Hence, when he asserted that the half-time was equal to its double, he was actually equating two *infinite sets*. If a modern mathematician wanted to establish this conclusion, he would probably argue as follows:

Suppose that $AB$ is twice as long as $CD$ and that both these lengths consist of infinitely many points. We say that $AB$ is 'equal' to $CD$ if there is a one to one correspondence between the points on $AB$ and those on $CD$. The existence of such a correspondence can be demonstrated by drawing the lines $AC$ and $BD$ and extending them until they intersect at a point $E$ (see Fig. 11). Now, let $x$ be an arbitrary point on $AB$; we correlate with $x$ the point $\xi$ at which the line $Ex$ intersects $CD$. Similarly, with each point $\mu$ on $CD$ we correlate the point $m$ obtained by extending the line $E\mu$ until it intersects $AB$. In this way, every point on $CD$ can be assigned to a unique point on $AB$ and vice versa. Considering the fact that $CD$ is shorter than $AB$, this situation appears to be somewhat paradoxical. It is described in the language of set theory by saying that $AB$ and $CD$ have the *same cardinality* even though one *(CD)* is a proper subset of the other *(AB)*.

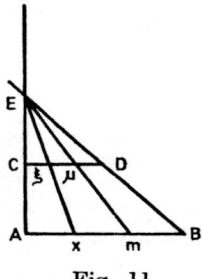

Fig. 11

[227] Aristotle, *Physics*, Z239b30.

[228] If we compare *Physics* Z9.239 and 2.223a21, it is apparent that Zeno thought not only of time as composed of infinitely many 'nows' (moments without durations) but also of space as composed of infinitely many points,

So it seems that Zeno was right; if time is made up of infinitely many instants (nows) and 'equality' is interpreted in a set theoretic sense (i.e. as cardinal equivalence), then the "half-time *is* equal to its double". No doubt, his main purpose in devising this paradoxical argument was to show the inconceivability (or inconsistency) of 'motion', 'space' and 'time'. But he also succeeded in demonstrating that our *intuitive concept* of 'equality' can only be applied to finite sets.

Aristotle's account suggests that Zeno was concerned exclusively with the nature of time. In my opinion, however, the Eleatics must have generalized the above argument to cover the case of geometrical *lengths* as well, for it is clear that the letters used to denote bodies serve at the same time as a measure of *distance*. Zeno's argument depends upon showing that the amount of time taken to travel a certain fixed distance is 'half of a length' or 'twice that half' according to whether the distance is measured by bodies at rest or bodies in motion. Nevertheless, the conclusion which he drew from it is valid for space as well as for time because the structure of the one is similar to that of the other. (Points in space correspond to instants of time.) In other words, the argument also establishes that the half-*length* (or distance) is 'equal' to its double.

There are three facts about the foundations of Euclidean geometry which indicate that the basic principles of this science were laid down expressly to avoid the paradox stated above:

(1) Euclid's first geometrical definition reads "A point is that which has no part." This means that *points* in geometry were exactly analogous to Zeno's *instants (νῦν)* of time.

(2) We quoted earlier a passage from Proclus[229] in which he defines "the line as *the flowing of a point* and . . . the straight line as *its uniform and undeviating flowing*". These definitions, however, are not to be found in the *Elements*. They give the impression that a 'line' is composed of *points* just as a 'number' is composed of *units*, and Zeno's paradox applies to such lines or lengths (i.e. to *infinite sets of points*). Euclid, who wanted to use the concept of 'infinity' as little as possible in geometry,[229a] gave a different definition of the relationship between

[229] See n. 187 above.

[229a] It is worth recalling Frajese's interpretation of *Philebus* 25a–b ('Platone e la matematica nel mondo antico'): "Si può seguire, almeno da un importante

points and lines: he asserted (Definition I.3) that "the extremities of a line are points".

(3) Axiom 8 states an empirical fact which was obtained by abstracting from *experience with finite sets*. Others of Euclid's axioms also seem to be concerned with laying down properties of *'equality'* which hold in general only for *finite sets*.

### 3.25 A COMPLEX OF AXIOMS

In the previous chapter it was argued that the formulation of Axiom 8 should be regarded as a response to Zeno's paradox about 'equality'. We now want to examine a few more of Euclid's axioms.

Heiberg, as is well known, rejected three of the first eight axioms (the fourth, fifth and sixth) as spurious. His authority for rejecting Axioms 5 and 6 was Proclus,[230] who considered these principles to be superfluous.[231] Indeed, Heiberg even thought that he had discovered interpolations in I.37 and I.38 because these propositions contain an almost literal restatement of Axiom 6.

Proclus and Heiberg are certainly right in maintaining that the axioms which they reject are not required by Euclid; the whole of his geometry can be built up without assuming these principles. Rather than discussing the question of their authenticity, however, let us first attempt to find out what led to their incorporation into the *Elements*.

Tannery believed that all of Euclid's principles were abstracted from the text of the *Elements as a kind of afterthought;*[232] but his view seems to be completely untenable. Much of what we find amongst the foundations of the *Elements* is the legacy of an earlier tradition and must already have been obsolete by Euclid's time. This remark may very well apply to Axioms 5 and 6, which state that equal magnitudes remain equal to one another when they are *doubled* or *halved*.

---

punto di vista, lo sviluppo della matematica greca come il succedersi di episodi di una lotta contro l'infinito. In questo passo Platon riafferma che la matematica è il dominio del finito."

[230] 196. 25ff. (in Friedlein's edition).

[231] See Euclides: *Elementa* (ed. I. L. Heiberg, Lipsiae 1883–1916), Vol. 1, p. 91.

[232] P. Tannery, *Mémoires scientifiques* **2**, 53.

As far as I know, the scientific literature of antiquity mentions only one thing which could possibly have prompted mathematicians to lay down these axioms. I am referring, of course, to Zeno's paradoxical claim that "the *half*-time is equal to its *double*". We saw in the last chapter that this assertion was illustrated by means of *lengths* (viewed as infinite sets of points); hence Zeno's argument also established that a *half*-length could be 'equal' to its *double*.

The mathematicians of antiquity must surely have attempted to refute Zeno's claim. Eleatic doctrine, however, prevented them from appealing to everyday experience or to the evidence of the senses. Furthermore, there is nothing wrong with Zeno's argument, so they could not have presented a counter-argument to show that his reasoning was fallacious. A 'refutation' such as Aristotle and his pupils gave, which simply disregarded Zeno's fundamental insight into the nature of 'infinite sets', would not even have been considered by anyone who had the slightest understanding of Eleatic dialectic. Therefore, it seems, that the paradox could only have been avoided by laying down axioms designed to block Zeno's argument. Of course this is at present nothing more than a conjecture on my part; nevertheless, it is worth taking seriously because it sheds new light on at least *four* of Euclid's axioms. The principles in question are listed below; if Zeno's paradox is kept in mind, they seem to form a coherent group of statements whose common purpose is to deny that a *half* can ever be equal to its *double*.

Axiom 5: "Things which are *double* of the same thing are equal to one another."

Axiom 6: "Things which are *halves* of the same thing are equal to one another."

Axiom 7: "Things which *coincide* with one another are equal to one another." (The lengths involved in Zeno's argument did not coincide; nevertheless, they were said to be 'equal'.)

Axiom 8: "The *whole* is greater than the *part*."

Of course, Axiom 8 without the other three suffices for the development of Euclidean geometry. Hardly any use is made of Axiom 7; hence it could be rejected on much the same grounds as Axioms 5 and 6.[233] My opinion is that Euclid included these four axioms because

[233] Cf. K. von Fritz, op. cit. (n. 2 above), pp. 76ff.

it was *traditional* to do so; this also explains why he gave definitions of 'straight line' and 'plane surface' and introduced such terms as ἑτερόμηκες, ῥόμβος, ῥομβοειδές, τρίπλευρα and πολύπλευρα (all of which are superfluous as far as his work is concerned).[234]

*

Our study of Euclid's axioms has led us to the following conclusions:

The axioms for equality which appear in the *Elements* are empirical assertions based upon experience with *finite sets*. Their validity is guaranteed only by *sensory* evidence; hence they could not have been accepted by the Eleatics, who required that all knowledge be obtained by purely intellectual means and without appealing to the senses. These principles were originally called *demands* (ἀξιώματα) because the other party in a dialectical debate had reservations about accepting them as a basis for further inquiry or, in other words, because their acceptance could *only* be *demanded*. (Euclid's axioms were no less incompatible with Eleatic teaching than these postulates were.)

After Plato's time, however, the essentials of Eleatic dialectic were no longer very well understood; hence the ancient term ἀξίωμα acquired a new meaning. Since it had always been used to refer to a group of principles which, from the viewpoint of common sense, were evidently valid, it came now to denote those statements whose truth was "*accepted as a matter of course.*" (Efforts were even made to justify the new meaning on etymological grounds.) Aristotle's dismissal of Zeno's paradoxes as mere sophistries undoubtedly helped to make this change possible. As a matter of fact, Aristotle seems to have been the ancient world's leading proponent of the idea that mathematics has to be based on "evidently true, indisputable and simple foundations".

We can now see why the old name ἀξιώματα was replaced in texts of the *Elements* by the phrase κοιναὶ ἔννοιαι. Tannery believed that it became necessary to make this substitution when the Stoics began using the word ἀξίωμα to mean any declarative statement.[235] In my opinion, however, he failed to mention the major reason why *axiomata*

---

[234] Cf. P. Tannery, *Mém. scient.* **2**, 540–4 and 48–63.
[235] Ibid., **2**, 62–3. See also Chapter 3.21 above.

gave way to *koinai ennoiai*. The latter term was preferred to the former mostly because it better described the new (Aristotelian) conception of 'axiom'; one would be even less inclined to doubt the 'evident truth' of axioms if they were called *common notions (κοιναὶ ἔννοιαι)*. Thus, in the course of time, the dialectical origins of both the term ἀξίωμα and the kind of principle which it had been used to denote were almost completely forgotten.

### 3.26 THE DIFFERENCE BETWEEN POSTULATES AND AXIOMS

One of the facts which has emerged from our discussion in the last few chapters is that *aitemata* and *axiomata* were synonymous. Furthermore, the manner in which they were used by every ancient mathematician except Euclid suggests that these terms were interchangeable with a number of other words having to do with the foundations of mathematics. In the later literature, however, attempts were made to distinguish between the various sorts of unproved principles upon which mathematics was based. For example, Book I of Archimedes' *On the Sphere and Cylinder* begins with *two* lists of such principles. The assertions in the first list are called ἀξιώματα and are obviously *definitions*, whereas those in the second are called λαμβανόμενα and seem to be related to Euclid's *postulates*. (For example, Archimedes' first λαμβανόμενον reads: "Of all lines which have the same extremities the straight line is the least." λαμβάνω δὲ τῶν τὰ αὐτὰ πέρατα ἐχουσῶν γραμμῶν ἐλαχίστην εἶναι τὴν εὐθεῖαν.)

Of course, no distinction of the kind described above was ever observed consistently in antiquity. Even Euclid's threefold division of principles into *definitions*, *postulates* and *axioms* was not adopted by any other ancient mathematician. It is true that Aristotle devised a system of *archai* which embodied his own views about what distinguished different kinds of basic principles. In my opinion, however, the historian of ancient mathematics has little to learn from Aristotle's treatment of this matter.

Proclus made several attempts to explain the difference between *aitemata* and *axiomata*, but none of these is entirely satisfactory from a historical point of view. Without going into the details of his explanations, I would like to mention two of them here.

(1) One of Proclus' remarks which seems to bear on Euclid's classification of first principles is the following: " . . . postulates are peculiar to geometry, while axioms are common to all sciences that deal with quantity and magnitude".[236] Euclid's *postulates (aitemata)* are indeed purely geometrical assertions. Despite what Proclus says, however, the *axioms* which are found in the *Elements* do not all apply to any science dealing with quantity and magnitude. There is at least *one* of them which only makes sense within the context of (plane) geometry.[237] I am referring to Axiom 7, "Things which coincide with one another are equal to one another." So, if the above distinction is taken seriously, some of Euclid's *axiomata* are not really axioms at all. But this means that Proclus' words describe a more or less accidental feature of the principles found in the *Elements;* they do not explain Euclid's system of classification. Perhaps Euclid did not use the terms *aitemata* and *axiomata* in the way that Proclus' remark suggests, and it was not until later that the equality axioms (or at least most of them) were seen to be a foundation for arithmetic as well.[238] Of course, nowadays, arithmetic is based upon a more comprehensive collection of such axioms. (For example, Euclid does not state that equality has to be reflexive and transitive, although modern mathematicians usually do so.) This, however, is not to say that Euclid's equality axioms played any role outside geometry when they were first formulated. As a matter of fact, I can find no traces of the conscious use of these principles in early Pythagorean arithmetic (Book VII of the *Elements*, for example). Before concluding this discussion of Proclus' remark, I would like to mention one further problem which it raises. The remark suggests that Euclid may have classified his principles according to their degree of generality. Had he done so, however, he would undoubtedly have placed the more general *axiomata* before the purely geometrical *aitemata;* yet in the *Elements* this order is reversed.

(2) Another explanation offered by Proclus is that the difference between *axiomata* and *aitemata* corresponds to the one between theorems

---

[236] Proclus (ed. Friedlein), 182.6ff; see also 58.7ff.

[237] It may be that Axiom 9, which is valid only for geometry, is spurious

[238] This was recognized already by Aristotle (*Posterior Analytics* I.10.76a37ff); cf. K. von Fritz, op. cit. (n. 2 above), p. 25.

*(θεωρήματα)* and problems *(προβλήματα)*. He enlarges upon this idea in the following passage:[239]

"A *postulate* prescribes that we construct or provide some simple or easily grasped object for the exhibition of a character, while an *axiom* asserts some inherent attribute that is known at once to one's auditors – such as that fire is hot, or some other quite evident truth about which we say that they who are in doubt lack sense organs or must be prodded to use them."

These words are instructive because they reveal the kind of distinction that was drawn between *aitemata* and *axiomata* after Euclid's time. However, they also reveal that at this late date the true origins of these two groups of principles had been forgotten.

From our previous discussion we see that Euclid's *postulates* differ against the Eleatic doctrine of the impossibility of *motion*, and they served to guarantee the *existence* (i.e. the *constructibility* by means of ruler and compasses) of certain geometrical forms. The axioms, on the other hand, were designed to avoid Zeno's paradoxes, and they laid down empirical facts about 'equality' which had been learned from experience with finite sets. Most of Euclid's fundamental principles were probably taken over unchanged from his precedessors (i.e. from those earlier mathematicians who had themselves compiled books of 'Elements'). So, acting in accordance with *tradition*, he retained the distinction between *axiomata* and *aitemata* even though it had lost much of its original importance by his time.

### 3.27 ARITHMETIC AND GEOMETRY

What we have learned so far in Part 3 of this book can be summarized as follows.

Eleatic philosophy had a decisive influence on the development of deductive mathematics. It was the teaching of Parmenides and Zeno which (in the first half of the fifth century, most probably) induced

---

[239] Proclus (ed. Friedlein), 181.5ff; see also 178.12ff. (The translation is by Morrow, Princeton 1970.) Cf. A. Frenkian, *Le postulat chez Euclide et chez les modernes* (Paris 1940), p. 19.

mathematicians to turn away from empiricism and to adopt the view that mathematical objects *exist only in the mind*. The method of indirect proof, so typical of mathematical reasoning, and the theoretical *foundations* of mathematics had their origins in Eleatic dialectic. Furthermore, all the terms which were used to denote fundamental mathematical principles came from there as well. Indeed, as we have seen, Euclid's very first *arithmetical* definition (VII. 1) presupposes the 'theory of the One' (*Republic* VII.525), which is in a sense identical with Parmenides' doctrine of 'Being'.

The definition of 'number' (VII.2) as "a multitude composed of units" marked a departure from Eleatic teaching. It is true that arithmeticians treated 'numbers' in much the same way that the Eleatics treated 'Being'; they emphasized that numbers were ideal entities which did not have visible or tangible bodies and could only be apprehended by the understanding. Arithmetic, however, required the existence of a 'plurality' (or, at least, of an 'ideal plurality'), whereas Eleatic philosophy admitted only the existence of the 'One'. In an earlier chapter we saw that the Eleatic problem of 'divisibility' took on a new meaning when numbers came to be regarded as multiples of the 'One'. This seems to have given rise to one of the most important and basic problems of pre-Euclidean arithmetic, namely the problem of divisibility for numbers.

When the foundations of *geometry* were laid, the gap between mathematics and Eleatic philosophy widened. The geometers of pre-Euclidean times had to formulate *postulates* which contradicted the Eleatic doctrine of the 'impossibility of motion' and *axioms* which characterized the relation of equality between finite sets (i.e. which avoided Zeno's paradoxes). If they had not done so, it would have been impossible to give a systematic development of geometry.

So there is an important difference between arithmetic and geometry. From an Eleatic point of view, the theoretical formulations of the former are much simpler than those of the latter. This is the reason why the ancients thought that arithmetic took precedence over geometry. As Proclus wrote:[240] "That geometry is a part of general mathe-

---

[240] Proclus (ed. Friedlein) 48.9ff: ὅτι μὲν οὖν ἡ γεωμετρία τῆς ὅλης ἐστὶ μαθηματικῆς μέρος καὶ ὅτι δευτέραν ἔχει τάξιν μετὰ τὴν ἀριθμητικὴν . . . εἴρηται τοῖς παλαιοῖς καὶ οὐ πολλοῦ δεῖται λόγου πρὸς τὸ παρόν. (The translation is by Morrow, Princeton 1970.)

matics and occupies a place second to arithmetic . . . has been asserted
by the ancients and needs no lengthy argument here."

It is relatively easy to maintain that numbers are ideal entities
which have no bodies. Any given number (23, for example) is readily
seen to be an abstraction once it has been distinguished from the
objects which it counts. A higher level of abstraction is reached when
properties (or sets) of numbers are considered instead of individual
ones. When it comes to *geometrical figures*, however, there is a case
to be made for saying that they are less abstract than numbers. Proclus
seems to have been fully aware of this, as the following remark of
his indicates:[241] "That numbers are purer and more immaterial than
magnitudes and that the starting point of numbers is simpler than
that of magnitudes are clear to everyone."

There is an obvious difference between a number and the letter,
numeral or straight line which is used to represent it. It is less clear,
however, that the 'ideal square' differs from a figure drawn to illustrate
it. Plato emphasized, of course, that geometers[242] "use the visible squares
and figures, and make their arguments about them, though they are
not thinking about them, but about those things of which the visible
are images. Their arguments concern the real square and a real diagonal, not the diagonal which they draw, and so with everything". Nevertheless, it was not so easy to distinguish the 'real' from the 'visible'
in geometry. Plato even found it necessary to reproach geometers
for confusing the two. He wrote:[243] "they talk in most ridiculous and
beggarly fashion; for they speak like men of business, and as though
all their demonstrations have a practical aim, with their talk of

[241] Ibid., 95.23ff: ὅτι μὲν οὖν ἀριθμοὶ τῶν μεγέθων αὐλότεροι καὶ καθαρώτεροι,
καὶ ὅτι τῶν ἀριθμῶν ἡ ἀρχὴ τῆς τῶν μεγέθων ἐστιν ἁπλουστέρα, παντὶ καταφανές.
(The translation is by Morrow, Princeton 1970.)

[242] Plato, *Republic* 510d: τοῖς ὁρωμένοις εἴδεσι προσχρῶνται καὶ τοὺς λόγους
περὶ αὐτῶν ποιοῦνται, οὐ περὶ τούτων διανοούμενοι, ἀλλ' ἐκείνων πέρι οἷς ταῦτα ἔοικε,
τοῦ τετραγώνου αὐτοῦ ἕνεκα τοὺς λόγους ποιούμενοι καὶ τῆς διαμέτρου αὐτῆς, ἀλλ' οὐ
ταύτης ἣν γράφουσιν. (The translation is by A. D. Lindsay, Everyman's Library
1950.)

[243] Ibid., 527: λέγουσι μέν που μάλα γελοίως τε καὶ ἀναγκαίως· ὡς γὰρ πράττοντές
τε καὶ πράξεως ἕνεκα πάντας τοὺς λόγους ποιούμενοι λέγουσιν τετραγωνίζειν τε καὶ
παρατείνειν καὶ προστιθέναι καὶ πάντα οὕτω φθεγγόμενοι, τὸ δ'ἔστι πού πᾶν τὸ
μάθημα γνώσεως ἕνεκα ἐπιτηδευόμενον. (The translation is by A. D. Lindsay,
Everyman's Library 1950.)

squaring *(τετραγωνίζειν)* and applying and adding, and so on. But surely the whole study is carried on for the sake of knowledge."

We can see that the foundations of geometry must have seemed much more problematic than those of arithmetic to the mathematicians of antiquity. In the next chapter, therefore, we shall examine the efforts which were made to reconcile geometry with the teaching of the Eleatics.


### 3.28 THE SCIENCE OF SPACE

On the face of it, the possibility of applying the basic tenets of Eleatic philosophy to geometry seems non-existent. The Eleatics flatly denied that there was such a thing as *space*.[244] Zeno, in particular, devised innumerable arguments to show that the concept of 'space' was inconsistent.[245]

If there is no space, however, there can be no science of space (i.e. there can be no geometry). Because they regarded space as contradictory, the Eleatics thought that it was a phenomenon like 'motion' which could not be apprehended by the understanding. Even they had to admit that motion could take place *in practice*. By denying its existence, the Eleatics were simply asserting that the concept of motion could not be grasped by consistent logical reasoning; they never tried to deny that our senses perceive movement everywhere. Similarly, when they asserted that there was no space, they meant that the notion of *space* was incoherent. In other words, the Eleatics thought that a 'science of space' (like a 'science of motion') could only be *based on sense perception*.

It appears that the earliest mathematicians adopted this view of geometry (the science of space). Iamblichus tells us that Pythagoras used to refer to geometry as *ίστορίη*.[246] Now the Pythagoreans were proud of their scientific achievements (especially the theory of num-

---

[244] *Theaetetus* 180e.

[245] Cf. W. Capelle, *Die Vorsokratiker*, Leipzig 1935, pp. 172f.

[246] Iamblichus, *Vita Pythagorica* 89. See also my paper 'Deiknymi, als mathematischer Terminus für beweisen' (referred to in n. 6 above) and Frenkian's paper in *Maia*, N. S. 11 (1959), 243–5.

bers) and called them μαθήματα.[247] The word ἱστορίη, on the other hand, was reserved for empirical knowledge which had been acquired by observation.[248] We can conclude from what Iamblichus says that geometry was at first considered to be a practical and experimental science (ἱστορίη rather than a true *mathema*).

Proclus, in the passage quoted above,[249] also seems to be saying that geometry did not become part of general mathematics until relatively late, and that even then it took second place to arithmetic. Plato agreed with him about the relative positions of geometry and arithmetic.[250] In fact, Plato's works enable us to show conclusively that it was the Eleatics who first made this assessment of geometry and posed the problems of 'space' and a 'science of space'. To see this we will need to recall Plato's classification of the various kinds of knowledge.[251]

\*

It is well known that Plato distinguished between the world of becoming or the perceptible world (ὁρατόν) and the world of being (νοητόν). In the *Republic* this distinction is used to develop an interesting theory of *knowledge*, *ignorance*, and *belief*, which Plato explains in the following way.[252]

He who has knowledge knows something (ὁ γιγνώσκων γιγνώσκει τι); furthermore, he knows "something which is (ὄν). For how could that which is not be known (πῶς γὰρ ἂν μὴ ὂν γέ τι γνωσθείη)? So *knowledge* is associated with what is (ἐπὶ μὲν τῷ ὄντι γνῶσις ἦν), and *ignorance* with what is not (ἀγνωσία δ'ἐξ ἀνάγκης ἐπὶ μὴ ὄντι). For that which lies between what is and what is not (i.e. for that which inhabits the

---

[247] B. L. van der Waerden, *Math. Ann.* **120** (1947/49), 127. Aristotle (Metaphysics A5) explicitly states that the Pythagoreans were the first people to concern themselves with μαθήματα. Cf. also K. Reidemeister, *Das exakte Denken der Griechen*, Hamburg 1949, p. 52.

[248] B. Snell, *Die Ausdrücke für den Begriff des Wissens in der vorplatonischen Philosophie*, Berlin 1924, pp. 59–71; A. Frekian, *Revue des Études Indoeuropéennes*, Bucarest–Paris 1938, pp. 468–74.

[249] See n. 240 above.

[250] Plato, *Epinomis* 990c–d.

[251] Cf. my paper 'Eleatica' (see n. 33 above), pp. 98ff.

[252] *Republic*, V 476e–477b.

world of *becoming* and *passing away*) there has to be discovered a corresponding intermediate between knowledge and ignorance *(ἐπὶ δὲ τῷ μεταξὺ τούτῳ μεταξύ τι καὶ ζητητέον ἀγνοίας τε καὶ ἐπιστήμης)*. This intermediate is called *belief (δόξα)*; it is "darker than knowledge and brighter than ignorance".

In my opinion, Plato's distinction between *knowledge, ignorance* and *belief* certainly has its roots in Eleatic philosophy. Parmenides says the following about his 'second way of inquiry':[253]

"The other, that *it is not*, and that *it is* bound *not to be:* this I tell you is a path that cannot be explored, for *you could neither recognize that which is not (οὔτε γὰρ ἂν γνοίης τό γε μὴ ἐόν)*, nor express it.[254]"

If, as Parmenides claims, what is not can be neither recognized nor expressed, it obviously cannot be an object of *knowledge;* only what *is* can be known and recognized. When Plato connects *knowledge* with Being and *ignorance* with Not-Being, he is drawing on one of Parmenides' ideas.

It can also be shown that Plato's account of *belief (δόξα)* derives from Parmenides. The latter discusses his 'third way of inquiry' (i.e. the *beliefs* of mortals – *βροτῶν δόξαι*) in the following terms:[255]

"But next I debar you from that way along which wander mortals knowing nothing, two-headed, for perplexity in their bosoms steers their intelligence astray, and they are carried along as deaf as they are blind, amazed, uncritical hordes, by whom To Be and Not To Be *(τὸ πέλειν τε καὶ οὐκ εἶναι)* are regarded as the same and not the same."

I have argued elsewhere[256] that Parmenides rejected *δόξα* (appearance, illusory knowledge, belief) because he considered it to be inconsistent.

[253] See my paper 'Zur Geschichte der Dialektik des Denkens', pp. 54ff. referred to in n. 33 above).

[254] Diels and Kranz, *Fragmente*, I.28b, fragment 2.5–8. The translation is by Kathleen Freeman.

[255] Fragment 6.4–9.

[256] See 'Zum Verständnis der Eleaten', pp. 262–5 (referred to in n. 33 above).

The contradiction which he saw in the concept of δοκεῖν involved the notions of Being and Not-Being (εἶναι καὶ οὐκ εἶναι). Plato associates *belief* with those things which lie between what is and what is not[257] (in the world of *becoming* and *passing away*): accordingly, he places it between *knowledge* and *ignorance*.

It seems fair to say that Plato, in the passage discussed above, is merely recapitulating Parmenides' view of *knowledge, ignorance* and *belief*.

*

My reason for giving this brief account of some features of Plato's epistemology is that these have to be kept in mind, if we are to make sense of two important passages in which he talks about *geometry*.

From what has been said above, it should be apparent that Plato related *knowledge* and *belief* to the world of Being (νοητόν) and the perceptible world (ὁρατόν) respectively. *Belief, δόξα,* was associated with sense perception and the world of *becoming* and *passing away*, whereas the domain of *true knowledge* was something *invisible* and accessible only to *reason*. The latter was often called *Being* by both Parmenides and Plato. Not only were these two philosophers in basic agreement about the nature of knowledge, but they also shared a common terminology. To substantiate this claim, I would need to quote some long passages from the *Timaeus* in the original Greek. Instead of doing this, however, I shall content myself with quoting the following translation of a small part of Plato's text:[258]

"There is one kind of being which is always the same, uncreated and indestructible (ἀγέννητον καὶ ἀνώλεθρον), never receiving anything into itself from without, nor itself going out to any other (οὔτε εἰς ἑαυτὸ εἰσδεχόμενον ἄλλοθεν, οὔτε αὐτὸ εἰς ἄλλο ποι ἰόν), invisible and imperceptible by any sense (ἀόρατον δὲ καὶ ἄλλως ἀναίσθητον), and of which the contemplation is granted to intelligence only (ὃ δὴ νόησις εἴληχεν ἐπισκοπεῖν). And there is another nature of the

---

[257] *Republic*, V 478d 5–9.

[258] *Timaeus* 52a–b; the translation is by Benjamin Jowett (*The Dialogues of Plato*, 3rd edn, Oxford 1892).

same name with it, and like to it, perceived by sense, created always in motion, becoming in place and again vanishing out of place, which is apprehended by δόξα and sense (δόξῃ μετ'αἰσθήσεως περιληπτόν)."

Taken out of context, this passage seems to contain nothing more than a restatement of the differences between the objects of *knowledge* and those of *belief*. When it is read together with its continuation, however, it provides us with some valuable new information.

In the *Timaeus* Plato is concerned not only with the *perceptible (horaton)* and *ideal (noeton)* worlds but also with a *third* world,[259] which he calls 'space'. Like the world of Being *(noeton)*, space is eternal and indestructible, nevertheless motion, generation and destruction take place in it.[260] Unlike the other two worlds, however, space is apprehensible by a kind of 'bastard reasoning' (μετ' ἀναισθησίας ἁπτὸν λογισμῷ τινι νόθῳ)[261] and not by pure intelligence or sense perception. Other passages in Plato's work, especially two which are to be found in the *Republic*,[262] attest to the fact that he identifies 'bastard reasoning' about space with geometry. The two passages in question contain references to the faculty employed by geometers; this is called διάνοια, and it is said to lie somewhere between mere *doxa* (i.e. perception by means of the senses) and the purely intellectual power of Reason *(νοῦς)*.

I believe that Plato's ideas about geometrical 'reasoning' arose as a result of the following historical circumstances.

The Eleatics flatly denied the existence of *space*. They showed that all phenomena belonging to the perceptible world, especially those which involved such concepts as 'motion', 'change', 'generation' and 'destruction', were contradictory. This led them to the realization that *space* and *time* were also inconsistent, for these two concepts are closely related to the others. (In particular, their arguments against 'motion' forced them to conclude that the notion of 'space' was

---

[259] *Timaeus* 50c–d: ἐν δ'οὖν τῷ παρόντι χρὴ γένη διανοηθῆναι τρίττα, τὸ μὲν γιγνόμενον, τὸ δ'ἐν ᾧ γίγνεται, τὸ δ'ὅθεν ἀφομοιούμενον φύεται τὸ γιγνόμενον.

[260] Frenkian (*Le Postulat chez Euclide et chez les modernes*, Paris 1940 p. 25) discusses the expressions used to describe "space" and refers to Zeller's *Die Philosophie der Griechen* II.1 (4th ed. 1889), note 722.

[261] *Timaeus*, 52b.

[262] *Republic*, VI 511d–e and VII 533e–534a.

incoherent.) As far as the Eleatics were concerned, however, anything contradictory was inconceivable and could not *be;* hence the very existence of 'space' was called in question.

Plato's conception of space evolved from a later and more refined version of the train of thought sketched in the preceding paragraph. At his time, space was no longer regarded as something intermediate between Being and Not-Being, a mere phenomenon which could be generated and destroyed. Although it was described as the 'receptacle' or 'nurse' of generation,[263] it was also considered to be no less indestructible *(φθοϱὰν οὐ πϱοσδεχόμενον)*[264] than the 'ideal' *(νοητόν).*

Thus 'space' was thought to have a kind of *dual* nature; on the one hand it was eternal and indestructible, and on the other it was inextricably bound up with the phenomena of the perceptible world. In view of this, we should not be surprised at the fact that the epithet 'bastard' was applied to geometry, 'the science of space'.

Something about what Plato means by 'bastard reasoning' can be learned from a passage in the *Republic*[265] where he points out that the nature of geometry is in flat contradiction to the language used by geometers. Plato has in mind such locutions as *squaring, applying* and *adding,* which, according to him, give the erroneous impression that the geometer is engaged in a practical activity. He goes on to say that the aim of geometry is to gain knowledge of eternal reality; it is not to effect the construction of ephemeral figures.

### 3.29 THE FOUNDATIONS OF GEOMETRY

In the previous chapter we discussed briefly some of Plato's remarks about space and the science of space. I believe that what he has to say about these subjects helps us to understand the process which led to the formulation of theoretical foundations for geometry.

The first step in this process necessarily involved a *revision* of Eleatic views about 'space'. There could be no science of space as long as the very existence of space was 'denied'. (The Eleatics, it will be recalled, rejected space, together with all phenomena belonging

---

[263] ὑποδοχὴ γενέσεως or τιθήνη τῆς γενέσεως; see n. 260 above.
[264] *Timaeus,* 52a.
[265] See n. 243 above.

to the perceptible world, as contradictory.) Geometry was at first nothing more than a kind of ἱστορίη; i.e. it was a body of empirical knowledge obtained by means of the senses and based principally on visual evidence.

Mathematicians must soon have realized, however, that, although awareness of space is invariably associated with awareness of those things which are located in space (and which move, change, are generated and destroyed in it), one can abstract from one's experiences and think of *space* itself as independent of these phenomena. Considered in this purely abstract way, it appears to be somewhat similar to the 'ideal' *(noeton)*.

In fact, the attempt to develop an *extrasensory* conception of space seems to have marked the beginning of theoretical geometry. As we have seen, Plato says that space is apprehensible not by sense perception but by a kind of 'bastard reasoning'.[266] After starting out as a kind of ἱστορίη whose chief tool was the sense of *sight*, geometry became the science of *space itself*. Less emphasis was placed on the role of sense perception in this science, since space was now regarded as something unchanging, indestructible and purely abstract, which was distinct from the objects located in it.

The renunciation of sense perception as a means of acquiring geometrical knowledge manifested itself in a number of interesting ways. For example, every effort was made to exclude the concept of *motion* from the definitions in Book I of the *Elements*.[267] We know from Proclus that a line was sometimes defined as 'the flowing of a point' and a straight line as 'its uniform and undeviating flowing'. However, these, are not the definitions given by Euclid; his are formulated in such a way as to avoid any mention of motion. As a matter of fact, the so-called 'geometry of motion' does not seem to have achieved respectability until after Plato's time.[268] It has even been argued that efforts were made to banish references to *visible* properties from geometry. Heath, for example, believed that Euclid's definition of a straight line was an attempt to explain *without any appeal to sight* what it meant to be straight.[269]

[266] *Timaeus*, 52b.
[267] J. E. Hofmann, *Geschichte der Mathematik*, vol. 1 (Berlin 1953), p. 32.
[268] Cf. A. D. Steele, op. cit. (n. 183 above), p. 308.
[269] T. L. Heath, *A History of Greek Mathematics* (Oxford 1921), Vol. 1, p. 373.

Inevitably, however, all the attempts which were made to treat 'space' as something close to the 'ideal' *(noeton)* ended in failure. There were certain seemingly contradictory facts about space which simply could not be handled within the framework of Eleatic philosophy.

The mere fact that *space* presents itself to the human mind as something *infinitely divisible* posed a considerable problem.[270] It led the Greeks to think that there were no 'least magnitudes' in geometry; every 'number' could be built up out of indecomposable *units*, but the same could not be said of space. In other words, they came to the conclusion that there was nothing in geometry which corresponded to the *unit* in arithmetic. As Proclus wrote:[271] "in geometry a *least magnitude* has no place at all. Peculiar to geometry are the propositions ... about irrationals (for the irrational *(τὸ ἄλογον)* has a place only where infinite divisibility is possible)."

Arithmetic was based on *numbers* and the *One*. These were ideal and completely abstract forms; hence they were immaterial and contained no contradictions. There was, however, no such simple starting-point for geometry. Proclus described the situation in the following way:[272] "That numbers are purer and more immaterial than magnitudes and that the starting-point of numbers is simpler than that of magnitudes are clear to everyone."

These difficulties effectively prevented Euclid from formulating the most basic geometrical definitions in a wholly consistent manner. He asserted that *"A point is that which has no part"* and *"A line is breadthless length"*. An Eleatic philosopher, however, would have regarded these statements as grounds for denying the existence of space and, hence, for doubting the possibility of a true science of geometry.

The infinite divisibility of space was by no means the only difficulty which stood in the way of those who wanted to lay the theoretical foundations of geometry. However abstract the concept of space may have been, it was inseparable from the inconsistent concept of *motion*. Geometrical 'points' or 'least magnitudes' could not be obtained without 'infinite division', a procedure characterized as irrational *(ἄλογον)*. On the other hand, any attempt to obtain 'lines' from 'points'

---

[270] See A. A. Fränkel, *Abstract Set Theory* (Amsterdam 1953), p. 9–13.
[271] Proclus (ed. Friedlein), 60.11ff.
[272] Ibid., 95.23ff.

(in the same way that 'numbers' were obtained from the 'One' by multiplying it) necessarily involved 'motion'. Furthermore, to study space it was necessary to consider practical questions (such as whether or not two finite sets were equal) which could be answered only with the help of the senses. The Eleatics, however, repudiated the idea that true knowledge could be based on sense perception.

In my opinion, the Greeks may very well have been led to devise *axiomatic foundations* (initially just for geometry) as a response to the difficulties described above. They must have felt the need to lay down certain empirical facts which were indispensable to the construction of a science of space, even though these facts did not satisfy the Eleatic requirement that knowledge be acquired in a purely intellectual way. Their first task was to make it clear that the forms of geometry (lines, points of intersection, angles, figures, etc.) *were not at all the same as those perceived by the senses;*[273] true geometrical forms were *ideal* entities (like numbers) whose visible counterparts served only to represent them. The definitions were intended to eliminate as many sensible features as possible from geometry, and the *axioms* and *postulates* discussed earlier were also aimed at making the foundations of this science purely abstract.

It is noteworthy, however, that in antiquity Euclid's *geometrical* definitions, postulates and axioms were never considered to be a completely satisfactory foundation. Plato observed that *geometers* did not bother to give any further account of the assumptions on which their science was based.[274] Furthermore, they were even accused of basing their investigations on *false* suppositions. Aristotle, who attempted to defend geometers against this charge, treated it as a senseless and uninteresting objection. He wrote:[275]

"Nor are the geometer's hypotheses false, as some have held, urging that one must not employ falsehood and that the geometer is uttering falsehood in stating that the line which he draws is a foot long or straight, when it is actually neither. The truth is that the geometer does not draw any conclusion from the being of the par-

[273] *Republic,* VI 510d.
[274] Ibid.
[275] *Posterior Analytics* I.10; the translation is by G. R. G. Mure (Oxford University Press 1928).

ticular line of which he speaks, but from what his diagrams symbolize."

As described by Aristotle, the objection raised against the methods of geometry was hardly worth refuting. So there is some reason to think that he may have misunderstood it. This seems even more likely to have been the case, when we consider the fact that his refutations of Zeno's arguments ignored completely the ingenious idea which lay behind them.

What seems to be the same objection is raised in a different form by the neo-Platonist philosopher Proclus, who believed that geometry had to be led *"out of Calypso's arms*, so to speak, to more perfect intellectual insight". He went so far as to call upon mathematicians to undertake this task in the future.[276] On the basis of the passage in which this view is expressed, Hartmann argued that Proclus should be credited with anticipating the discovery of analytic geometry (an event which did not take place until the time of Descartes).[277]

*

I would like to conclude this chapter by recalling some of the points which were made earlier.

At the beginning of Part 3, I stated the conjecture that new techniques of proof, which appealed to something other than visible evidence, were developed as a result of a *geometrical* discovery. I asserted that Greek science underwent a decisive change which was very closely connected with the discovery of linear incommensurability. This change manifested itself in two ways; on the one hand, *purely theoretical methods of proof* were introduced into mathematics, and, on the other, *empiricism and the visual were rejected*. When Greek mathematicians first came across geometrical lengths for which they could find no common measure, they were compelled to seek new techniques of proof. Although the discovery of such lengths undoubtedly prepared the way for the discovery of incommensurability, the *existence of*

[276] Proclus (ed. Friedlein), pp. 54–5.
[277] N. Hartmann, *Des Proklus Diadochus philosophische Anfangsgründe der Mathematik* (Giessen 1909), p. 44. See also A. Speiser, *Die mathematische Denkweise* (Basel–Stuttgart 1952), pp. 64f.

incommensurability could not be conclusively proved by practical or empirical methods. Hence a complex of problems associated with incommensurability made it necessary to adopt Eleatic techniques of proof in geometry.

The application of Eleatic methods to particular geometrical problems may well have suggested the idea that an Eleatic approach could be taken to the whole of geometry. It turned out, however, that the teachings of the Eleatics were not altogether compatible with this branch of mathematics and, as a result, it was found necessary to provide Euclidean geometry with theoretical foundations.

My claim is that the construction of mathematics as a deductive system came about because of certain problems encountered in *geometry*. It is true that Eleatic doctrine can be applied more easily to arithmetic than to *geometry* and that the Greeks therefore regarded arithmetic as the superior science; however, this ranking was only a *theoretical* one. Euclid's mathematics is predominantly geometrical in character; even his arithmetic takes a geometrical form. This should not surprise us in view of the fact that the problems which caused mathematicians to break with Eleatic philosophy came principally from *geometry* and the outcome of this break was a theoretical foundation for *geometry*.

### 3.30 A RECONSIDERATION OF SOME PROBLEMS RELATING TO EARLY GREEK MATHEMATICS

Part 3 of this book has been devoted to an explanation of the origins of Euclidean axiomatics and the way in which deductive mathematics came into being. The conclusions which we have reached may very well have some bearing on other historical problems concerning early Greek mathematics. I do not want to go into details about any of these here, but I would like to discuss two of them briefly.

I

An important question in the history of ancient mathematics is how *mathematical existence* came to be a problem. As far as I can see, there is no adequate formulation of this question in the literature. Many scholars have addressed themselves to it, basing their research on observations which are at least partially correct, but their efforts have served mainly to *obscure* the true nature of the problem.

It is generally acknowledged that the noted Danish scholar, H. G. Zeuthen, was the first person to clarify the principles used by the ancients in their *proofs of existential assertions*.[278] Becker, for example, wrote:[279]

"According to his [Zeuthen's] research, the one and only method of proving existence was *construction*, furthermore, since ancient mathematics consisted only of geometry (both arithmetic and algebra were presented in a geometrical form), this meant construction of *figures*. All such constructions were based on two fundamental ones, the drawing of a straight line between two given points and of a circle with given radius and centre. That these constructions are *possible* or, what amounts to the same thing, that the figures yielded by them *exist* is precisely what is asserted by Euclid's first two postulates ..."

The individual statements in this quotation may seem at first sight to be both clear and convincing, but there is an important weakness in the position which they represent. Although the Greeks appear to have put most of their mathematics (including the arithmetic found in the *Elements*) in geometrical form, it is extremely doubtful whether the one and only method of proving existence in arithmetic was *construction*. If we keep in mind that Zeuthen meant by 'construction' the 'construction of figures', the weakness of his position becomes even more apparent.

In my opinion, it is quite clear that Zeuthen's view cannot possibly be the correct one. Consider, for example, the following two propositions of Euclidean arithmetic.

*Elements* VII.31: "Any composite number is measured by some prime number."

*Elements* IX.20: "Prime numbers are more than any assigned multitude of prime numbers."

[278] H. G. Zeuthen, *Math. Ann.* **47** (1896), pp. 222–8.
[279] O. Becker, *Mathematische Existenz, Untersuchungen zur Logik und Ontologie mathematischer Phänomene* (Halle 1927), p. 130.

The proofs of both these propositions are undoubtedly *existential*. The first one shows that, given any composite number $a$, there *exists* a prime number which divides it. Similarly, to prove the second theorem, it has to be shown that no assigned (i.e. finite) multitude of prime numbers can contain all of them; in other words, given any (finite) set of primes, there *exists* at least one prime number which is not a member of it.

There is nothing which even resembles a *geometrical construction* in Euclid's proofs of these two propositions. (In fact, it is debatable whether the proofs should be described as constructions of any kind.) If we are to maintain the view that *construction* was the only method used by the Greeks to prove existence, the notion of construction cannot be limited to geometry; it must be given a much broader interpretation despite what is asserted in the passage quoted above.

Frajese has pointed out other weaknesses in Zeuthen's position.[280] One of his objections, which deserves special mention, is the following: There is general agreement amongst both ancients[281] and moderns[282] on the fact that Euclid was a 'Platonist'. But the idea that geometrical constructions can serve as proofs of existence was completely foreign to Plato.[283] In the final analysis, such constructions are carried out in the domain of sensible objects, and Plato thought that 'existence' was to be found elsewhere. (According to his view, what *existed (τὸ ὄν)* was not in the world of the *horaton* but in that of the *noeton*.) – There is no doubt that these remarks of Frajese's are very much to the point.

In the light of our investigations, the historical problem of *mathematical existence* can be treated in an entirely new way, for we know how the Eleatics regarded existence and, of course, how their philosophy influenced Plato's view of it. Furthermore, we have seen that deductive mathematics would never have come into being, had it not been for the Eleatics. Hence it seems reasonable to attempt an explanation of the problem of mathematical existence which starts out from the problem of 'existence' in Eleatic philosophy.

[280] A. Frajese, *La matematica nel mondo antico* (Roma 1951), p. 92.
[281] Proclus (ed. Friedlein), 68.20ff.
[282] A. Frajese (op. cit. (n. 280 above), p. 95) mentions as examples U. von Willamowitz-Moellendorff (*Platon* 3rd edn, Berlin 1929, p. 754) and E. Sachs (*Die fünf platonischen Körper*, Berlin 1917, p. 159).
[283] A. Frajese, op. cit. (n. 280 above), p. 95.

The Eleatics never doubted that *Being (τὸ ὄν)*, or *the One (τὸ ἕν)*, existed, but they did not establish this fact directly. Instead, Parmenides gave an indirect proof of it. He started out by supposing that *Being did not exist* and then went on to show that this supposition could not possibly be true because it was inconsistent. Thus a proof by contradiction was thought sufficient to establish the existence of *Being*.

Existential proofs in arithmetic were similar to the argument sketched above except that they involved an abstract notion of *many* (i.e. number) which was obtained by 'multiplying' (in an ideal sense) the Eleatic *One*. The Eleatics had originally denied the existence of *plurality* and had thus ruled out the possibility of arithmetic. Therefore, it was found necessary to base arithmetic on the following definition (*Elements* VII.2) of number: "A number is a multitude composed of units." This definition is really an unproved proposition concerning *the existence of numbers;* it was intended to provide a firm foundation for the whole of arithmetic, and, in addition, it enabled arithmeticians to make use of Parmenides' indirect method in their proofs of existence.

One of the techniques which Euclid employed to establish the existence of arithmetical forms (i.e. numbers) was proof by contradiction; he did not use constructions. Proposition VII.31, for example, asserts that every compound number has a prime divisor and is proved by showing that a contradiction follows from its negation. (Any compound number which has no prime divisor must have an infinite decreasing sequence of divisors, and this is impossible in the domain of numbers because it contradicts Definition VII.2.) Thus, given any compound number, the existence of a prime number which divides it is established indirectly. (Some modern intuitionists would deny that a proof of this kind establishes the existence of a number,[284] but their views need not concern us here.) Proposition IX.20 is proved in a similar manner; given a finite set of primes, the existence of a prime number which lies outside it is established by showing that, if no such number exists, a contradiction follows.

---

[284] A. Robinson (in *Problems in the Philosophy of Mathematics* ed. by I. Lakatos, Amsterdam 1967, p. 11) has pointed out that "A contemporary intuitionist would reject this proof as it stands, since it tries to establish the proof of a positive, existential statement."

It is apparent, therefore, that *construction* was *not* the one and only method used in Greek mathematics to prove existence. Arithmetical proofs of existence, even when they had a geometrical form, were not really constructions at all; they depended upon purely logical considerations and were frequently indirect.

The problem of *mathematical existence* took on a different meaning in geometry, the science of space. The Eleatics, whose ideas figured so prominently in deductive mathematics, did not believe in space. Unlike *numbers* and the *One*, *space* was thought to lie outside what Parmenides called the sphere of *Being* and Plato described as the world of *noeton*. Both Plato and the Eleatics had some doubts about *the existence of space* and were certainly less sure about it than they were about the existence of *numbers* and the *One*. Geometrical existence was more a matter of perception than reflection; it was more empirical than ideal. This explains why Euclid's geometrical definitions are not all of the same kind. He defines point and line, for example, in a purely *speculative* manner ('A point is that which has no part' and 'A line is length without breadth'), whereas his definitions of angle and various geometrical figures simply describe the *appearances* of these things.

Even though they were frowned on by Plato and the Eleatics, practical and empirical considerations had to be granted a role in geometry. For this reason geometrical existence had to be established by a practical and empirical method, namely *construction*.

It seems that Zeuthen was right in maintaining that construction was an important (perhaps even the most important) method of proving existence in *geometry*. Euclid's first and third postulates are *existential statements* designed to ensure that two fundamental constructions can be carried out. Yet the fact that these statements were called *aitemata* indicates to me that the Eleatics at one time *doubted the existence* of those forms which could be constructed on the basis of the postulates.

These doubts were shared by later philosophers who had been subjected to the influence of Plato and the Eleatics. It remained questionable whether geometrical figures actually 'existed' or whether existence was a mere 'appearance', the object of a 'belief'. (The Greek word δόξα means both 'appearance' and 'belief'.) The existence of these figures was thought to involve a contradiction. According to

Eleatic/Platonic philosophy, no visible or perceptible object (least of all one which was obtained as the result of a construction, a kind of *motion*) could be regarded as consistent; such objects were assigned a position 'between Being and Not-Being' similar to the one between *knowledge* and *ignorance* which was occupied by '*doxa*'.

As theoretical geometry became more and more independent of Eleatic philosophy, however, philosophical questions about the existence of those geometrical forms which were obtained by construction receded into the background.

(One further question, which ought at least to be mentioned here, is whether Euclid's first three postulates had at one time the kind of special significance which it is often claimed that they did. Becker has written:[285] "It appears that in the school of Plato constructions were carried out, wherever possible, solely by means of ruler and compasses, whereas previously the use of verging *(νεῦσις)* had also been allowed."

We may ask whether it was in fact the Platonists who wanted to limit themselves to the first three postulates and, if so, how such a restriction fits in with the *teachings of Plato*. Certainly, the above quotation gives the impression that the use of just a ruler and compasses was an *innovation* made by the Platonists. This seems unlikely to have been the case, however. On the basis of Becker's own research, it can be shown that Euclid's first three postulates probably date back to the fifth century B.C. It might reasonably be argued, therefore, that by Plato's time observance of the restriction in question was already an *ancient custom* to which his followers may only have adhered for the sake of *tradition*. – However, I do not want to suggest that I can give a definitive solution to this problem. It is mentioned only as an example of the kind of fundamental question about the history of mathematics which still remains to be investigated.)

II

I would now like to turn to another interesting question which is worth reconsidering in the light of what we have learned in this book. It concerns the relationship between the philosophy of Plato

[285] O. Becker, *Das mathematische Denken der Antike* (Göttingen 1957), p. 20.

on the one hand and deductive mathematics on the other. Let me begin by outlining the kind of treatment which this question has been given in the literature (and, as a matter of fact, which it still receives from some modern scholars).

In 1913 Zeuthen published an important paper under the title 'Sur les connaissances géométriques des Grecs avant *la réforme platonicienne.*'[286] All that interests us here is this title, for it suggests that at some stage Greek geometry underwent a so-called 'Platonic reform'. In fact, Zeuthen was of the opinion that Plato himself (if not directly, then at least through his pupils) exercised a very important influence on geometry, which resulted in a transformation of the subject.

Twelve years after the appearance of Zeuthen's paper, Töplitz gave the following account of how most historians explained the relationship between Plato and mathematics.[287]

"Plato, of course, did not make any mathematical discoveries; the tradition which attributes the discovery of the dodecahedron to him is not to be taken seriously. *He did, however, influence the general direction of mathematics. The axiomatic structure of the 'Elements', the requirement that only ruler and compasses be used for constructions, the analytical method, these are all Plato's work. Theaetetus and Eudoxus, the greatest mathematicians in his circle, created Euclidean mathematics under his influence.*"

We can see from the words in italics that the 'Platonic reform' which Zeuthen regarded as a milestone in the history of ancient science was thought to have given rise to the whole body of systematic and deductive mathematics produced by the Greeks before and during the time of Euclid. In 1927, Becker, expressed his agreement with Zeuthen's views and asserted that Plato's achievements marked the beginning of a new epoch in the history of mathematics. He wrote:[288]

"Broadly speaking, mathematics before Plato can be said to have concerned itself with visual shapes (Zeuthen described it as a 'geom-

[286] *Oversigt det kgl. Dansk Videnskabernes Selskabs Forhandlinger*, 1913, No. 6, pp. 431–73.
[287] Töplitz, 'Mathematik und Antike' *Die Antike* 1 (1925), 175–203.
[288] *Mathematische Existenz*, p. 250 (see n. 279 above).

etry based on perception'). The properties of various striking figures (symmetrical ones and the like) were investigated in a desultory manner. Furthermore there were no strict rules governing 'constructions'. ('Verging' and all kinds of kinematic constructions were permitted; Hippias of Elis, for example, made use of such techniques to construct the quadratrix). *Plato carried out a thorough-going reform of mathematics, which gave us the axiomatic method and the definition of mathematical existence in terms of constructibility.*"

In the last few decades, Zeuthen's theory about a supposed 'Platonic reform' of mathematics has gained such wide acceptance that very few dissenting voices have been heard. The slight modification of it which was proposed by Töplitz is therefore very significant.

Töplitz argued that Zeuthen's theory allowed the development of Greek geometry to be divided into two stages, an empirical (pre-reform) stage and a theoretical (post-reform) one. He then went on to say:[289]

"Nevertheless, we can insert a third stage between the other two; at this intermediate stage proofs were already in use, but the question of how many unprovable axioms were required to ensure their validity had not yet been systematically investigated. The way in which mathematics is often taught to children indicates that there could very well have been a stage of this kind."

Of course, the addition of such an intermediate stage lessens considerably the significance of Plato's alleged 'reform'. This brings us to the second part of the proposed modification:[290]

"It is possible that the great mathematicians, including those who were members of the Academy, *were led to undertake this task not by a suggestion of Plato's but by considerations which were essentially mathematical in nature; perhaps it was Plato who learnt from them and incorporated the principles of their method into a general theory of knowledge.*"

---

[289] Op. cit. (n. 287 above), p. 201.
[290] Op. cit. (n. 287 above), pp. 201–2.

Becker gave Töplitz's ideas rather a cool reception, as we can see from the following passage in which he discusses them:[291]

"This hypothesis can be neither confirmed nor refuted on the basis of the few records which have come down to us from that time. We can at least be sure of one thing, however: *Plato was the first person to obtain a clear understanding of the strictly methodical technique of elementary construction and, in so doing, he contributed greatly to the development of positive mathematical research.*"

We should not be astonished to find Becker, in 1927, lamenting the scarcity of records from pre-Platonic times and writing as if this scarcity presented an insurmountable obstacle to learning more about the mathematics of the period. It was not until a few years later that he succeeded in reconstructing the original version of the so-called theory of even and odd. (This theory is the oldest piece of deductive mathematics known to us[292] and may well date back to the first half of the fifth century B.C.) What is surprising, however, is that Becker seems never to have abandoned completely his belief in Zeuthen's theory about a 'Platonic reform' of mathematics. In 1951 he responded in the following way to van der Waerden's conjecture that Book VII of the *Elements* originated in the fifth century B.C. "Before being put in its final form, *Book VII may perhaps have been revised by the mathematicians of the Academy.*"[293] Becker himself must have realized that this was an inadequate reply, for he added:

"It should be mentioned, however, that the author (van der Waerden) raises an important argument against this; he points out that it is impossible to distinguish between the content and form of a strictly logical investigation of the kind undertaken in Book VII."

The suggestion that "the mathematicians of the Academy" may have revised Book VII serves only to show how reluctant Becker was to abandon Zeuthen's position.

[291] *Mathematische Existenz* (Halle 1927), p. 250, note 2.
[292] *Quellen und Studien* ... B, **3** (1936), 533–53. Cf. also O. Becker, *Die Grundlagen der Mathematik in geschichtlicher Entwicklung*, Freiburg–Munich 1954, pp. 38ff.
[293] *Gnomon* 23 (1951), 299.

I feel sure that my attitude towards Plato's 'reform' of mathematics can be inferred from the other parts of this book, so I will not discuss it in any more detail here. Instead, I would like to give a brief account of how it came about that Plato was thought to have had such a decisive influence on mathematics.

Plato's close connection with the mathematics of his time is mentioned repeatedly by the authors of antiquity. Proclus, for example, wrote the following about it.[294]

"Plato, who appeared after them, greatly advanced mathematics in general and geometry in particular because of his zeal for these studies. It is well known that his writings are thickly sprinkled with mathematical terms and that he everywhere tries to arouse admiration for mathematics among students of philosophy."

Moreover, the great mathematicians of that time, whose names are listed by Proclus, were Plato's friends, his mathematics teachers or his students in philosophy. It cannot be denied that Plato may well have had a significant effect upon mathematics. Neither his mathematical knowledge nor the close relationship between his method of reasoning and mathematical argument can be called in question. Furthermore, my theory about the origins of deductive mathematics, which was developed in Part 3, depends heavily upon the evidence furnished by Plato's writings.

None of the above, however, gives us any reason to think that Plato's teaching inspired mathematicians to undertake a *thorough-going reform* of their subject. In fact, as far as I can see, there is no evidence at all to support this idea. The claim that "Plato was the first person to obtain a clear understanding of the strictly methodical technique of elementary construction and, in so doing, he contributed greatly to the development of positive mathematical research" seems to me to be nothing more than an unfounded and rather arbitrary modern conjecture.

Even those who do not accept my thesis *that Eleatic philosophy is the common root of both Plato's dialectic and the method of early Greek mathe-*

[294] Proclus (ed. Friedlein), 66.8ff; the translation is by Glenn R. Morrow., Princeton 1970.

*matics* will not find the idea of a Platonic 'reform' of mathematics very plausible. They are more likely to agree with Reidemeister[295] that the method of indirect proof, the most important tool of Platonic dialectic, was borrowed from mathematics.[296]

We are now faced with the task of explaining what led scholars to think that Plato instigated a 'reform' of mathematics when there was no hard evidence to support this view. (Proclus' references to the great zeal which Plato showed for the study of mathematics and the great advances which he made in this subject have, of course, some bearing on the matter, but they are far from being sufficient to justify the conclusion that Plato reformed mathematics.)

I believe that this curious modern idea arose in part as a result of the fact that earlier scholars (for no good reason) mistrusted Hippocrates of Chios, an outstanding mathematician of the fifth century B.C. This is not the place to discuss Hippocrates' quadratures of lunes, but I would like to mention some of the things which have been said about this achievement of his. The remarks in question are worth recalling because they help us to understand why a 'Platonic reform' was thought to have taken place in the history of Greek mathematics.

Let me begin by mentioning that Hippocrates was not credited with using indirect methods of proof by earlier scholars.[297] This opinion about Hippocrates' knowledge and capabilities led to certain textual emendations being made. Passages in Simplicius which contained references to indirect proofs were freely excised on the grounds that they could not possibly have been the work of Hippocrates and must have been added later by Eudemus.[298] Not content with this, scholars proceeded to go far beyond the emended text in their interpretations and made Hippocrates out to be a true sophist.

For example, Simplicius tells us that Hippocrates proved the following theorem: "Circles have the same ratio as the squares on the diameters", but he fails to say how this result was established. Con-

[295] K. Reidemeister, *Das exacte Denken der Griechen*, Hamburg 1949, pp. 44 and 65.

[296] B. L. van der Waerden, *Erwachende Wissenschaft*, p. 247.

[297] See H. G. Zeuthen, 'Sur les connaissances géométriques des Grecs', p. 448 (a full reference is given in n. 286 above).

[298] Ibid., p. 452.

sequently, some commentators were led to ask whether Hippocrates was able to give a convincing proof of it. With amazing self-confidence, they argued that he could not have known the proof which appears in Euclid (Proposition XII. 2) since this was *most likely* due to Eudoxus. Töplitz (in the paper mentioned above) therefore attempted to reconstruct a 'less complete proof' of the theorem in question; in other words, he was prepared to credit Hippocrates with having discovered a kind of 'quasi-proof'.

According to Töplitz, Hippocrates began by considering regular polygons (4, 8, 16, 32, . . . ) and showed that the areas of any two such polygons *(A* and *a)* stood in the same ratio as the squares on the radii of the corresponding circles *(R* and *r)*; that is, he showed that $A : a = = R^2 : r^2$. From this, Töplitz conjectured, Hippocrates concluded that the areas of the circles themselves *(K* and *k)* also stood in this ratio *(K : k = R^2 : r^2)* because he realized that, as the number of sides was increased, the areas of the inscribed polygons *obviously* approached those of the circles.

Having made up this naive 'proof' for Hippocrates out of his own head (he certainly did not look for traces of it in the tradition), Töplitz went on to say the following:[299]

"The keen mind of Eudoxus perceived that an infinite process was involved here, something which could not be handled within the framework of purely deductive *apodeixis*. The inference from *n*-sided polygons to circles is invalid; however plausible it may be, it contains a logical gap which cannot be filled in by any axiom. *The present case was not the only one in which such gaps occurred. Similar ones were to be found in dozens of other propositions* (?) *which had been formulated by the Sophists.*"

He then gave a brief account of the Eudoxian axiom which is required in the proof of Hippocrates' theorem;[300] this enabled him to make an effective comparison between the Sophists and the 'true Platonist scholars':[301]

---

[299] *Die Antike* 1 (1925), 182–3.
[300] See n. 112 above.
[301] *Die Antike* 1 (1925), 192.

*"The sophistical masters,* on the one hand, who executed elegantly the mysterious leap into the infinite, and *the Academicians* on the other, who exposed all flaws in reasoning by means of *their ingenious methodology,* kept the framework of finite geometry intact and made everything follow *more geometrico* from a single new axiom discovered by Eudoxus."

It is apparent from these last two quotations that we are dealing with a *twofold historical construction.* An interesting body of 'sophistical mathematics' needs to be constructed[302] in order to establish thaf Plato's 'reform' of mathematics actually took place; the 'reform' itselt can then be interpreted as a great advance over this earlier approach to science. I need hardly say that neither part of this historical construction will stand up to much critical scrutiny.

There is another question which is sometimes asked in the context of Plato's alleged 'reform of mathematics' and which can, in my opinion, now be answered with some certainty. Töplitz, at the end of his paper, formulates it in the following way: "Was there ever a time when philosophy played a decisive role in mathematics and helped to form the distinctive character of this subject, or did it (i.e. mathematics) evolve entirely on its own?"

We know from our earlier investigations, however, that deductive mathematics started out as a branch of philosophy (more precisely, as a branch of Eleatic dialectic). It was only when the theoretical foundations of *geometry* were laid that Greek mathematics found itself in opposition to philosophy and was gradually transformed into a separate science.

[302] Cf. B. L. van der Waerden, *Math. Ann.* **120** (1947/49) 139–40 and *Erwachende Wissenschaft,* p. 214.

# POSTSCRIPT

After reading this book, the non-mathematician might welcome some remarks about the ways in which Euclid's axiom system differs from modern axiomatizations of mathematics. He might also like to see how the various ancient scientific concepts which we have discussed, are treated in contemporary mathematics. For example, it would be instructive to compare Eudoxus' theory of incommensurables with Dedekind's theory of real numbers.

As a matter of fact, I thought seriously about the possibility of attempting such a comparison; finally however, I abandoned the idea. It became clear to me that Greek mathematics could not be placed in its proper historical perspective until a lot more research had been done. Let me illustrate this point with an example.

It is usual nowadays to distinguish between *rational* and *irrational* numbers in the following way. *Rational* numbers can be written as *fractions* or *ratios (m/n)*, whereas *irrational* ones cannot. (This is why the latter are said to be *irrational*.) Historians have been of the opinion that the Greeks must also have made this distinction. A *ratio* in Greek mathematics was called a λόγος *(a : b)* and so they reasoned, the word ἄλογος in Book X of the *Elements* must have had the same meaning as our mathematical term *irrational*. I do not want to comment upon this view here, but I do wish to point out that *the problem of mathematical ἄλογον has not even been touched on in this book*. I have ignored it completely since it does not bear on my main theme, the discovery of linear incommensurability. (In the quotation from Proclus on p. 153, the word ἄλογον has an entirely different meaning, of course.) Another topic which we did not really discuss (although it was mentioned in passing), is Eudoxus' theory of proportions. As I have already remarked in the introduction, we are still a long way from being able to give a definitive new picture of the early period of Greek science. Therefore, I am obliged to conclude this book without presenting a comparison of ancient and modern mathematical concepts.

There is a very informative book, *What is Mathematics* by Courant and Robbins (Oxford University Press 1969), which can be recommended to the non-mathematician who is nonetheless interested in these matters. Chapter 2 of that work ('The number system of mathematics') is especially relevant as far as rational and irrational numbers are concerned. Also, the sections on the Euclidean algorithm, axioms and axiomatics are well worth reading.

# APPENDIX

## HOW THE PYTHAGOREANS DISCOVERED
## PROPOSITION II. 5 OF THE *ELEMENTS**

### 1 THE PREVAILING VIEW

Nowadays most scholars would agree that Proposition II.5 forms part of *ancient Pythagorean mathematics* or, more precisely, that it is a theorem of the so-called '*geometrical algebra of the Pythagoreans*'. Since it is usually discussed together with II. 6, let me begin by quoting both of these propositions:

Proposition II. 5: "If a straight line be cut into equal and unequal segments, the rectangle contained by the unequal segments of the whole together with the square on the straight line between the points of section is equal to the square on the half."

Proposition II. 6: "If a straight line be bisected and a straight line be added to it in a straight line, the rectangle contained by the whole with the added straight line and the added straight line together with the square on the half is equal to the square on the straight line made up of the half and the added straight line."

According to the standard historical account of Greek mathematics, II. 5 can be regarded as a geometric equivalent of the algebraic formula

$$a^2 - b^2 = (a - b) \cdot (a + b)$$

(or $a^2 = (a - b) \cdot (a + b) + b^2$).[1] Since this can be said of II. 6 as well, it would appear that both propositions amount to the same thing. The explanation offered for this curious fact[2] is that II. 5 and II. 6 are not really propositions, but rather *the solutions to certain problems*. II. 5

---

* This paper was written (after the rest of the book had been completed) at the invitation of Professor Ph. M. Vassiliou, who presented it at a meeting of the Greek Academy of Sciences in Athens on 16 May 1968. It is reprinted here because it continues the work begun in the present book and points out a direction for future research.

[1] B. L. van der Waerden, *Erwachende Wissenschaft*, p. 196.

[2] Ibid., p. 198. This explanation goes back to Zeuthen and even to Tannery (see n. 6 below).

is concerned with the construction of two straight lines whose *sum and product* are given, whereas in II. 6 it is the *difference and product* of the two lines which is given. These problems are supposed to be algebraic in nature[3] and to have originated amongst the Babylonians who also devised procedures for solving them.[4] It is claimed that the Pythagoreans subsequently gave a geometrical formulation of this part of Babylonian mathematics and also supplied the necessary proofs.[5]

I disagree with the above interpretation on the following grounds:

(1) Even if we are convinced by Neugebauer's arguments and accept that there was such a thing as 'Babylonian algebra', this does not mean that the Greeks in pre-Euclidean times actually knew about it, let alone that they took it over and put it into geometric form; in fact, no one has yet succeeded in producing any concrete evidence to support the view that they did. (After all the Greeks never adopted the place-value system of notation for numbers from the Babylonians, even though it would have been much easier for them to have done this than to have reworked the whole of Babylonian algebra.)

(2) Ever since the time of Tannery[6] it has been customary to regard certain of Euclid's propositions as 'algebraic theorems in geometric form'. However, these propositions are algebraic only in the sense that we can very easily find algebraic results which are equivalent to them. It certainly cannot be maintained that they started out as *algebraic* theorems or as the solutions to *algebraic* problems, for all of them have purely *geometric* origins. One such proposition is II. 5 which, when interpreted from a modern point of view, can be compared to the solu-

---

[3] Heath (*Euclid's Elements*, Vol. 1, pp. 382ff) also points out that II.5 can be interpreted as a "geometrical solution of a quadratic equation". The difference between van der Waerden and Heath is that the latter makes no mention of 'Babylonian science'; Heath's translation of the *Elements* appeared before Neugebauer's research was published.

[4] The standard reference for this is O. Neugebauer, 'Zur geometrischen Algebra. Studien zur Geschichte der antiken Algebra III'. *Quellen und Studien . . .* B **3** (1936) 245–59.

[5] B. L. van der Waerden, *Erwachende Wissenschaft*, p. 203.

[6] P. Tannery, 'De la solution géométrique des problèmes du second degré avant Euclide', 1882 (reprinted in *Mémoires scientifiques* 1, 254–80). Neugebauer regarded Zeuthen as the discoverer of 'geometrical algebra', whereas it was in fact Tannery who originated this idea.

tion of an algebraic problem. Nonetheless, it is a proposition which originally had a geometric meaning, and we should not allow ourselves to forget this fact.


## 2 MY OWN VIEW

In my opinion, II. 5 is a purely geometrical lemma needed for the solution of a purely geometrical problem (or to prove the correctness of this solution). The problem is stated as Proposition II. 14; it is *"to construct a square equal to a given rectilineal figure"*.

It is obvious that II. 14 depends upon II. 5, both from modern commentaries and translations which almost always refer back to the latter proposition in their discussions of the former, and even more so from the original Greek text. To a large extent II. 5 is repeated word for word in the proof of II. 14, a sure sign that whoever composed the text intended to refer back to this proposition.

Indeed, the complicated and awkward language of II. 5 suggests that it was tailored especially to fit the needs of II. 14. The author appears to have formulated it in such a complicated way because he knew from the start that this seemingly awkward form of the proposition was the one best suited for use in the proof of II. 14. I shall show that this was in fact the case.

First, however, let me take this opportunity to point out that the kind of explanation offered above is equally convincing for other Euclidean propositions which up to now have been regarded as part of a 'geometrical algebra'. II. 6, for example, is a purely geometrical lemma which bears exactly the same relationship to Proposition II. 11 as II. 5 does to II. 14. The reason why it appears to be a special case of II. 5, is that II. 11 is basically nothing more than a special case of II. 14.

Yet another purely geometrical lemma which has been viewed as part of a 'geometrical algebra', is Proposition II. 10.[7] The writings of Proclus attest to the fact that it is in reality a lemma[8] needed for the proof of an ancient Pythagorean theorem. Furthermore, they have

---

[7] See B. L. van der Waerden, *Erwachende Wissenschaft*, p. 202.

[8] *In Platonis Rem Publicam Commentarii* (ed. W. Kroll), 1901 Chapter II, pp. 23 and 27.

enabled modern researchers to reconstruct this theorem even though it was never incorporated into the *Elements*.[9]

I now want to give a more detailed account of the genesis of Proposition II. 5. This will require some discussion of the interesting way in which the Pythagoreans dealt with the areas of parallelograms. The next chapter is devoted to an outline of their theory.

### 3 ELEMENTS OF A PYTHAGOREAN THEORY ABOUT THE AREAS OF PARALLELOGRAMS

Our starting point is the scene in Plato's *Meno* (82b–85e) where Socrates asks an uneducated slave how to find a square having twice the area of one with sides two feet long. In other words, he wants the slave to find *a number n* such that the sides of the required square will

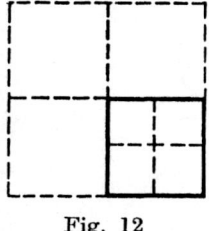

Fig. 12

have a length of $n$ feet. The slave's first thought is that the way to double the area of a square is to double the length of its sides; so he answers that the required length is *four feet*. Socrates, however, proceeds to show him that this would yield a square having four times the original area. (The area of such a square would consist of sixteen unit squares, whereas the area of the original one consisted of only four; see Fig. 12.) The slave's next suggestion is that a square with *sides three feet long* might have twice the original area. Socrates then draws a diagram (see Fig. 13) showing that this square will not do because its area consists of nine unit squares. In recounting the slave's attempts to answer Socrates' question, Plato seems to be hinting that, once the sides of a given square have been *assigned some number as their length*, the same *cannot* be done for the sides of a square having twice the area

---

[9] B. L. van der Waerden, *Erwachende Wissenschaft*, p. 206.

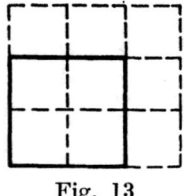

Fig. 13

of the first one. In fact, the two squares will have sides which are *incommensurable in length*. Strangely enough, however, the problem of incommensurability is never mentioned explicitly in this scene. Instead Socrates draws one last diagram (Fig. 14) which shows that a square is divided into two equal isosceles triangles by its *diagonal*, and hence that the square constructed upon the diagonal of the original one will have exactly twice its area.

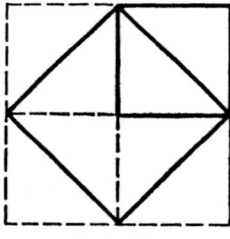

Fig. 14

I believe that this scene from the *Meno* gives us an idea of how the Greeks may have arrived at the discovery of linear incommensurability by way of an *arithmetical* problem which appeared in *geometry*. Given a square whose sides had been assigned a *number*, they wanted to find the *number* corresponding to the sides of a square with twice its area. Euclid's fundamental definition of 'number' (VII. 2) precluded the possibility of this problem having an *arithmetical* solution, nonetheless it remained of interest from a *geometric* point of view. The realization that any square could be doubled by constructing another one on its *diagonal*, together with their inability to ascertain the *numerical* ratio between the side of a square and its diagonal, must soon have led Greek mathematicians to conclude that these two segments were *incommensurable in length*. Of course, they must also have been aware that the two were *measurable in square* (since the square on the diagonal had exactly twice the area of the square on the side). Hence the discovery of *linear incommensurability* went hand in hand with a knowledge of quadratic commensurability and the difficulties presented by the former could be circumvented by means of the latter.

The historical interest of Socrates' conversation with the slave is heightened by the fact that the *arithmetical problem of doubling the square* is equivalent to the problem of finding *a mean proportional between two numbers, one of which is twice as large as the other*.[10] The Pythagoreans must have known from their study of arithmetic that the latter problem had no numerical solutions. This fact was certainly familiar to them, since it follows from a theorem (Proposition 3) in the *Sectio Canonis* which states: "There does not exist a mean proportional number between two numbers in a ratio superparticularis (i.e. in a ratio of the form $(n + 1) : n$)."

Bearing in mind this proposition and the scene from the *Meno* discussed above, we can reconstruct three interesting stages in the development of pre-Euclidean mathematics.

*(a)* The first stage saw the emergence of certain arithmetical problems whose solutions could not be found. On closer inspection, it turned out that some of these were *completely insoluble*, whereas others had *solutions only under certain conditions*. Included among the former was the problem of finding a mean proportional number, $x$, between some number, $a$, and $2a$. The discovery that there was no such $x$ for any $a$, represented of course a considerable achievement.

*(b)* Somewhat later it was realized that these same problems could easily be solved if they were given geometrical interpretations. For example, the mean proportional between $a$ and $2a$ was found by taking $a$ to be an arbitrary straight line instead of a number, and constructing a square on it. The diagonal, $d$, of this square was the required mean proportional since $a : d = d : 2a$. This equation can be proved with the

---

[10] It was Hippocrates of Chios who suggested that the problem of *doubling the cube* could be solved by finding *two* mean proportionals between a number and its double. (Cf. O. Becker, *Das mathematische Denken der Antike*, Göttingen 1957, p. 75). It is interesting to consider how he might have come upon this idea. He knew, of course, that the *planimetric problem* of doubling the square could be solved by finding *one* mean proportional between a number and its double. In my opinion, this probably led him to conclude that the related *stereometric problem* of doubling the cube would be solved if one could find *two* such mean proportionals.

help of similar right-angled triangles, *provided that the notion of 'being in the same ratio' is defined for incommensurable magnitudes, i.e. provided that something like the so-called Eudoxian definition of proportionality* (V. 5 in the *Elements*) *is available.*

So we see that at this stage of development, problems which had seemed to be *insoluble* from the *arithmetical* viewpoint of stage *(a)* were given geometrical solutions which depended both on a knowledge of *incommensurability* and on *Eudoxus' definition* of proportionality.

*(c)* It was found at an 'intermediate stage'[11] that the original problems could be *transformed* into essentially equivalent ones which were capable of being solved geometrically without taking problems of commensurability (or incommensurability) into account and without the use of proportions. For example, the problem treated above (to find the mean proportional between some *number* or *magnitude, a,* and *2a*) was transformed into one of finding *the straight line required to double a square with sides of length a.* The significance of this intermediate stage lies in the fact that questions of proportionality and incommensurability were, in a manner of speaking, eliminated. It was probably the discovery that this could be accomplished by a transformation of the problems concerned, which led the Pythagoreans to develop an interesting *'geometry of areas'.* This theory, which earlier scholars interpreted as 'geometrical algebra', will be discussed below.

*

First of all it should be noted that there are a great many examples to be found in the *Elements*, which illustrate the three stages of development outlined above in *(a), (b)* and *(c).* I will quote three of these here.

*Example 1*

*Stage (a)* Consider the following arithmetical propositions from the *Elements:*

---

[11] I do not want to claim that stage *(c)* must have preceded stage *(b)* in time. Questions of chronology are not my prime concern here. What does seem to me to be important is that stage *(c)* shows how the problems raised at stage *(a)* could be given solutions which made no mention of proportions or the problem of incommensurability.

IX. 16: *"If two numbers be prime to one another, the second will not be to any other number as the first is to the second."* (I.e. if $a$ and $b$ are relatively prime, then there is no $x$ such that $a : b = b : x$.)

IX. 18: *"Given two numbers, to investigate whether it is possible to find a third proportional to them."* (I.e. to find conditions on $a$ and $b$ which are necessary and sufficient to ensure the existence of an $x$ such that $a : b = b : x$.)

IX. 19: *"Given three numbers, to investigate whether it is possible to find a fourth proportional to them."* (I.e. to find conditions on $a$, $b$ and $c$ which are necessary and sufficient to ensure the existence of an $x$ such that $a : b = c : x$.)

These propositions were obviously written down at a time when it was known that the problem of finding a third proportional to two given numbers (or a fourth proportional to three given ones) could only be solved *under certain conditions.*

*Stage (b)* Now if $a$, $b$ and $c$ are taken to be arbitrary straight lines instead of numbers, *geometrical* solutions to the above problems can *always* be found. This is proved by Euclid in Propositions VI. 11 and VI. 12 where he shows how to construct the required third (or fourth proportional). His constructions, however, frequently yield *incommensurable magnitudes.* Hence it cannot be claimed that they always provide correct solutions, unless the *Eudoxian definition of proportionality* is assumed.

*Stage (c)* It is possible to solve these same problems without considering proportions or *incommensurability,* if they are first transformed in the following way. The problem of finding the third proportional (i.e. the $x$ such that $a : b = b : x$) becomes that of finding the rectangle *(ax)* which has one side of length $a$ and the same area as a given square, $b^2$. Similarly, instead of trying to find the fourth proportional (i.e. the $x$ such that $a : b = c : x$), one looks for the rectangle which has one side of length $a$ and the same area as a given rectangle, $bc$. Euclid's Proposition I. 44 provides the means for solving both problems. It runs as follows: *"To a given straight line to apply, in a given rectilineal angle, a parallelogram equal to a given triangle."*

(Euclid speaks of 'a given triangle' and a 'parallelogram in a given rectilineal angle', instead of dealing with rectangles, for the sake of greater generality.) The proof of this proposition uses *'applications of areas',* a method developed by the Pythagoreans which has nothing to do with incommensurability or proportions.

22*

*Example 2*

*Stage (a)* The next two of Euclid's propositions which we want to examine are:[12]

VIII. 18: "*Between two similar plane numbers there is one mean proportional number.*"

VIII. 20: "*If one mean proportional number falls between two numbers, the numbers will be similar plane numbers.*"

(Both propositions assume Definitions VII. 16 and VII. 21.)

So, according to these two propositions, a mean proportional number exists between two numbers *(a* and *b)* if, and only if, $a = cd$ and $b = ef$ for some *c*, *d*, *e* and *f* such that $c : e = d : f$. In other words, this conditon is necessary and sufficient for the problem of finding a mean proportional to have an *arithmetical solution*.

*Stage (b)* Now Proposition VI. 13 shows how to construct a mean proportional to *any* two straight lines, *a* and *b*. As in the previous example, however, the construction is one which depends upon *Eudoxus' definition of proportionality* because of the fact that it yields an *incommensurable magnitude* in many cases.

*Stage (c)* The above problem (to find an *x* such that $a : x = x : b$) is equivalent to that of finding a square $(x^2)$ which has the same area as a given rectangle, *ab*. Once it has been reformulated in this way, it can be solved *without using proportions and without bothering about incommensurability*. This is exactly what Euclid does in Proposition II. 14.


*Example 3*

*Stage (a)* As my third example I would like to take the so-called 'Theorem of Pythagoras'. It is well known that this proposition can be verified *arithmetically* only for right-angled triangles with rational sides. The pioneers of Greek mathematics seem to have been interested in this particular case. (They even went so far as to devise rules for generating triples of numbers which could be the sides of right triangles.[13]) So the verification of Pythagoras' theorem is yet another

---

[12] For the sake of simplicity, Proposition VIII.18 is quoted in an abbreviated form.

[13] Cf. O. Becker, *Das mathematische Denken der Antike*, Göttingen 1957, pp. 52ff.

problem which has an *arithmetical solution only under very restricted conditions*.

*Stage (b)* The usual proofs of the theorem make use of *proportions*, for they rely on the fact that the perpendicular from the right angle to the hypotenuse divides a right triangle into two smaller ones which are similar both to each other and to the original triangle. This means that the corresponding sides of these triangles are proportional, from which it is easy to derive the formula $a^2 + b^2 = c^2$. It must be admitted that no proof of this kind is to be found in Euclid. Nonetheless, it is reasonable to conjecture that one was known in pre-Euclidean times, since he himself uses *proportions* to prove a more general form of Pythagoras' theorem (VI. 31). Of course such a proof is not convincing unless proportionality has somehow been defined for incommensurable magnitudes. (This is accomplished by Definition V. 5 as far as Euclid is concerned.)

*Stage (c)* A kind of 'geometry of areas' is used to prove Pythagoras' theorem (I. 47) in the *Elements*. The proof, like those in our previous examples of stage *(c)*, is distinguished by the fact that *it avoids entirely both proportions and the problem of incommensurability*. Figures 15 and 16 are intended to illustrate Euclid's argument, which runs roughly as follows:

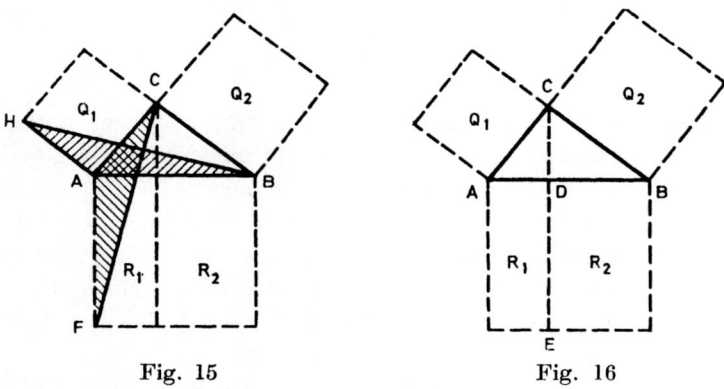

Fig. 15 Fig. 16

The perpendicular *(CD)* from the right angle to the hypotenuse of an arbitrary right triangle *(ACB)* can be extended (by *DE*) so as to divide the square on the hypotenuse, into two rectanges, $R_1$ and $R_2$ (see Fig. 15). Now $Q_1 = R_1$, and $Q_2 = R_2$. The shaded triangles $CAF$ and $HAB$ (see Fig. 16) are obviously congruent. Furthermore, $HAB$ has half the area of $Q_1$, and $CAF$ has half the area of $R_1$ (by Proposition I. 41). Hence $R_1$ and $Q_1$ must have the same area. In a similar manner one can prove that $R_2 = Q_2$. The proof is completed by combining these two equations.

The preceding examples together with my discussion of Socrates' mathematical demonstration in the *Meno* have, I hope, lent some plausibility to the claim that the kind of 'geometry of areas' used at stage *(c)* was designed to eliminate the application of proportions and the problem of incommensurability.

The most elementary portions of this Pythagorean 'geometry of areas' (or 'geometrical algebra', as Tannery called it)[14] are surveyed in (1), (2), and (3) below.

(1) Once again our starting point is that part of the *Meno* which deals with squares and their areas. The passage in question describes how Socrates, having drawn a square with *sides two feet long, bisects*

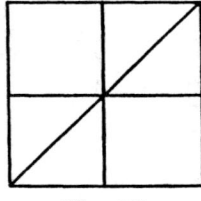

Fig. 17

these sides to show that it is made up of four unit squares, and how he goes on to indicate that the square's area is *halved* by its *diagonal*. (See Fig. 17. I should mention here the trivial fact that the diagonal and the lines which bisect the sides intersect *at a point*. This point of intersection will play an important role in the sequel.) In short, Socrates demonstrates two ways of splitting up the area of a square. Both of these are treated *in a more general form* by Euclid. For his Proposition I. 34, "*In parallelogrammic areas ... the diameter bisects the areas*", is just Socrates' remark about the diagonal of a square generalized to the case of parallelograms; whereas Proposition II. 4 discusses the way in which the area of a square can be split up when its side, instead of being bisected, is divided into two *arbitrary line segments*. This latter proposi-

---

[14] See n. 6 above.

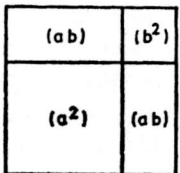

Fig. 18

tion, which is perhaps an even more interesting generalization than the former, runs as follows: "*If a straight line be cut at random, the square on the whole is equal to the squares on the segments and twice the rectangle contained by the segments.*"

So, when the side of a square is 'cut at random', its area is obtained as the sum of two distinct squares and two congruent rectangles (see Fig. 18). Although the above proposition is a geometric equivalent of the algebraic formula $(a + b)^2 = a^2 + 2ab + b^2$, this fact should not in my opinion be taken to mean that II. 4 is "a geometric version of what was originally an algebraic idea". In the first place there is no evidence that Greek mathematics contained any genuinely algebraic ideas either before or during Euclid's time, and in the second place such an interpretation would obliterate the fundamental difference between the algebraic equation and its geometric counterpart. In the final analysis the one is much more general than the other. (Euclid's proposition is one possible interpretation of the algebraic formula, but it is by no means the only one.)

(2) It can be shown that the Pythagoreans also generalized Proposition II. 4 to *parallelograms*. They succeeded in proving a theorem something like the following (see Fig. 19):

If two lines which are parallel to the sides of a given parallelogram are drawn through an arbitrary point of its diagonal, they will split the area of the parallelogram into four parts. Two of these parts, the 'parallelograms about the diameter' (left unshaded in Fig. 19), will be similar

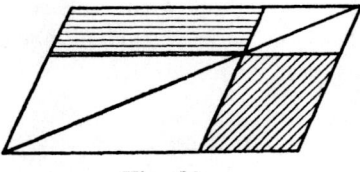

Fig. 19

to each other and to the original parallelogram, whereas the other two, the so-called παραπληρώματα (shaded in Fig. 19), will be equal.

I must admit that the above theorem is largely my own reconstruction. Euclid does not state it in this form, but treats it instead as two separate propositions. One, which discusses the *similarity* of the '*parallelograms about the diameter*', does not appear until Book VI; the other deals with the *equality* of their '*complements*' *(parapleromata)* and is included in Book I. This arrangement was forced upon Euclid by the fact that the proof of the former (Proposition VI. 24) depends in an essential way on *Eudoxus' definition* (V. 5) *of proportionality.* On the other hand, the latter (Proposition I. 43) is proved simply by subtracting areas; its proof has nothing to do with proportionality or incommensurability.

I should emphasize here that the theorem which I have reconstructed played a very important role in the Pythagorean 'geometry of areas' or, to use the unfortunate and misleading name which Tannery gave this theory, 'geometrical algebra'. If, for example, the Pythagoreans[15] had not known that the *parapleromata* were equal, they would not have been able to develop their method of *application of areas.* This method was mentioned in *Example 1* above, where it was pointed out that finding a fourth proportional to three given *numbers* or *magnitudes* (i.e. an $x$ such that $a : b = c : x$) could be construed as the problem of finding a rectangle with a given side *(a)* which has the same area as a given rectangle *(bc).*

Proposition I. 44 of the *Elements* provides the following solution to this problem (see Figs 20 and 21). If we think of the given rectangle *(bc)* as a *parapleroma* and the rectangle which the line $a$ makes with one of its sides *(c* or *b)* as a 'parallelogram about the diagonal', then the second 'parallelogram about the diagonal' *(bx* or *cx)* can be constructed by extending the side *(c* or *b)* of the original rectangle until it meets the continuation of the diagonal of *ac* (Fig. 20) or of *ab* (Fig. 21). Now these three rectangles together determine uniquely the second *parapleroma* and, in particular, its side $x$ which we set out to find. (This is sometimes called *parabolic* application of areas, because the given area *bc* is *applied* to the line $a$; cf. the word παραβάλλειν.)

---

[15] Proclus (ed. Friedlein), 419.15–420.23. See also T. L. Heath, *Euclid's Element*, Vol. 1, pp. 343ff.

It seems to me, however, that the importance of our theorem for Pythagorean geometry can be illustrated even more convincingly. Let us look once more at the theorem itself. It states that two lines drawn parallel to the sides of a parallelogram through an arbitrary point on its diagonal will split the area of the parallelogram into four parts; two of these will be equal, and the other two will be similar (both to each

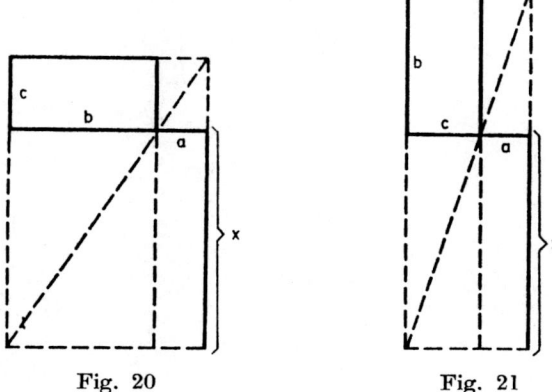

Fig. 20                    Fig. 21

other and to the original parallelogram). My claim is that this simple yet beautiful discovery led the way to a famous problem of the early Pythagoreans about which Plutarch wrote the following (*Symposium* VIII. 2.4):

> "Among the most geometrical theorems, or rather problems, is the following: given two figures, to apply a third equal to the one and similar to the other, on the strength of which discovery they say moreover that Pythagoras sacrificed. This is indeed unquestionably more subtle and scientific than the theorem which demonstrated that the square on the hypotenuse is equal to the squares on the sides about the right angle."[16]

(3) I cannot conclude this account of the elements of the Pythagorean 'geometry of areas' without mentioning Euclid's definition of the geometrical concept *gnomon:*

[16] Quoted by Heath, loc. cit. (*Euclid's Elements*, Vol. 1, pp. 343ff).

*Definition* II. 2: "*In any parallelogrammic area let any one whatever of the parallelograms about its diagonal with the two complements be called a gnomon.*"

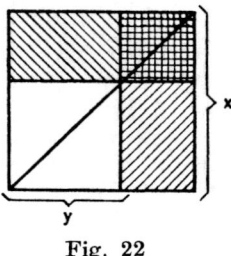

Fig. 22

As we can see, the definition of this concept presupposes that the area of the parallelogram has already been split into *four parts* in the manner described previously. A *gnomon* consists of *three* of these parts, namely the two *parapleromata* together with either one of the 'parallelograms about the diagonal'. The *gnomon* of a square is illustrated in Fig. 22, where its three components are indicated by different kinds of shading. The figure also shows that the original square *(x²)* can be obtained as the sum of a smaller square *(y²)* and the *gnomon*. (In other words, subtracting the smaller square from the greater leaves a *gnomon* as the remainder.)

This fact is important because the *gnomon* itself can be very easily transformed into a *rectangle*. There are two ways of carrying out the transformation; these are illustrated in Figs. 23 and 24 respectively.

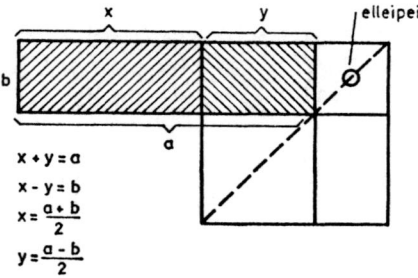

Fig. 23

In one case (Fig. 23) the area consisting of one of the 'parallelograms about the diagonal' (the smallest square in the illustration, which for convenience we will denote by 'c') and one of the *parapleromata* is, in a manner of speaking, joined onto the left-hand side of the second *parapleroma*; whereas in the other case (Fig. 24) it is a parapleroma which is joined onto the left-hand side of the area consisting of the smallest square and the other *parapleroma*. The resulting rectangle (the shaded part in Figs 23 and 24) is the same in both cases; it has sides *(x + y)* and *(x − y)* (these letters refer to the same lengths as in Fig. 22). The only difference between the two constructions has to do with *the square o, a component of the original gnomon.* In the first case, a rectangle

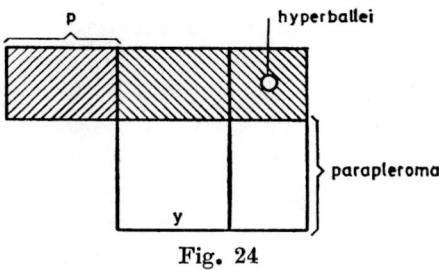

Fig. 24

having the same area as the gnomon is applied to the line 2*x* and *falls short (ἐλλείπειν)* by *O* (of the rectangle on this line); hence this is sometimes called *elliptic application of areas.* In the second case, the rectangle is applied to 2*y* and *exceeds (ὑπερβολή)* by *O* (the rectangle on this line); so this is sometimes referred to as *hyperbolic application of areas.* Now is not the time to go into any more detail about *elleipsis* and *hyperbola*,[17] suffice it to say that Euclid employs *elliptic application* in the proof of Proposition II. 5 and *hyperbolic application* in II. 6.

#### 4 HOW TO FIND A SQUARE WITH THE SAME AREA AS A GIVEN RECTANGLE

We are now in a position to return to II. 5, the proposition whose genesis we set out to explain.

[17] See the discussion in Heath's edition of the *Elements* (Vol. 1, pp. 343–4)

I have already mentioned (in Chapter 2 above) that it is a lemma which is needed to prove another important proposition, namely II. 14, and that it is quoted word for word in the proof of the latter. Let us begin by taking a closer look at II. 14, which is concerned with the following problem:

"*To construct a square equal to a given rectilineal figure.*" The proposition is really about *tetragonismes* the transformation of a rectangle into a square of the same area. It is stated for an arbitrary rectilinear figure only because Euclid always sought after the greatest possible generality. Therefore his proof begins with a reference to Proposition I. 45, which makes it possible for any such figure to be transformed into a rectangle.

In example 2 (of Chapter 3) it was observed that the transformation of a rectangle into a square of the same area is equivalent to the problem of *finding the mean proportional between two numbers or magnitudes.* Euclid discusses the construction of the mean proportional to two straight lines in Proposition VI. 13. It turns out that II. 14 and VI. 13 lead to the same result.[18] Our interest, however, is in the differences between these two propositions.

The surprising thing about the constructions in II. 14 and VI. 13 is that at first sight they appear to be identical in all important respects (see Fig. 25). In both cases the sides of the rectangle concerned (or the two lines, $a$ and $b$, to which the mean proportional is being sought) are *added* together to form one straight line. This line is then *bisected*, and a *semicircle* with radius $\dfrac{a + b}{2}$ is drawn around it. Finally a perpendicular is raised from the point at which $a$ and $b$ meet to the circumference of the semicircle. This perpendicular, $d$, is the side of a square having the same area as $ab$ (or the required mean proportional between $a$ and $b$).

The steps outlined above make it seem as though the identical construction is used in both propositions. Yet the significance of these steps in the two cases is as different as it could be. Completely different ideas lie behind II. 14 and VI. 13. Let us first interpret the above from the point of view of the latter proposition.

---

[18] See Aristotle, *De Anima* II.2.413a13–20 and *Metaphysics* B2. 996b18–21. These two passages convinced both Heiberg ('Mathematisches bei Aristoteles', *Abhandlungen zur Geschichte der mathematische Wissenschaften* 18 Heft, Leipzig 1904) and Heath (*Mathematics in Aristotle*, Oxford 1949, pp. 191–3) that VI.13 came before II.14.

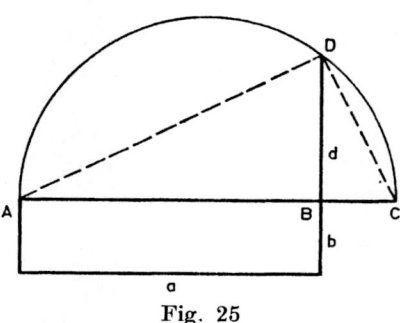

Fig. 25

The two segments $a$ and $b$ are *added* together in order to obtain the *hypotenuse* ($AC$ in Fig. 25) of a right-angled triangle (with the thought already in mind that $a$ will eventually form the *longer* side about the right angle of another triangle, and $b$ the *shorter* side about the right angle of a third one). The segment $AC$ is bisected and a semicircle is drawn around it, since this gives, by Thales' theorem, the locus of points which can be the third vertex of a right triangle with hypotenuse $AC$. Finally, $d$ is the required mean proportional between $a$ and $b$, because the triangles $ABD$ and $BCD$ are similar, and $d$ is the *shorter side* about the right angle in the former and the *longer side* about the right angle in the latter. This is undoubtedly the argument upon which the construction in VI. 13 is based. What appears to be the same construction in II. 14, however, rests upon an entirely different chain of reasoning, which we shall now attempt to reproduce.[19]

In order to transform a given rectangle, $ab$, into a square of the same area, an argument which runs something like the following is required. There is, so to speak, quite a distance to cover. Our starting point, the rectangle $ab$, is given, and our aim is to find a square, $d^2$, which has the same area. Since our goal is so far removed from our starting point, it would be desirable to find some way of bringing them closer together. If, for example, it could be shown that both the rectangle and the square can be obtained by performing *certain geometrical operations*, a link would have been established between them. We shall now take up the question of how this is to be accomplished in the case of the rectangle.

[19] Cf. G. Polya, *How to solve it?* 2nd edn., New York 1957. This book was my guide in the reconstruction which follows in the text.

Consider the subtraction illustrated in Figs 22 and 23. A square, $y^2$, is subtracted from a larger one, $x^2$, leaving a *gnomon* as the remainder; this *'gnomon'* can easily be transformed into the rectangle $(x + y) \cdot (x - y)$. Our present case differs from the preceding in that the sides $(a$ and $b)$ of the rectangle are given instead of the two squares. However, it is a simple matter to find $x^2$ and $y^2$ from $a$ and $b$. Let $x^2 = \left(\dfrac{a + b}{2}\right)^2$ and $y^2 = \left(\dfrac{a - b}{2}\right)^2$ (see Fig. 23), then the rectangle $ab$ can be regarded as the difference of two squares, since $ab = \left(\dfrac{a + b}{2}\right)^2 - \left(\dfrac{a - b}{2}\right)^2$ ; hence it can be described as the result of performing a geometrical operation. (This is actually the transformation which is accomplished by the lemma II. 5.)

Similarly the square $d^2$ can be viewed as the result of performing another geometrical operation, the operation being *subtraction* in this case as well. Given the right-angled triangle shown in Fig. 26, we know by Pythagoras' Theorem that $x^2 = y^2 + d^2$. So the square on one of the sides about the right-angle can be obtained by subtracting the square on the other from the square on the hypotenuse, and in particular $d^2 = x^2 - y^2$. The solution to our problem is now almost at hand.

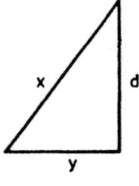

Fig. 26

Subtracting one square from another led us first to a *rectangle* and then to a *square*,[20] so we should be able to transform the rectangle $ab$ into a square of the same area by combining these two results. $\left(\dfrac{a + b}{2}\right)^2$ will be the square on the hypotenuse of a right-angled triangle, whereas

[20] The lettering used in Fig. 23 is intended to make the discussion easier to understand.

$\left(\dfrac{a-b}{2}\right)^2$ will be the square on one of the sides about the right angle. Since we know that the *hypotenuse* equals $\dfrac{a+b}{2}$ and one of the other *sides* equals $\dfrac{a-b}{2}$, we should be able to obtain the third side, $d$, of the triangle (see Fig. 27). The square on $d$ will then have the same area as the given rectangle $ab$. This is the argument which underlies the construction in Proposition II. 14.

So we see the *addition* of the two sides of the rectangle plays a different role in this construction (see Fig. 27) to the one which it plays in VI. 13 (for the latter see Fig. 25). The lines $a$ and $b$ are added together in VI. 13 because $a + b$ will form the hypotenuse of a right triangle; this hypotenuse then has to be *bisected* in order to apply Thales' theorem. In II. 14, on the other hand, $a$ is *added* to $b$, and their sum *bisected* because $\dfrac{a+b}{2}$ will itself be the hypotenuse of a right triangle (see Fig. 27).

There is one other interesting fact which I should like to mention here. It concerns the side of the square $\left(\dfrac{a-b}{2}\right)^2$. No special construction is required to obtain this side in the proof of II. 14, for *the line*

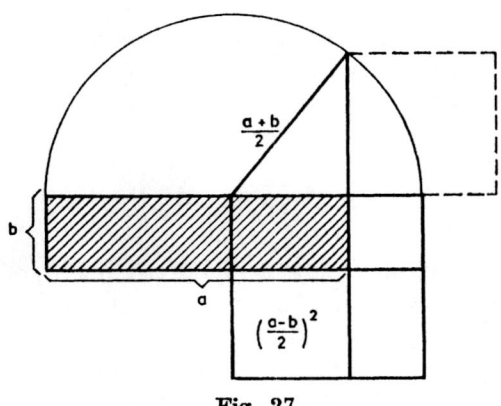

Fig. 27

*between the midpoint of $a + b$ and the point at which $a$ and $b$ meet is exactly* $\dfrac{a-b}{2} = y$.

In conclusion, I would like to say something more about my claim that II. 5 is a lemma which was especially tailored to suit the needs of Proposition II. 14.

First of all I should note that Euclid, in his explanation and proof of Proposition II. 5, makes use of a diagram which corresponds in all essential respects to my Fig. 23. This figure shows how the Pythagoreans obtained an arbitrary square as the sum of a *smaller square* and a *gnomon* (or as the sum of a *square* and a *rectangle*). Basically this is the subject matter of II. 5 as well. Of course Euclid does not start out with a square, but with a straight line which is *cut into equal and unequal segments*. Cutting the line into unequal segments is the reverse of *adding* the two sides of the rectangle in II. 14. The latter operation converts a rectangle into a straight line, whereas the former turns a straight line into a rectangle. *Bisecting* the line plays the same role in both propositions; it furnishes the side of the larger square (from which the smaller one is subtracted).

The following is even more interesting. As I have already emphasized, it is necessary to find *half the difference between the two sides of the rectangle* before Pythagoras' Theorem can be applied in II. 14. However, there is no need to construct this length since it is just *the line between the midpoint of a + b and the point at which a and b meet*. In II. 5 this same segment is described as *"the straight line between the points of section"*. Euclid's use of this clumsy and awkward phrase, in my opinion, proves conclusively that he had II. 14 in mind when he stated and proved Proposition II. 5.

## 5 CONCLUSION

It seems to me that the results of the preceding investigation are of some interest from the viewpoint of further research. They are summarized below.

(1) Proposition II. 5 is a purely geometrical lemma needed for the proof of II. 14. Although it is equivalent to 'the solution of an algebraic equation', it should not be interpreted in this way. Such an interpretation is misleading because it obscures the true geometric meaning of the proposition and suggests the false historical idea that the Greeks actually operated with algebraic equations in pre-Euclidean times.

(2) As was mentioned in Chapter II above, the other propositions which are usually regarded as part of 'Pythagorean geometrical algebra' can also be given a purely geometrical explanation. On the other hand, no traces of genuine algebraic ideas[21] have yet been discovered in the mathematical tradition which culminated in Euclid's *Elements*.

(3) There is however one point on which I agree with Tannery, the scholar whose ideas gave rise to later speculation about the supposed 'geometrical algebra of the Greeks'.[22] He was of the opinion that this kind of geometry owed its existence to the discovery of incommensurability. In fact, as I have tried to show, the 'Pythagorean geometry of areas' *eliminated* the problem of incommensurability and consequently *avoided* the use of proportions.

(4) The claim that the 'geometrical algebra of the Pythagoreans' resulted from the Greeks either taking over or developing further an idea of the Babylonians has no basis in fact. No connection has ever been established between this branch of mathematics and 'Babylonian science'. It seems much more likely that the Greeks were solely responsible for the creation of this 'geometry of areas'.

---

[21] Van der Waerden (in *Science Awakening*, p. 118) gives a genuinely algebraic interpretation of Proposition II.14, but this modern interpretation is foreign to the spirit of ancient mathematics.

[22] See n. 6 above.

23 Szabó

# INDEX OF NAMES

# SUBJECT INDEX

# SYNTHESE HISTORICAL LIBRARY

Texts and Studies
in the History of Logic and Philosophy

*Editors:*

N. KRETZMANN (Cornell University)
G. NUCHELMANS (University of Leyden)
L. M. DE RIJK (University of Leyden)

1. M. T. Beonio-Brocchieri Fumagalli, *The Logic of Abelard*. Translated from the Italian. 1969, IX + 101 pp.
2. Gottfried Wilhelm Leibnitz, *Philosophical Papers and Letters*. A selection translated and edited, with an introduction, by Leroy E. Loemker. 1969, XII + 736 pp.
3. Ernst Mally, *Logische Schriften*, ed. by Karl Wolf and Paul Weingartner. 1971, X + 340 pp.
4. Lewis White Beck (ed.), *Proceedings of the Third International Kant Congress*. 1972, XI + 718 pp.
5. Bernard Bolzano, *Theory of Science*, ed. by Jan Berg. 1973, XV + 398 pp.
6. J. M. E. Moravcsik (ed.), *Patterns in Plato's Thought. Papers Arising Out of the 1971 West Coast Greek Philosophy Conference*. 1973, VIII + 212 pp.
7. Nabil Shehaby, *The Propositional Logic of Avicenna: A Translation from al-Shifā: al-Qiyās*, with Introduction, Commentary and Glossary. 1973, XIII + 296 pp.
8. Desmond Paul Henry, *Commentary on De Grammatico: The Historical-Logical Dimensions of a Dialogue of St. Anselm's*. 1974, IX + 345 pp.
9. John Corcoran, *Ancient Logic and Its Modern Interpretations*. 1974, X + 208 pp.
10. E. M. Barth, *The Logic of the Articles in Traditional Philosophy*. 1974, XXVII + 533 pp.
11. Jaakko Hintikka, *Knowledge and the Known. Historical Perspectives in Epistemology*. 1974, XII + 243 pp.
12. E. J. Ashworth, *Language and Logic in the Post-Medieval Period*. 1974, XIII + 304 pp.
13. Aristotle, *The Nicomachean Ethics*. Translated with Commentaries and Glossary by Hypocrates G. Apostle. 1975, XXI + 372 pp.
14. R. M. Dancy, *Sense and Contradiction: A Study in Aristotle*. 1975, XII + 184 pp.
15. Wilbur Richard Knorr, *The Evolution of the Euclidean Elements. A Study of the Theory of Incommensurable Magnitudes and Its Significance for Early Greek Geometry*. 1975, IX + 374 pp.
16. Augustine, *De Dialectica*. Translated with Introduction and Notes by B. Darrell Jackson. 1975, XI + 151 pp.

# SYNTHESE LIBRARY

Studies in Epistemology, Logic, Methodology,
and Philosophy of Science

*Managing Editor:*
JAAKKO HINTIKKA (Academy of Finland and Stanford University)

*Editors:*

ROBERT S. COHEN (Boston University)
DONALD DAVIDSON (University of Chicago)
GABRIËL NUCHELMANS (University of Leyden)
WESLEY C. SALMON (University of Arizona)

1. J. M. Bocheński, *A Precis of Mathematical Logic.* 1959, X + 100 pp.
2. P. L. Guiraud, *Problèmes et méthodes de la statistique linguistique.* 1960, VI + 146 pp.
3. Hans Freudenthal (ed.), *The Concept and the Role of the Model in Mathematics and Natural and Social Sciences. Proceedings of a Colloquium held at Utrecht, The Netherlands, January 1960.* 1961, VI + 194 pp.
4. Evert W. Beth, *Formal Methods. An Introduction to Symbolic Logic and the Study of Effective Operations in Arithmetic and Logic.* 1962, XIV + 170 pp.
5. B. H. Kazemier and D. Vuysje (eds.), *Logic and Language. Studies Dedicated to Professor Rudolf Carnap on the Occasion of His Seventieth Birthday.* 1962, VI + 256 pp.
6. Marx W. Wartofsky (ed.), *Proceedings of the Boston Colloquium for the Philosophy of Science, 1961–1962,* Boston Studies in the Philosophy of Science (ed. by Robert S. Cohen and Marx W. Wartofsky), Volume I. 1973, VIII + 212 pp.
7. A. A. Zinov'ev, *Philosophical Problems of Many-Valued Logic.* 1963, XIV + 155 pp.
8. Georges Gurvitch, *The Spectrum of Social Time.* 1964, XXIV + 152 pp.
9. Paul Lorenzen, *Formal Logic.* 1965, VIII + 123 pp.
10. Robert S. Cohen and Marx W. Wartofsky (eds.), *In Honor of Philipp Frank,* Boston Studies in the Philosophy of Science (ed. by Robert S. Cohen and Marx W. Wartofsky), Volume II. 1965, XXXIV + 475 pp.
11. Evert W. Beth, *Mathematical Thought. An Introduction to the Philosophy of Mathematics.* 1965, XII + 208 pp.
12. Evert W. Beth and Jean Piaget, *Mathematical Epistemology and Psychology.* 1966, XII + 326 pp.
13. Guido Küng, *Ontology and the Logistic Analysis of Language. An Enquiry into the Contemporary Views on Universals.* 1967, XI + 210 pp.

14. Robert S. Cohen and Marx W. Wartofsky (eds.), *Proceedings of the Boston Colloquium for the Philosophy of Science 1964–1966, in Memory of Norwood Russell Hanson*, Boston Studies in the Philosophy of Science (ed. by Robert S. Cohen and Marx W. Wartofsky), Volume III. 1967, XLIX + 489 pp.

15. C. D. Broad, *Induction, Probability, and Causation. Selected Papers*. 1968. XI + 296 pp.

16. Günther Patzig, *Aristotle's Theory of the Syllogism. A Logical-Philosophical Study of Book A of the Prior Analytics*. 1968, XVII + 215 pp.

17. Nicholas Rescher, *Topics in Philosophical Logic*. 1968, XIV + 347 pp.

18. Robert S. Cohen and Marx W. Wartofsky (eds.), *Proceedings of the Boston Colloquium for the Philosophy of Science 1966–1968*, Boston Studies in the Philosophy of Science (ed. by Robert S. Cohen and Marx W. Wartofsky), Volume IV. 1969, VIII + 537 pp.

19. Robert S. Cohen and Marx W. Wartofsky (eds.), *Proceedings of the Boston Colloquium for the Philosophy of Science 1966–1968*, Boston Studies in the Philosophy of Science (ed. by Robert S. Cohen and Marx W. Wartofsky), Volume V. 1969, VIII + 482 pp.

20. J. W. Davis, D. J. Hockney, and W. K. Wilson (eds.), *Philosophical Logic*. 1969, VIII + 277 pp.

21. D. Davidson and J. Hintikka (eds.), *Words and Objections: Essays on the Work of W. V. Quine*. 1969, VIII + 366 pp.

22. Patrick Suppes, *Studies in the Methodology and Foundations of Science. Selected Papers from 1911 to 1969*. 1969, XII + 473 pp.

23. Jaakko Hintikka, *Models for Modalities. Selected Essays*. 1969, IX + 220 pp.

24. Nicholas Rescher *et al.* (eds.), *Essays in Honor of Carl G. Hempel. A Tribute on the Occasion of His Sixty-Fifth Birthday*. 1969, VII + 272 pp.

25. P. V. Tavanec (ed.), *Problems of the Logic of Scientific Knowledge*. 1969, XII + 429 pp.

26. Marshall Swain (ed.), *Induction, Acceptance, and Rational Belief*. 1970, VII + 232 pp.

27. Robert S. Cohen and Raymond J. Seeger (eds.), *Ernst Mach: Physicist and Philosopher*, Boston Studies in the Philosophy of Science (ed. by Robert S. Cohen and Marx W. Wartofsky), Volume VI. 1970, VIII + 295 pp.

28. Jaakko Hintikka and Patrick Suppes, *Information and Inference*. 1970, X + 336 pp.

29. Karel Lambert, *Philosophical Problems in Logic. Some Recent Developments*. 1970, VII + 176 pp.

30. Rolf A. Eberle, *Nominalistic Systems*. 1970, IX + 217 pp.

31. Paul Weingartner and Gerhard Zecha (eds.), *Induction, Physics, and Ethics: Proceedings and Discussions of the 1968 Salzburg Colloquium in the Philosophy of Science*. 1970, X + 382 pp.

32. Evert W. Beth, *Aspects of Modern Logic*. 1970, XI + 176 pp.

33. Risto Hilpinen (ed.), *Deontic Logic: Introductory and Systematic Readings*. 1971, VII + 182 pp.

34. Jean-Louis Krivine, *Introduction to Axiomatic Set Theory*. 1971. VII + 98 pp.
35. Joseph D. Sneed, *The Logical Structure of Mathematical Physics*. 1971, XV + 311 pp.
36. Carl R. Kordig, *The Justification of Scientific Change*. 1971, XIV + 119 pp.
37. Milič Čapek, *Bergson and Modern Physics*, Boston Studies in the Philosophy of Science (ed. by Robert S. Cohen and Marx W. Wartofsky), Volume VII. 1971, XV + 414 pp.
38. Norwood Russell Hanson, *What I Do Not Believe, and Other Essays* (ed. by Stephen Toulmin and Harry Woolf). 1971, XII + 390 pp.
39. Roger C. Buck and Robert S. Cohen (eds.), *PSA 1970. In Memory of Rudolf Carnap*, Boston Studies in the Philosophy of Science (ed. by Robert S. Cohen and Marx W. Wartofsky), Volume VIII. 1971, LXVI + 615 pp. Also available as paperback.
40. Donald Davidson and Gilbert Harman (eds.), *Semantics of Natural Language*. 1972, X + 769 pp. Also available as paperback.
41. Yehoshua Bar-Hillel (ed.), *Pragmatics of Natural Languages*. 1971, VII + 231 pp.
42. Sören Stenlund, *Combinators, λ-Terms and Proof Theory*. 1971, 184 pp.
43. Martin Strauss, *Modern Physics and Its Philosophy. Selected Papers in the Logic, History, and Philosophy of Science*. 1972, X + 297 pp.
44. Mario Bunge, *Method, Model and Matter*. 1973, VII + 196 pp.
45. Mario Bunge, *Philosophy of Physics*. 1973, IX + 248 pp.
46. A. A. Zinov'ev, *Foundations of the Logical Theory of Scientific Knowledge (Complex Logic)*, Boston Studies in the Philosophy of Science (ed. by Robert S. Cohen and Marx W. Wartofsky), Volume IX. Revised and enlarged English edition with an appendix, by G. A. Smirnov, E. A. Sidorenka, A. M. Fedina, and L. A. Bobrova. 1973, XXII + 301 pp. Also available as paperback.
47. Ladislav Tondl, *Scientific Procedures*, Boston Studies in the Philosophy of Science (ed. by Robert S. Cohen and Marx W. Wartofsky), Volume X. 1973, XII + 268 pp. Also available as paperback.
48. Norwood Russell Hanson, *Constellations and Conjectures* (ed. by Willard C. Humphreys, Jr.). 1973, X + 282 pp.
49. K. J. J. Hintikka, J. M. E. Moravcsik, and P. Suppes (eds.), *Approaches to Natural Language. Proceedings of the 1970 Stanford Workshop on Grammar and Semantics*. 1973, VIII + 526 pp. Also available as paperback.
50. Mario Bunge (ed.), *Exact Philosophy — Problems, Tools, and Goals*. 1973, X + 214 pp.
51. Radu J. Bogdan and Ilkka Niiniluoto (eds.), *Logic, Language, and Probability. A Selection of Papers Contributed to Sections IV, VI, and XI of the Fourth International Congress for Logic, Methodology, and Philosophy of Science, Bucharest, September 1971*. 1973, X + 323 pp.
52. Glenn Pearce and Patrick Maynard (eds.), *Conceptual Chance*. 1973, XII + 282 pp.

53. Ilkka Niiniluoto and Raimo Tuomela, *Theoretical Concepts and Hypothetico-Inductive Inference*. 1973, VII + 264 pp.
54. Roland Fraïssé, *Course of Mathematical Logic* — Volume 1: *Relation and Logical Formula*. 1973, XVI + 186 pp. Also available as paperback.
55. Adolf Grünbaum, *Philosophical Problems of Space and Time*. Second, enlarged edition, Boston Studies in the Philosophy of Science (ed. by Robert S. Cohen and Marx W. Wartofsky), Volume XII. 1973, XXIII + 884 pp. Also available as paperback.
56. Patrick Suppes (ed.), *Space, Time, and Geometry*. 1973, XI + 424 pp.
57. Hans Kelsen, *Essays in Legal and Moral Philosophy*, selected and introduced by Ota Weinberger. 1973, XXVIII + 300 pp.
58. R. J. Seeger and Robert S. Cohen (eds.), *Philosophical Foundations of Science. Proceedings of an AAAS Program, 1969*, Boston Studies in the Philosophy of Science (ed. by Robert S. Cohen and Marx W. Wartofsky), Volume XI. 1974, X + 545 pp. Also available as paperback.
59. Robert S. Cohen and Marx W. Wartofsky (eds.), *Logical and Epistemological Studies in Contemporary Physics*, Boston Studies in the Philosophy of Science (ed. by Robert S. Cohen and Marx W. Wartofsky), Volume XIII. 1973, VIII + 462 pp. Also available as paperback.
60. Robert S. Cohen and Marx W. Wartofsky (eds.), *Methodological and Historical Essays in the Natural and Social Sciences. Proceedings of the Boston Colloquium for the Philosophy of Science, 1969–1972*, Boston Studies in the Philosophy of Science (ed. by Robert S. Cohen and Marx W. Wartofsky), Volume XIV. 1974, VIII + 405 pp. Also available as paperback.
61. Robert S. Cohen, J. J. Stachel and Marx W. Wartofsky (eds.), *For Dirk Struik. Scientific, Historical and Political Essays in Honor of Dirk J. Struik*, Boston Studies in the Philosophy of Science (ed. by Robert S. Cohen and Marx W. Wartofsky), Volume XV. 1974, XXVII + 652 pp. Also available as paperback.
62. Kazimierz Ajdukiewicz, *Pragmatic Logic*, transl. from the Polish by Olgierd Wojtasiewicz. 1974, XV + 460 pp.
63 Sören Stenlund (ed.), *Logical Theory and Semantic Analysis. Essays Dedicated to Stig Kanger on His Fiftieth Birthday*. 1974, V + 217 pp.
64. Kenneth F. Schaffner and Robert S. Cohen (eds.), *Proceedings of the 1972 Biennial Meeting, Philosophy of Science Association*, Boston Studies in the Philosophy of Science (ed. by Robert S. Cohen and Marx W. Wartofsky), Volume XX. 1974, IX + 444 pp. Also available as paperback.
65. Henry E. Kyburg, Jr., *The Logical Foundations of Statistical Inference*. 1974, IX + 421 pp.
66. Marjorie Grene, *The Understanding of Nature: Essays in the Philosophy of Biology*, Boston Studies in the Philosophy of Science (ed. by Robert S. Cohen and Marx W. Wartofsky), Volume XXIII. 1974, XII + 360 pp. Also available as paperback.
67. Jan M. Broekman, *Structuralism: Moscow, Prague, Paris*. 1974, IX + 117 pp.

68. Norman Geschwind, *Selected Papers on Language and the Brain*, Boston Studies in the Philosophy of Science (ed. by Robert S. Cohen and Marx W. Wartofsky), Volume XVI. 1974, XII + 549 pp. Also available as paperback.

69. Roland Fraïssé, *Course of Mathematical Logic* — Volume 2: *Model Theory*. 1974, XIX + 192 pp.

70. Andrzej Grzegorczyk, *An Outline of Mathematical Logic. Fundamental Results and Notions Explained with All Details*. 1974, X + 596 pp.

71. Franz von Kutschera, *Philosophy of Language*. 1975, VII + 305 pp.

72. Juha Manninen and Raimo Tuomela (eds.), *Essays on Explanation and Understanding. Studies in the Foundations of Humanities and Social Sciences*. 1976, VII + 440 pp.

73. Jaakko Hintikka (ed.), *Rudolf Carnap, Logical Empiricist. Materials and Perspectives*. 1975, LXVIII + 400 pp.

74. Milič Čapek (ed.), *The Concepts of Space and Time. Their Structure and Their Development*, Boston Studies in the Philosophy of Science (ed. by Robert S. Cohen and Marx W. Wartofsky), Volume XXII. 1976, LVI + 570 pp. Also available as paperback.

75. Jaakko Hintikka and Unto Remes, *The Method of Analysis. Its Geometrical Origin and Its General Significance*, Boston Studies in the Philosophy of Science (ed. by Robert S. Cohen and Marx W. Wartofsky), Volume XXV. 1974, XVIII + 144 pp. Also available as paperback.

76. John Emery Murdoch and Edith Dudley Sylla, *The Cultural Context of Medieval Learning. Proceedings of the First International Colloquium on Philosophy, Science, and Theology in the Middle Ages — September 1973*, Boston Studies in the Philosophy of Science (ed. by Robert S. Cohen and Marx W. Wartofsky), Volume XXVI. 1975, X + 566 pp. Also available as paperback.

77. Stefan Amsterdamski, *Between Experience and Metaphysics. Philosophical Problems of the Evolution of Science*, Boston Studies in the Philosophy of Science (ed. by Robert S. Cohen and Marx W. Wartofsky), Volume XXXV. 1975, XVIII + 193 pp. Also available as paperback.

78. Patrick Suppes (ed.), *Logic and Probability in Quantum Mechanics*. 1976, XV + 541 pp.

79. H. von Helmholtz, *Epistemological Writings*. (A New Selection Based upon the 1921 Volume edited by Paul Hertz and Moritz Schlick, Newly Translated and Edited by R. S. Cohen and Y. Elkana), Boston Studies in the Philosophy of Science, Volume XXXVII. 1977, XXXVII + 205 pp.

80. Joseph Agassi, *Science in Flux*, Boston Studies in the Philosophy of Science (ed. by Robert S. Cohen and Marx W. Wartofsky), Volume XXVIII. 1975, XXVI + 553 pp. Also available as paperback.

81. Sandra G. Harding (ed.), *Can Theories Be Refuted? Essays on the Duhem-Quine Thesis*. 1976, XXI + 318 pp. Also available as paperback.

82. Stefan Nowak, *Methodology of Sociological Research: General Problems*. 1977, XVIII + 504 pp.

83. Jean Piaget, Jean-Blaise Grize, Alina Szeminska, and Vinh Bang, *Epistemology and Psychology of Functions*. 1977, XIV + 205 pp.

84. Marjorie Grene and Everett Mendelsohn (eds.), *Topics in the Philosophy of Biology*, Boston Studies in the Philosophy of Science (ed. by Robert S. Cohen and Marx W. Wartofsky), Volume XXVII. 1976, XIII + 454 pp. Also available as paperback.

85. E. Fischbein, *The Intuitive Sources of Probabilistic Thinking in Children*. 1975, XIII + 204 pp.

86. Ernest W. Adams, *The Logic of Conditionals. An Application of Probability to Deductive Logic*. 1975, XIII + 156 pp.

87. Marian Przełęcki and Ryszard Wójcicki (eds.), *Twenty-Five Years of Logical Methodology in Poland*. 1977, VIII + 803 pp.

88. J. Topolski, *The Methodology of History*. 1976, X + 673 pp.

89. A. Kasher (ed.), *Language in Focus: Foundations, Methods and Systems. Essays Dedicated to Yehoshua Bar-Hillel*, Boston Studies in the Philosophy of Science (ed. by Robert S. Cohen and Marx W. Wartofsky), Volume XLIII. 1976. XXVIII + 679 pp. Also available as paperback.

90. Jaakko Hintikka, *The Intentions of Intentionality and Other New Models for Modalities*. 1975, XVIII + 262 pp. Also available as paperback.

91. Wolfgang Stegmüller, *Collected Papers on Epistemology, Philosophy of Science and History of Philosophy*, 2 Volumes, 1977, XXVI + 525 pp.

92. Dov M. Gabbay, *Investigations in Modal and Tense Logics with Applications to Problems in Philosophy and Linguistics*, 1976, XI + 306 pp.

93. Radu J. Bogdan, *Local Induction*. 1976, XVI + 340 pp.

94. Stefan Nowak, *Understanding and Prediction: Essays in the Methodology of Social and Behavioral Theories*. 1976, XIX + 482 pp.

95. Peter Mittelstaedt, *Philosophical Problems of Modern Physics*, Boston Studies in the Philosophy of Science (ed. by Robert S. Cohen and Marx W. Wartofsky), Volume XVIII. 1976, X + 211 pp. Also available as paperback.

96. Gerald Holton and William Blanpied (eds.), *Science and Its Public: The Changing Relationship*, Boston Studies in the Philosophy of Science (ed. by Robert S. Cohen and Marx W. Wartofsky), Volume XXXIII. 1976, XXV + 289 pp. Also available as paperback.

97. Myles Brand and Douglas Walton (eds.), *Action Theory. Proceedings of the Winnipeg Conference on Human Action, Held at Winnipeg, Manitoba, Canada, 9–11 May 1975*. 1976, VI + 345 pp.

98. Risto Hilpinen, *Knowledge and Rational Belief*. 1979 (forthcoming).

99. R. S. Cohen, P. K. Feyerabend, and M. W. Wartofsky (eds.), *Essays in Memory of Imre Lakatos*, Boston Studies in the Philosophy of Science (ed. by Robert S. Cohen and Marx W. Wartofsky), Volume XXXIX. 1976, XI + 762 pp. Also available as paperback.

100. R. S. Cohen and J. Stachel (eds.), *Léon Rosenfeld, Selected Papers*. Boston Studies in the Philosophy of Science (ed. by Robert S. Cohen and Marx W. Wartofsky), Volume XXI. 1978, XXX + 927 pp.

101. R. S. Cohen, C. A. Hooker, A. C. Michalos and J. W. van Evra (eds.), *PSA 1974: Proceedings of the 1974 Biennial Meeting of the Philosophy of Science Association*, Boston Studies in the Philosophy of Science (ed. by Robert S. Cohen and Marx W. Wartofsky), Volume XXXII. 1976, XIII + 734 pp. Also available as paperback.

102. Yehuda Fried and Joseph Agassi, *Paranoia: A Study in Diagnosis*, Boston Studies in the Philosophy of Science (ed. by Robert S. Cohen and Marx W. Wartofsky), Volume L. 1976. XV + 212 pp. Also available as paperback.

103. Marian Przełęcki, Klemens Szaniawski, and Ryszard Wójcicki (eds.), *Formal Methods in the Methodology of Empirical Sciences*. 1976, 455 pp.

104. John M. Vickers, *Belief and Probability*. 1976, VIII + 202 pp.

105. Kurt H. Wolff, *Surrender and Catch: Experience and Inquiry Today*, Boston Studies in the Philosophy of Science (ed. by Robert S. Cohen and Max W. Wartofsky), Volume LI. 1976, XII + 410 pp. Also available as paperback.

106. Karel Kosík, *Dialectics of the Concrete*, Boston Studies in the Philosophy of Science (ed. by Robert S. Cohen and Marx W. Wartofsky), Volume LII. 1976, VIII + 158 pp. Also available as paperback.

107. Nelson Goodman, *The Structure of Appearance*, Boston Studies in the Philosophy of Science (ed. by Robert S. Cohen and Marx W. Wartofsky), Volume L. Third edition. 1977, L + 285 pp.

108. Jerzy Giedymin (ed.), *Kazimierz Ajdukiewicz: Scientific World-Perspective and Other Essays, 1930–1963*. 1977, LIII + 378 pp.

109. Robert L. Causey, *Unity of Science*. 1977, VIII + 185 pp.

110. Richard Grandy, *Advanced Logic for Applications*. 1977, XIII + 168 pp.

111. Robert P. McArthur, *Tense Logic*. 1976, VII + 84 pp.

112. Lars Lindahl, *Position and Change: A Study in Law and Logic*. 1977, IX + 299 pp.

113. Raimo Tuomela, *Dispositions*. 1977, X + 450 pp.

114. Herbert A. Simon, *Models of Discovery and Other Topics in the Methods of Science*, Boston Studies in the Philosophy of Science (ed. by Robert S. Cohen and Marx W. Wartofsky), Volume LIV. 1977, XX + 456 pp.

115. Roger D. Rosenkrantz, *Inference, Method and Decision*. 1977, XV + 262 pp.

116. Raimo Tuomela, *Human Action and Its Explanation. A Study on the Philosophical Foundations of Psychology*. 1977, XII + 426 pp.

117. Morris Lazerowitz, *The Language of Philosophy*, Boston Studies in the Philosophy of Science (ed. by Robert S. Cohen and Marx W. Wartofsky), Volume LV. 1977, XV + 2C9 pp.

118. Tran Duc Thao, *Origins of Language and Consciousness*, Boston Studies in the Philosophy of Science (ed. by Robert S. Cohen and Marx W. Wartofsky), Volume LVI. 1979 (forthcoming).

203 common measure    19-20 Tennon — amateur
208 C + fr.

Printed in the United Kingdom
by Lightning Source UK Ltd.
108506UKS00001B/38

9 789027 708199